T0319499

**Computational Fractional
Dynamical Systems**

Computational Fractional Dynamical Systems

Fractional Differential Equations and Applications

Snehashish Chakraverty, Rajarama Mohan Jena, and Subrat Kumar Jena

National Institute of Technology
Rourkela, Odisha, IN

This edition first published 2023
© 2023 John Wiley & Sons, Inc.

The right of Snehashish Chakraverty, Rajarama Mohan Jena, and Subrat Kumar Jena to be identified as the authors of this work has been asserted in accordance with law.

Registered Office
John Wiley & Sons, Inc., 111 River Street, Hoboken, NJ 07030, USA

Editorial Office
111 River Street, Hoboken, NJ 07030, USA

For details of our global editorial offices, customer services, and more information about Wiley products visit us at www.wiley.com.

Wiley also publishes its books in a variety of electronic formats and by print-on-demand. Some content that appears in standard print versions of this book may not be available in other formats.

Limit of Liability/Disclaimer of Warranty
In view of ongoing research, equipment modifications, changes in governmental regulations, and the constant flow of information relating to the use of experimental reagents, equipment, and devices, the reader is urged to review and evaluate the information provided in the package insert or instructions for each chemical, piece of equipment, reagent, or device for, among other things, any changes in the instructions or indication of usage and for added warnings and precautions. While the publisher and authors have used their best efforts in preparing this work, they make no representations or warranties with respect to the accuracy or completeness of the contents of this work and specifically disclaim all warranties, including without limitation any implied warranties of merchantability or fitness for a particular purpose. No warranty may be created or extended by sales representatives, written sales materials or promotional statements for this work. The fact that an organization, website, or product is referred to in this work as a citation and/or potential source of further information does not mean that the publisher and authors endorse the information or services the organization, website, or product may provide or recommendations it may make. This work is sold with the understanding that the publisher is not engaged in rendering professional services. The advice and strategies contained herein may not be suitable for your situation. You should consult with a specialist where appropriate. Further, readers should be aware that websites listed in this work may have changed or disappeared between when this work was written and when it is read. Neither the publisher nor authors shall be liable for any loss of profit or any other commercial damages, including but not limited to special, incidental, consequential, or other damages.

Library of Congress Cataloging-in-Publication Data
Names: Chakraverty, Snehashish, author. | Jena, Rajarama Mohan, author.
   | Jena, Subrat Kumar, author.
Title: Computational fractional dynamical systems : fractional differential
   equations and applications / Snehashish Chakraverty, Rajarama Mohan
   Jena, Subrat Kumar Jena.
Description: Hoboken, NJ : Wiley, 2023. | Includes index.
Identifiers: LCCN 2022023201 (print) | LCCN 2022023202 (ebook) | ISBN
   9781119696957 (cloth) | ISBN 9781119696834 (adobe pdf) | ISBN
   9781119696995 (epub)
Subjects: LCSH: Fractional differential equations.
Classification: LCC QA372 .C42528 2023 (print) | LCC QA372 (ebook) | DDC
   515/.35–dc23/eng20220826
LC record available at https://lccn.loc.gov/2022023201
LC ebook record available at https://lccn.loc.gov/2022023202

Cover Design: Wiley
Cover Image: © local_doctor/Shutterstock

Set in 9.5/12.5pt STIXTwoText by Straive, Pondicherry, India

Contents

Preface

Subject of fractional calculus has gained considerable popularity and importance during the past three decades, mainly due to its validated applications in various fields of science and engineering. It deals with the differential and integral operators with nonintegral powers. Fractional differential equations are the pillar of various systems occurring in a wide range of science and engineering applications, namely physics, chemical engineering, mathematical biology, financial mathematics, structural mechanics, control theory, circuit analysis, and biomechanics. The fractional derivative has also been used in various other physical problems, such as frequency-dependent damping behavior of structures, the motion of a plate in a Newtonian fluid, and $PI^\lambda D^\mu$ controller for the control of dynamical systems. The mathematical models in electromagnetics, rheology, viscoelasticity, electrochemistry, control theory, Brownian motion, signal and image processing, fluid dynamics, financial mathematics, and material science are well defined by fractional-order differential equations. Generally, these physical models are demonstrated either by ordinary or partial differential equations. But, modeling these problems by fractional differential equations sometimes makes the physics of the systems more practical. In order to know the behavior of these systems, we need to study the solutions of the governing fractional equations. The exact solution of fractional differential equations may not always be possible using known classical methods. Generally, the physical models occurring in nature comprise complex phenomena. So, it is challenging to get the solution (both analytical and numerical) of nonlinear differential equations of fractional order. For the last few decades, a great deal of attention has been directed toward the solution to these kinds of problems. Researchers throughout the globe are trying to develop various efficient methods to handle these problems. Although there exist a variety of standard books related to the solution of fractional differential equations and related methods. But, the existing books are method specific or subject specific or sometimes may not be efficient. Few existing books deal with basic numerical and analytical methods for solving the fractional differential equations, whereas some other books may be found related to particular semi-analytical methods only. But, as per the authors' knowledge, books covering the basic concepts of the computationally efficient and variety of advanced methods in one place in a systematic manner are rare.

As such, the authors realized the need for a book that contains traditional as well as recent numerical and analytic methods with simple example problems. With respect to student-friendly, straightforward, and easy understanding of the methods, this book may be a benchmark for the teaching/research courses for students, teachers, and industry. The present book consists of 25 chapters giving basic knowledge of various recent and challenging procedures with respect to semi-analytical and expansion methods. The best part of the book is that it discusses different computationally efficient and recently developed methods for solving linear and nonlinear fractional problems for better understanding. Before we address the details of the book, the authors consider that the readers have essential knowledge of calculus, differential equations, partial differential equation, fractional calculus, functional analysis, real analysis, and linear algebra.

Accordingly, Chapter 1 addresses the preliminaries on fractional calculus in which various important functions related to the fractional calculus and popular differential and integral operator of fractional order have been included. Chapter 2 deals with the various mathematical models arising in day to day life. It is worth mentioning that semi-analytical techniques based on perturbation parameters also exist and have broad applicability. As such, the Adomian decomposition method (ADM) for solving linear and nonlinear fractional differential equations has been presented in Chapter 3. Chapter 4 discusses the four hybrid methods, which are the coupling of various transform methods and ADM. Examples of simple linear and nonlinear fractional differential equations have been deliberated to understand the methodologies of these four methods. In this regard, another well-known semi-analytical technique is the homotopy perturbation method (HPM). The HPM is easy to implement for handling various types of fractional differential equations. As such, a detailed procedure of the HPM is described and applied to linear and nonlinear fractional problems in Chapter 5. Chapter 6 deals with the four hybrid methods that combine four transform methods and HPM. These four methods are getting more popular due to their widespread application for solving various fractional problems. Another important method, namely the fractional differential transform method (FDTM), has been presented in Chapter 7. Due to some difficulties/complexity arising for solving fractional problems in DTM, an advanced version of this method has been developed called the fractional reduced differential transform method (FRDTM) given in Chapter 8. The main benefit of this method is that it does not require any assumption, perturbation, and discretization for solving the fractional dynamical model. Also, less computation time is needed as compared to other techniques.

Further, Chapter 9 deals with a semi-analytical method, viz., variational iteration method (VIM) for finding the approximate series solution of linear and nonlinear fractional differential equations. It may be worth mentioning that the methods, namely ADM, HPM, and VIM discussed, respectively, in Chapters 4, 5, and 9, yield approximate solutions and may produce exact solutions depending upon the considered problem. Another powerful approximation technique, namely the weighted residual method (WRM), is addressed in Chapter 10 for finding solutions of fractional differential equations subject to boundary conditions. In this regard, this chapter is organized such that various WRMs, viz., collocation, least-square, and Galerkin methods are applied for solving boundary value problems. A new challenging technique, viz., the use of boundary characteristic orthogonal polynomials (BCOPs) in well-known methods like Galerkin, collocation, etc., have also been introduced in Chapter 11. In Chapter 12, we have discussed the residual power series method (RPSM). Some main characteristics of this method are (i) this technique obtains expansions of the solutions in the form of polynomials, (ii) the solutions and all their derivatives are applicable for each arbitrary point in the given interval, and (iii) it does not require any modification while switching from the first order to the higher order. So this technique can be applied directly to the given system by choosing an appropriate value for the initial guess approximations. This technique needs small computational cost with high precision and less time. Further, Chapter 13 confers the homotopy analysis method (HAM), which is based on the coupling of the traditional perturbation method and homotopy in topology. Generally, the HAM involves a control parameter that controls the convergent region and rate of convergence of the solution. Four transform methods coupled with HAM have been discussed in Chapter 14. Similarly, the q-homotopy analysis method (q-HAM) and four transform methods coupled with q-HAM have also been presented along with simple examples in Chapters 15 and 16, respectively.

Further, in the present day, expansion methods are getting more attention of researchers for obtaining the exact solution of the fractional nonlinear partial differential equations. In this regard, we have discussed three reliable and efficient methods, namely (G'/G)-expansion method, (G'/G^2)-expansion method, and $(G'/G,1/G)$-expansion method in Chapters 17, 18, and 19, respectively. In Chapters 20, 21, and 22, we have addressed the procedure and implementation of the modified simple equation method and Sine-Cosine and Tanh methods, respectively, to obtain the traveling wave solution of fractional differential equations. Fractional subequation method, Exp-function method, and $\text{Exp}(-\varphi(\xi))$-expansion method and their applications to nonlinear fractional partial differential equations have been illustrated in Chapters 23, 24, and 25, respectively.

In view of the above, this book aims to provide basic concepts of fractional-order differential equations with various numerical example problems as well as important applications in science and engineering systems along with the recently developed methods in a systematic manner. The book will certainly find an important source for graduate and postgraduate students, teachers, and researchers in colleges, universities/institutes, and industries in various sciences and engineering fields, wherever one wants to model and analyze their physical problems.

Finally, we believe that the book may represent a new vista because it demonstrates how the most current, advanced, and novel mathematical and computational techniques given in a series of 25 chapters can be put to effective use of fractional calculus in fractional-order differential equations.

Acknowledgments

Our journey to write this book would not have been possible without the support, encouragement, and motivation we received from the people around us who helped us through our hardships and gave us the confidence to pursue our dreams.

As such, the first author would like to thank his parents for their blessings and love that they have bestowed on my life, whose memory always inspire me at each and every step. Next, he would like to thank his wife Shewli and daughters Shreyati and Susprihaa for their support and source of inspiration during this project. The support of all the PhD students of the first author and the NIT Rourkela facilities are also gratefully acknowledged.

The second and third authors would like to express their sincere gratitude to their family members, especially Sh. Ullash Chandra Jena, Sh. Durga Prasad Jena, Sh. Laxmidhara Jena, Smt. Urbasi Jena, Smt. Renu Bala Jena, Smt. Arati Jena, sisters Jyotrimayee, Truptimayee, and Nirupama for their enormous love, continuous motivation, support, and blessings. Further, the second author, Dr. Rajarama Mohan Jena, would like to acknowledge the Department of Science and Technology, Govt. of India for providing INSPIRE fellowship (IF170207) to carry out the present work. The second author will forever be thankful to his brother Dr. Subrat Kumar Jena for his help, encouragement, and support during his stay at NIT and for making it a memorable experience in his life. Last but not least, Dr. Rajarama Mohan Jena also wants to thank Dr. Sujata Swain, Department of Physics and Astronomy, National Institute of Technology Rourkela, for showing her endless love, moral support, generous spirit, faith, and encouragement, which at all times rejuvenated his vigour for research and motivated him to have achievements beyond his expectations.

Also, the second and third authors greatly appreciate the inspiration of the first author for his support and inspiration. This work would not have been possible without his guidance, support, and encouragement.

Further, the authors sincerely acknowledge the reviewers' fruitful suggestions and appreciation. All the authors do appreciate the support and help of the whole team of Wiley. Finally, the authors are greatly indebted to all the authors/researchers mentioned in the bibliography/reference sections at the end of each chapter.

Snehashish Chakraverty
Rajarama Mohan Jena
Subrat Kumar Jena

About the Authors

Dr. Rajarama Mohan Jena is currently working as a Senior Researcher at the Department of Mathematics, National Institute of Technology Rourkela. He has defended his PhD thesis at Department of Mathematics, National Institute of Technology Rourkela. Dr. Rajarama does research in fractional calculus, partial differential equations, numerical analysis, mathematical modeling, and uncertainty modeling. He has published 30 research papers in peer-reviewed international journals, one international conference paper, three book chapters, and four books (one published in Springer, one in press of Wiley, and two books ongoing in Elsevier). One of his papers published in *ZAMM – Journal of Applied Mathematics and Mechanic (Wiley)* is among the top-cited papers for the year 2020–2021. He has been awarded an *IOP Publishing Top Cited Paper Award 2021* from India, published across the entire IOP Publishing journal portfolio within the recent three years (2018–2020). He was awarded *DST-INSPIRE Scholarship for Higher Education* for five years (2011–2016). Again, he was conferred with *DST-INSPIRE Fellowship* for doctoral research. Till date, he is having 696 citations with H-index:16 as per Google Scholar. He also serves as a reviewer for many prestigious international journals. Further, he is continuing collaborative research works with renowned researchers from Italy, Turkey, Iran, Poland, Canada, Egypt, China, South Africa, Saudi Arabia, etc.

Prof. Snehashish Chakraverty has 30 years of experience as a researcher and teacher. Presently, he is working in the Department of Mathematics (Applied Mathematics Group), National Institute of Technology Rourkela, Odisha, as a senior (Higher Administrative Grade) professor. Prior to this, he was with CSIR-Central Building Research Institute, Roorkee, India. After completing graduation from St. Columba's College (Ranchi University), his career started from the University of Roorkee (now, Indian Institute of Technology Roorkee) and he did MSc (mathematics) and MPhil (computer applications) from there securing first positions in the university. Dr. Chakraverty received his PhD from IIT Roorkee in 1993. Thereafter, he did his postdoctoral research at the Institute of Sound and Vibration Research (ISVR), University of Southampton, UK, and at the Faculty of Engineering and Computer Science, Concordia University, Canada. He was also a visiting professor at Concordia and McGill universities, Canada, during 1997–1999 and visiting professor of University of Johannesburg, South Africa, during 2011–2014. He has authored/co-authored/edited 29 books, published 416 research papers (till date) in journals and conferences, and two books are ongoing. He is in the editorial boards of various international journals, book series, and conferences. Prof. Chakraverty is the chief editor of "International Journal of Fuzzy Computation and Modelling" (IJFCM), Inderscience Publisher, Switzerland (http://www.inderscience.com/ijfcm); associate editor of "Computational Methods in Structural Engineering, Frontiers in Built Environment" and "Curved and Layered Structures (De Gruyter)"; and happens to be the editorial board member of "Springer Nature Applied Sciences," "IGI Research Insights Books," "Springer Book Series of Modeling and Optimization in Science and Technologies," "Coupled Systems Mechanics (Techno Press)," "Journal of Composites Science (MDPI)," "Engineering Research Express (IOP)," and "Applications and Applied Mathematics: An International Journal." He is also the reviewer of around 50 national and international journals of repute, and he was the president of the Section of Mathematical Sciences (including Statistics) of "Indian Science Congress" (2015–2016) and was the vice president of "Orissa Mathematical Society" (2011–2013). Prof. Chakraverty is a recipient of prestigious awards, viz. Indian National Science Academy (INSA) nomination under International Collaboration/Bilateral Exchange Program (with the Czech Republic), Platinum Jubilee ISCA Lecture Award (2014), CSIR Young Scientist Award (1997), BOYSCAST Fellow. (DST), UCOST Young Scientist Award (2007, 2008), Golden Jubilee Director's (CBRI) Award (2001), INSA International Bilateral Exchange Award ([2010–11 (selected but could not undertake), 2015 (selected)], and Roorkee University Gold Medals (1987, 1988) for first positions in MSc and MPhil (computer applications). He is in the list of 2% world scientists recently (2020, 2021) in Artificial Intelligence and Image Processing category based on an independent study done by Stanford University scientists. He is the recipient IOP Publishing Top Cited Paper Award for one of the most cited articles from India, published across the entire IOP Publishing journal portfolio within the recent three years (2018–2020). It also features in the top 1% of most cited papers in the materials subject category. This data is from the citations recorded in Web of Science. One of our papers has been awarded an IOP Publishing Top Cited Paper Award 2022 from India, published across the entire IOP Publishing journal portfolio within the past three years (2019–2021), and the paper was among the top 1% of most cited papers in the materials subject category. He has guided 19 PhD students and 12 are ongoing. Prof. Chakraverty has undertaken around 16 research projects as principal investigator funded by international and national agencies totaling about Rs.1.5 crores. He has hoisted around eight international students with different international/national fellowships to work in his group as postdoctoral fellow, PhD, and visiting researchers for different periods. A good number of international and national conferences, workshops, and training programs have also been organized by him. His present research area includes differential equations (ordinary, partial, and fractional), numerical analysis and computational methods, structural dynamics (FGM, nano) and fluid dynamics, mathematical and uncertainty modeling, soft computing, and machine intelligence (artificial neural network; fuzzy, interval, and affine computations).

Dr. Subrat Kumar Jena is currently working as an honorary postdoctoral researcher at Nonlinear Multifunctional Composites-Analysis & Design Lab, Indian Institute of Science (IISC) Bangalore. He has defended his PhD thesis at Department of Mathematics, National Institute of Technology Rourkela. Dr. Subrat does research in structural dynamics, applied mathematics, mathematical modeling, uncertainty modeling, etc. Till now, he has authored 32 research papers in peer-reviewed international journals, 2 international conference papers, 8 book chapters, and 6 books (1 book is published, 2 books are in press, and 3 books are ongoing). He has been awarded an IOP Publishing Top Cited Paper Award 2021 and 2022 in succession from India, published across the entire IOP Publishing journal portfolio within the recent three years, i.e. (2018–2020) and (2019–2021), respectively, and the papers were among the top 1% of most cited papers in the materials subject category. One of his paper published in *ZAMM – Journal of Applied Mathematics and Mechanic* (Wiley) is among top cited papers for the year 2020–2021. He was also featured in Stanford University's Top 2% Most Influential Scientists List in 2021 for the Year 2020. Till date, he is having 732 citations with H-index: 20 as per Google Scholar. Additionally, he has been listed in Shell Buckling website as "Shell Buckling People" for his substantial contributions to the fields of shell buckling and static or dynamic analysis of general structures. He is serving as reviewer/review editor for good number of prestigious international journals and has reviewed over 50 manuscripts so far.

1

Introduction to Fractional Calculus

1.1 Introduction

Fractional calculus is a generalization of ordinary differentiation and integration to arbitrary (non-integer) order. It is also an area of mathematics that investigates the possibilities of using real or even complex numbers as powers of the differential operator. This area is three centuries old compared to conventional calculus, but initially, it was not very popular. Fractional derivatives and integrals are not local in nature, so the nonlocal distributed effects are considered. The subject of fractional calculus has gained considerable popularity and importance during the past three decades, mainly due to its validated applications, dynamic nature, and comprehensive representation of complex nonlinear phenomena in various fields of science and engineering. The mathematical models in electromagnetics, rheology, viscoelasticity, electrochemistry, control theory, fluid dynamics, financial mathematics, and material science are well defined by fractional-order differential equations.

1.2 Birth of Fractional Calculus

In a letter to L'Hospital in 1695, Leibniz asked the following question: "Can the meaning of integer-order derivatives be generalized to non-integer-order derivatives?" L'Hospital was very curious about that question and replied to Leibniz by asking what would happen to the term $\frac{d^n \psi(x)}{dx^n}$ if $n = \frac{1}{2}$. In order to explain the answer to the query raised by L'Hospital, Leibniz wrote a letter dated 30 September 1695, known as the birthday of fractional calculus, which mentioned that "It will lead to a paradox, from which one-day useful consequences will be drawn." This was the beginning of fractional calculus. Many famous mathematicians, namely Liouville, Riemann, Weyl, Fourier, Abel, Lacroix, Leibniz, Grünwald, and Letnikov, contributed to fractional calculus over the years. Recently, various types of fractional differential and integral operators have been developed, namely the Riemann–Liouville fractional integral and derivative, Caputo fractional derivative, Grünwald–Letnikov fractional derivative, Riesz fractional derivative, modified Riemann–Liouville derivative, and local fractional derivative, which are all discussed in this chapter.

1.3 Useful Mathematical Functions

In order to understand various types of fractional derivatives and integrals arising in fractional calculus, we need first to understand some necessary preliminaries and related functions used in fractional calculus. These functions include the gamma function, the Euler psi function, incomplete gamma function, beta function, incomplete beta function, Mittag-Leffler functions (MLFs), Mellin-Ross function, the Wright function, the error function, the hypergeometric functions (Gauss, Kummer, and generalized hypergeometric functions), and the H-function.

1.3.1 The Gamma Function

In this section, the definitions and some properties of the gamma function have been covered. The fundamental characteristic of the gamma function is just an extension of the factorial for all real numbers. It can also be defined in terms of a complex number.

Computational Fractional Dynamical Systems: Fractional Differential Equations and Applications, First Edition.
Snehashish Chakraverty, Rajarama Mohan Jena, and Subrat Kumar Jena.
© 2023 John Wiley & Sons, Inc. Published 2023 by John Wiley & Sons, Inc.

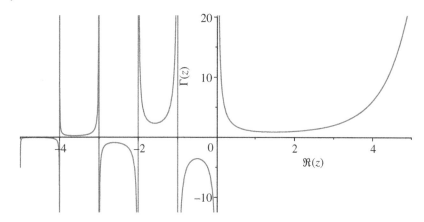

Figure 1.1 The graph of gamma function in the real axis.

Definition 1.1 The gamma function is most important in the fractional-order calculus, and it is written as (Baleanu et al. 2012; Chakraverty et al. 2020; Das 2011; Kilbas et al. 2006; Kiryakova 1993; Miller and Ross 1993; Oldham and Spanier 1974; Podlubny 1999; Samko et al. 2002):

$$\Gamma(z) = \int_0^\infty e^{-x}x^{z-1}dx, \quad \Re(z) > 0, \tag{1.1}$$

where $\Re(z)$ is the real part of the complex number $z \in C$. Equation (1.1) is convergent for all complex numbers $z \in C$ ($\Re(z) > 0$). The gamma function is defined everywhere on the real axis except its singular points, viz. 0, −1, −2, ... As a result, the domain of the gamma function is $... \cup (-2,-1) \cup (-1,0) \cup (0,+\infty)$. The graph of the gamma function is depicted in the Figure 1.1.

Some properties of the gamma function are as follows (Chakraverty et al. 2020; Miller and Ross 1993; Podlubny 1999; Samko et al. 2002):

i) $\Gamma(z+1) = z\Gamma(z), \quad$ for $z \in R^+$.

ii) $\Gamma(z+1) = z!, \quad \Gamma(1) = 1 \quad$ and $\quad \Gamma\left(\frac{1}{2}\right) = \sqrt{\pi}$.

iii) $\Gamma\left(\frac{1}{2} - z\right) = \dfrac{z!(-4)^z}{(2z)!}\sqrt{\pi} \quad$ and $\quad \Gamma\left(\frac{1}{2} + z\right) = \dfrac{(2z)!}{(4)^z z!}\sqrt{\pi}$.

iv) $\Gamma(z)\Gamma(-z) = \dfrac{-\pi}{n\sin(\pi z)}, \quad \Gamma(z)\,\Gamma(1-z) = \dfrac{\pi}{\sin(\pi z)}, \quad z \notin \aleph, \ \Re(z) < 1$.

Definition 1.2   (Euler psi Function)
The Euler psi function is the logarithmic derivative of the gamma function, which is defined as (Kilbas et al. 2006):

$$\psi(z) = \frac{d}{dz}\log\Gamma(z) = \frac{\Gamma'(z)}{\Gamma(z)}, \quad z \in C, \tag{1.2}$$

with the following property:

$$\psi(z+n) = \psi(z) + \sum_{k=0}^{n-1}\frac{1}{z+k}, \quad z \in C, n \in \aleph. \tag{1.3}$$

Definition 1.3   (Incomplete Gamma Function)
The incomplete gamma function (Herrmann 2011; Kilbas et al. 2006) is derived from Eq. (1.1) by decomposing into an integral from 0 to ω and another from ω to ∞ as:

$$\gamma(z,\omega) = \int_0^\omega e^{-x}x^{z-1}dx, \quad z,\omega \in C, \Re(z) > 0. \tag{1.4}$$

$$\Gamma(z,\omega) = \int_\omega^\infty e^{-x}x^{z-1}dx, \quad |\arg(\omega)| < \pi, \Re(z) > 0. \tag{1.5}$$

The incomplete gamma functions have the following properties (Herrmann2011; Kilbas et al. 2006):

i) $\gamma(z, \infty) = \Gamma(z, 0) = \Gamma(z)$,
ii) $\gamma(z, \omega) + \Gamma(z, \omega) = \Gamma(z), \quad \Re(z) > 0$.

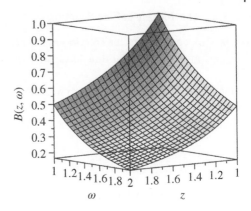

Figure 1.2 The graph of beta function.

1.3.2 The Beta Function

Definition 1.4   The beta function is defined as (Kilbas et al. 2006; Miller and Ross 1993; Podlubny 1999; Samko et al. 2002):

$$B(z,\omega) = \int_0^1 x^{z-1}(1-x)^{\omega-1}dx, \quad \Re(\omega), \ \Re(z) > 0. \tag{1.6}$$

3D plot of the beta function Eq. (1.6) has been illustrated in Figure 1.2.

Some properties of the beta function are given as follows (Kilbas et al. 2006; Miller and Ross 1993; Podlubny 1999; Samko et al. 2002):

i) $B(z, \omega) = B(\omega, z)$,

ii) $B(z, \omega) = 2\int_0^{\frac{\pi}{2}} (\sin\theta)^{2z-1}(\cos\theta)^{2\omega-1}d\theta, \quad \Re(\omega), \Re(z) > 0$.

iii) $B(z, \omega) = \int_0^\infty \frac{x^{z-1}}{(1+x)^{z+\omega}}dx, \quad \Re(\omega), \Re(z) > 0$.

iv) $B(z, \omega) = B(z, \omega+1) + B(z+1, \omega)$,

v) $B(z, \omega+1) = B(z, \omega)\frac{\omega}{z+\omega}$,

vi) $B(z+1, \omega) = B(z, \omega)\frac{z}{z+\omega}$,

vii) $B(z, \omega)B(z+\omega, 1-\omega) = \frac{\pi}{z\sin(\pi\omega)}$.

Note 1.1   The relationship between gamma and beta functions is written as:

$$B(z,\omega) = \frac{\Gamma(z)\Gamma(\omega)}{\Gamma(z+\omega)}.$$

Definition 1.5   (Incomplete Beta Function)

The generalized form beta function is known as incomplete beta function, which is given as:

$$B(z; a, b) = \int_0^z x^{a-1}(1-x)^{b-1}dx, \quad \Re(z) > 0. \tag{1.7}$$

It is worth mentioning that when $z = 1$, the incomplete beta function transforms into the beta function, which has several applications in physics, functional analysis, and integral calculus.

1.3.3 The Mittag-Leffler Function

The MLF comes from the solution of fractional-order differential equations or fractional-order integral equations. It is an extension of exponential functions that may be expressed as a power series.

Definition 1.6 (One-Parametric Mittag-Leffler Function)
One-parameter MLF is defined as (Baleanu et al. 2012; Chakraverty et al. 2020; Das 2011; Kilbas et al. 2006; Kiryakova 1993; Miller and Ross 1993; Oldham and Spanier 1974; Podlubny 1999; Samko et al. 2002):

$$E_\alpha(z) = \sum_{n=0}^\infty \frac{z^n}{\Gamma(1+n\alpha)}, \quad \text{for } z \in C \text{ and } \alpha > 0. \tag{1.8}$$

If we put $\alpha = 1$ in Eq. (1.8), we obtain

$$E_1(z) = \sum_{n=0}^\infty \frac{z^n}{\Gamma(1+n)}, \quad \text{for } z \in C. \tag{1.9}$$

which is the summation form of the exponential function e^z. So, MLF is an extension of the exponential function in one parameter.

Definition 1.7 (Two-Parametric Mittag-Leffler Function)
Two-parameter representation of the MLF may be written as (Miller and Ross 1993; Podlubny 1999; Samko et al. 2002):

$$E_{\alpha,\beta}(z) = \sum_{n=0}^\infty \frac{z^n}{\Gamma(\beta+n\alpha)}, \quad \text{for } z \in C \text{ and } \alpha,\beta > 0. \tag{1.10}$$

Definition 1.8 (Generalized Mittag-Leffler Function)
The generalized MLF can be defined as (Haubold et al. 2011; Kilbas et al. 2006; Kurulay and Bayram 2012):

$$E_{\alpha,\beta}^\gamma(z) = \sum_{k=0}^\infty \frac{(\gamma)_n}{\Gamma(\beta+k\alpha)}\frac{z^k}{k!} \quad \text{for } \Re(\alpha),\Re(\beta),\Re(\gamma) > 0 \text{ and } z,\alpha,\beta \text{ and } \gamma \in C. \tag{1.11}$$

where $(\gamma)_n$ is the Pochhammer symbol and is defined as

$$(\gamma)_n = \frac{\Gamma(\gamma+n)}{\Gamma(\gamma)} = \begin{cases} 1, & n=0,\gamma \neq 0, \\ (\gamma+n-1)\cdots(\gamma+2)(\gamma+1)\gamma, & n \in \aleph, \gamma \in C. \end{cases} \tag{1.12}$$

Note 1.2 The derivative of the two-parametric MLF can be expressed in the form of generalized MLF as (Kilbas et al. 2006):

$$\frac{d^n}{dz^n}\left[E_{\alpha,\beta}(z)\right] = n!E_{\alpha,\beta+\alpha n}^{n+1}(z), \quad n \in \aleph, z \in C. \tag{1.13}$$

Some properties of the MLF are given as follows (Mathai and Haubold 2008):

i) $E_{\alpha,\beta}(z) = \frac{1}{\Gamma(\beta)} + z\,E_{\alpha,\alpha+\beta}(z),$
ii) $E_{\alpha,\beta}(z) = \beta\,E_{\alpha,\beta+1}(z) + \alpha z\frac{d}{dz}E_{\alpha,\beta+1}(z),$
iii) $\left(\frac{d}{dz}\right)^m\left[z^{\beta-1}E_{\alpha,\beta}(z^\alpha)\right] = z^{\beta-m-1}E_{\alpha,\beta-m}(z^\alpha), \quad \Re(\beta-m) > 0, m = 0,\ 1,\ 2,\ldots$

1.3.4 The Mellin-Ross Function

The Mellin-Ross function $E_t(\nu, a)$ is obtained while evaluating the fractional derivative of an exponential function e^{at}. Both the incomplete gamma and MLFs are closely related to this function.

Definition 1.9   The Mellin-Ross function is defined as (Mathai and Haubold 2008):

$$E_t(\nu, a) = t^\nu \sum_{k=0}^{\infty} \frac{(at)^k}{\Gamma(k + \nu + 1)} = t^\nu E_{1, \nu + 1}(at). \tag{1.14}$$

1.3.5   The Wright Function

Definition 1.10   The Wright function was proposed by Wright (1993) in 1933, which is denoted by $W(z; \alpha, \beta)$ and is defined as (Kilbas et al. 2006; Miller and Ross 1993; Podlubny 1999; Samko et al. 2002):

$$W(z; \ \alpha, \beta) = \sum_{k=0}^{\infty} \frac{z^k}{k! \Gamma(\alpha k + \beta)}, \quad \alpha > -1 \text{ and } z, \beta \in C. \tag{1.15}$$

Note 1.3   The Wright function can be expressed with the help of the MLF as (Kurulay and Bayram 2012):

$$\frac{d^n}{dz^n} \left[z^{\beta - 1} E^\gamma_{\alpha, \beta + r\alpha}(z) \right] = \sum_{k=0}^{\infty} z^{\beta - 1 - n} (\gamma)_k W(z; \alpha, \beta + r\alpha - n). \tag{1.16}$$

1.3.6   The Error Function

Definition 1.11   The error function which is denoted by erf (z) and is defined as (Kilbas et al. 2006; Miller and Ross 1993; Podlubny 1999; Samko et al. 2002):

$$\text{erf}(z) = \frac{2}{\sqrt{\pi}} \int_0^z e^{-t^2} dt, z \in C. \tag{1.17}$$

Some properties of the error function are given as (Miller and Ross 1993; Podlubny 1999; Samko et al. 2002):

i) (Complementary error function) $\text{erfc}(z) = \dfrac{2}{\sqrt{\pi}} \displaystyle\int_z^{\infty} e^{-t^2} dt = 1 - \text{erf}(z),$

ii) $\text{erf}(-z) = -\text{erf}(z),$

iii) $E_{1/2}(z^{1/2}) = e^z (1 + \text{erf}(z^{1/2})).$

1.3.7   The Hypergeometric Function

This section discusses the definitions and characteristics of Gauss, Kummer, and generalized hypergeometric functions. When $\alpha > 0$, the following lemma gives the integral representation of $E_\alpha(z)$ as a Mellin–Barnes contour integral.

Lemma 1.1   For $\alpha > 0$ and $z \in C(|\arg(z)| < \pi)$, the following relations hold

$$E_\alpha(z) = \frac{1}{2\pi i} \int_{\gamma - i\infty}^{\gamma + i\infty} \frac{\Gamma(s) \Gamma(1 - s)}{\Gamma(1 - \alpha s)} (-z)^{-s} ds, \tag{1.18}$$

where the integration path separates all the poles $s = -k \, (k \in \aleph_0)$ to the left and all poles $s = n + 1 \, (n \in \aleph_0)$ to the right.

Definition 1.12   (The Gauss Hypergeometric Function)

The Gauss hypergeometric function $_2F_1(a, b; c; z)$ in the unit disk is defined as the sum of hypergeometric series provided by (Kilbas et al. 2006):

$$_2F_1(a, b; c; z) = \sum_{k=0}^{\infty} \frac{(a)_k (b)_k}{(c)_k} \frac{z^k}{k!}, \tag{1.19}$$

where $|z| < 1$, $a, b \in C, c \in C \backslash Z_0^- := \{0, -1, -2, \ldots\}$ and $(a)_k$ is the Pochhammer symbol defined in Eq. (1.12).

Note 1.4   The Euler integral representation of Gauss hypergeometric function is written as:

$$_2F_1(a,b;c;z) = \frac{\Gamma(c)}{\Gamma(b)\Gamma(c-b)}\int_0^1 t^{b-1}(1-t)^{c-b-1}(1-zt)^{-a}dt, \quad 0 < \Re(b) < \Re(c), |\arg(1-z)| < \pi. \tag{1.20}$$

Note 1.5   If $c \notin Z_0^-$, then $_2F_1(a,b;c;z)$ has another integral representation in terms of the Mellin–Barnes contour integral, which is given as (Kilbas et al. 2006; Podlubny 1999):

$$_2F_1(a,b;c;z) = \frac{1}{2\pi i}\frac{\Gamma(c)}{\Gamma(a)\Gamma(b)}\int_{\gamma-i\infty}^{\gamma+i\infty}\frac{\Gamma(a-s)\Gamma(b-s)}{\Gamma(c-s)}(-z)^{-s}\Gamma(s)ds, \tag{1.21}$$

where $|\arg(-z)| < \pi$ and the path of integration starts at the point $\gamma - i\infty$ ($\gamma \in R$) and terminates at the point $\gamma + i\infty$, separating all the poles $s = -k$ ($k \in \aleph_0$) to the left and all poles $s = a + n$ ($n \in \aleph_0$) and $s = b + m$ ($m \in \aleph_0$) to the right.

Some properties of the Gauss hypergeometric function are given as (Kilbas et al. 2006):

i) $_2F_1(a,b;c;z) = {}_2F_1(b,a;c;z)$,

ii) $_2F_1(a,b;c;0) = {}_2F_1(0,b;c;z)$,

iii) $_2F_1(a,b;b;z) = (1-z)^{-a}$,

iv) $_2F_1(a,b;c;1) = \dfrac{\Gamma(c)\Gamma(c-a-b)}{\Gamma(c-a)\Gamma(c-b)}, \quad \Re(c-a-b) > 0,$

v) (Euler transformation formula) $_2F_1(a,b;c;z) = (1-z)^{c-a-b}\,{}_2F_1(c-a,c-b;c;z)$,

vi) $\left(\dfrac{d}{dz}\right)^n {}_2F_1(a,b;c;z) = \dfrac{(a)_n(b)_n}{(c)_n}\,{}_2F_1(a+n,b+n;c+n;z), \quad n \in \aleph,$

vii) $\left(\dfrac{d}{dz}\right)^n \left[z^{a+n-1}\,{}_2F_1(a,b;c;z)\right] = (a)_n z^{a-1}\,{}_2F_1(a+n,b;c;z), \quad n \in \aleph.$

Definition 1.13   (The Kummer Hypergeometric Function)

The Kummer hypergeometric function is defined as (Kilbas et al. 2006):

$$\Phi(a;c;z) = {}_1F_1(a;c;z) = \sum_{k=0}^{\infty}\frac{(a)_k}{(c)_k}\frac{z^k}{k!}, \tag{1.22}$$

where $z, a \in C$ and $c \in C\backslash Z_0^-$.

Note 1.5   The Euler integral representation of Kummer hypergeometric function is given as:

$$\Phi(a;c;z) = \frac{\Gamma(c)}{\Gamma(a)\Gamma(c-a)}\int_0^1 t^{a-1}(1-t)^{c-a-1}e^{zt}dt, \quad 0 < \Re(a) < \Re(c). \tag{1.23}$$

Note 1.6   The function $\Phi(a;c;z)$ can be represented in terms of the Mellin–Barnes contour integral as follows:

$$\Phi(a;c;z) = \frac{1}{2\pi i}\frac{\Gamma(c)}{\Gamma(a)}\int_{\gamma-i\infty}^{\gamma+i\infty}\frac{\Gamma(a-s)}{\Gamma(c-s)}(-z)^{-s}\Gamma(s)ds, \tag{1.24}$$

where $|\arg(-z)| < \pi$ and the path of integration separates all the poles $s = -k$ ($k \in \aleph_0$) to the left and all poles $s = a + n$ ($n \in \aleph_0$) to the right.

Definition 1.14   (The Generalized Hypergeometric Function)

The Gauss hypergeometric series Eq. (1.19) and the Kummer hypergeometric series Eq. (1.22) are extended to the generalized hypergeometric function, which is defined as (Kilbas et al. 2006):

$$_pF_q\left(a_1, ..., a_p; b_1, ..., b_q; z\right) = \sum_{k=0}^{\infty} \frac{(a_1)_k ... (a_p)_k}{(b_1)_k ... (b_q)_k} \frac{z^k}{k!}, \tag{1.25}$$

where $a_i, b_j \in C$ and $b_j \neq 0, \ -1, \ -2, ...(i = 1, ..., p; j = 1, ..., q)$.

Note 1.7   If $b_j \notin Z_0^- \{j = 1, ..., q\}$, then the generalized hypergeometric function has another integral representation in terms of the Mellin–Barnes contour integral as presented in Eqs. (1.21) and (1.24), which is given as:

$$_pF_q\left(a_1, ..., a_p; b_1, ..., b_q; z\right) = \frac{1}{2\pi i} \frac{\prod_{j=1}^{q} \Gamma(b_j)}{\prod_{i=1}^{p} \Gamma(a_i)} \int_{\gamma-i\infty}^{\gamma+i\infty} \frac{\prod_{i=1}^{p} \Gamma(a_i - s)}{\prod_{j=1}^{q} \Gamma(b_j - s)} (-z)^{-s} \Gamma(s) ds, \tag{1.26}$$

where $|\arg(-z)| < \pi$ and the path of integration separates all the poles $s = -k \, (k \in \aleph_0)$ to the left and all poles $s = a_j + n \, (n \in \aleph_0; j = 1, 2, ..., p)$ to the right.

Note 1.8   The property of a generalized hypergeometric function is given as (Kilbas et al. 2006):

$$\left(\frac{d}{dz}\right)^n {_pF_q}\left(a_1, ..., a_p; b_1, ..., b_q; z\right) = \frac{(a_1)_n ... (a_p)_n}{(b_1)_n ... (b_q)_n} {_pF_q}\left(a_1 + n, ..., a_p + n; b_1 + n, ..., b_q + n; z\right), \quad n \in \aleph$$

1.3.8   The *H*-Function

In this section, we have presented the definitions and properties of *H*-function.

Definition 1.15   (The *H*-Function)

For integers m, n, p, q such that $0 \leq m \leq q$ and $0 \leq n \leq p$, for $a_i, b_j \in C$ and for $\alpha_i, \beta_j \in R^+ (i = 1, \ 2, ..., p, j = 1, \ 2, ..., q)$, the *H*-function is defined as (Fox 1962; Kilbas et al. 2006):

$$
\begin{aligned}
H_{p,q}^{m,n}(z) &= H_{p,q}^{m,n}\left[z \, \middle| \, \begin{matrix} (a_i, \alpha_i)_{1,p} \\ (b_j, \beta_j)_{1,q} \end{matrix}\right] \\
&= H_{p,q}^{m,n}\left[z \, \middle| \, \begin{matrix} (a_1, \alpha_1), ..., (a_p, \alpha_p) \\ (b_1, \beta_1), ..., (b_q, \beta_q) \end{matrix}\right] \\
&= \frac{1}{2\pi i} \int_C H_{p,q}^{m,n}(s) z^{-s} ds,
\end{aligned}
\tag{1.27}
$$

where

$$H_{p,q}^{m,n} = \frac{\prod_{j=1}^{m} \Gamma\left(b_j + \beta_j s\right) \prod_{i=1}^{n} \Gamma(1 - a_i - \alpha_i s)}{\prod_{i=n+1}^{r} \Gamma(a_i + \alpha_i s) \prod_{j=m+1}^{m} \Gamma\left(1 - b_j - \beta_j s\right)}. \tag{1.28}$$

Note 1.9   The relationship between the *H*-function and the generalized MLF may be described as (Mathai and Haubold 2008):

$$E_{\alpha,\beta}^{\gamma}(z) = \frac{1}{\Gamma(\gamma)} H_{1,2}^{1,1}\left[-z \, \middle| \, \begin{matrix} (1-\gamma, \ 1) \\ (0, \ 1), \ (1-\beta, \ \alpha) \end{matrix}\right].$$

There are various ways of defining fractional derivatives. As such, preliminaries of some definitions are incorporated in the following section for the sake of completeness.

1.4 Riemann–Liouville (R-L) Fractional Integral and Derivative

Different definitions have been developed by various researchers for the fractional (non-integer) order integral or derivative. The Riemann–Liouville definition is one of the most well-known forms of fractional calculus. Riemann was the first to introduce the Riemann–Liouville definition, derived from Abel's integral.

Definition 1.16   A real function $\psi(t)$, $t > 0$ is said to be in space C_γ, $\gamma \in R$ if there exists a real number $p \, (>\gamma)$, such that $\psi(t) = t^p \psi_1(t)$, where $\psi_1(t) \in C[0, \infty]$, and it is said to be in the space C_γ^m if $\psi^{(m)} \in C_m$, $m \in \aleph$.

Definition 1.17   (Riemann–Liouville Fractional Integral)
The Riemann–Liouville fractional integral operator J_t^α of an order α of a function $\psi \in C_\gamma$, $\gamma \geq -1$ is defined as (Podlubny 1999; Samko et al. 2002):

$$J_t^\alpha \psi(t) = \frac{1}{\Gamma(\alpha)} \int_0^t (t - \xi)^{\alpha - 1} \psi(\xi) d\xi, \quad t > 0 \text{ and } \alpha \in R^+ . \tag{1.29}$$

Proof
Let us take the first integral of a function $\psi(t)$ as

$$J_t^1 \psi(t) = \int_0^t \psi(u) du, \quad t > 0. \tag{1.30}$$

Integrating Eq. (1.30) once more, we have

$$J_t^2 \psi(t) = \int_0^t \int_0^u \psi(v) dv du, \quad t > 0. \tag{1.31}$$

Successive integration of $\psi(t)$ for n-times (n, integer) is

$$J_t^n \psi(t) = \int_0^t \int_0^u \cdots \int_0^w \psi(v) dv dw \ldots du, \quad t > 0. \tag{1.32}$$

The closed-form formula of Eq. (1.32), which is given by Cauchy, is as follows (https://en.wikipedia.org/wiki/Cauchy_formula_for_repeated_integration n.d.):

$$J_t^n \psi(t) = \frac{1}{(n-1)!} \int_0^t (t - u)^{n-1} \psi(u) du, \quad t > 0. \tag{1.33}$$

By replacing n by α and factorial by gamma function, one may have the desired result Eq. (1.29).

Definition 1.18   (Riemann–Liouville Left-Hand Side and Right-Hand Side Integrals)
The left- and right-hand sides Riemann–Liouville fractional integral of a function $\psi \in C_\gamma$, $\gamma \geq -1$ are defined, respectively, as (Chakraverty et al. 2020; Miller and Ross 1993; Podlubny 1999; Samko et al. 2002):

$$_{-\infty} J_t^\alpha \psi(t) = \frac{1}{\Gamma(m - \alpha)} \int_{-\infty}^t (t - \xi)^{m - \alpha - 1} \psi(\xi) d\xi, \quad m - 1 < \alpha < m, m \in \aleph, \tag{1.34}$$

and

$$_tJ_\infty^\alpha \psi(t) = \frac{(-1)^m}{\Gamma(m-\alpha)} \int\limits_t^\infty (\xi-t)^{m-\alpha-1}\psi(\xi)d\xi, \quad m-1 < \alpha < m, \ m \in \aleph. \tag{1.35}$$

Proof

Hints: Substituting $m - \alpha$ in place of α in Eq. (1.29) and taking integration over $-\infty$ to t, we get left-hand side R-L integral. Similarly, the right-hand side R-L integral is obtained using the same technique, but the $(-1)^m$ term arises due to a change of order of integration $\left(\text{i.e. } \int_a^b f(x)dx = -\int_b^a f(x)dx \right)$. For this concept, one may see the detailed derivations of the aforementioned definitions in any standard fractional calculus book.

From Podlubny (1999), we have the following result:

$$J_t^\alpha t^n = \frac{\Gamma(n+1)}{\Gamma(n+\alpha+1)} t^{n+\alpha}, \quad n > -1, \alpha > -1-n. \tag{1.36}$$

Proof

We know that mth time integration of t^n is defined as:

$$J_t^m t^n = \frac{\Gamma(n+1)}{\Gamma(n+m+1)} t^{n+m}, \quad n > -1. \tag{1.37}$$

In place of $m - \alpha$, if we substitute α in Eq. (1.37), then we obtain α times integration of t^n, which is defined as in Eq. (1.36).

Example 1.1 Suppose we want to calculate 1/2-times integration of $\psi(t) = t$, then we have

$$J_t^{0.5} t = \frac{\Gamma(1+1)}{(1+0.5+1)} t^{1+0.5} = \frac{1}{\Gamma(2.5)} t^{1.5} = \frac{3}{4}\sqrt{\pi} t^{1.5}. \tag{1.38}$$

Definition 1.19 **(The Riemann–Liouville Fractional Derivative)**

The fractional-order R-L derivative of order α is defined as (Kilbas et al. 2006; Podlubny 1999; Samko et al. 2002):

$$D_t^\alpha \psi(t) = \begin{cases} \dfrac{1}{\Gamma(m-\alpha)} \dfrac{d^m}{dt^m} \displaystyle\int_0^t (t-\xi)^{m-\alpha-1}\psi(\xi)d\xi, & m-1 < \alpha < m, m \in \aleph, \\[4mm] \dfrac{d^m}{dt^m}\psi(t), & \alpha = m, m \in \aleph. \end{cases} \tag{1.39}$$

Proof

Hints: we know that

$$\begin{aligned} \frac{d^\alpha \psi(t)}{dt^\alpha} &= D_t^\alpha \psi(t) = \left(D_t^m D_t^{\alpha-m}\right)\psi(t) \\ D_t^\alpha \psi(t) &= \left(D_t^m D_t^{-(m-\alpha)}\right)\psi(t) = \left(D_t^m J_t^{(m-\alpha)}\right)\psi(t) \end{aligned} \tag{1.40}$$

Now, using Eq. (1.29), Eq. (1.40) yields

$$\begin{aligned} D_t^\alpha \psi(t) &= D_t^m \frac{1}{\Gamma(m-\alpha)} \int_0^t (t-\xi)^{m-\alpha-1}\psi(\xi)d\xi \\[4mm] D_t^\alpha \psi(t) &= \frac{1}{\Gamma(m-\alpha)} \frac{d^m}{dt^m} \int_0^t (t-\xi)^{m-\alpha-1}\psi(\xi)d\xi \end{aligned} \tag{1.41}$$

Definition 1.20 (Riemann–Liouville Left-Hand and Right-Hand Sides Derivatives)
The left-hand and right-hand sides R-L fractional derivative of order α can be defined, respectively, as follows (Podlubny 1999; Samko et al. 2002):

$$_{-\infty}D_t^\alpha \psi(t) = \frac{1}{\Gamma(m-\alpha)} \frac{d^m}{dt^m} \int_{-\infty}^{t} (t-\xi)^{m-\alpha-1} \psi(\xi)d\xi, \quad m-1 < \alpha < m, m \in \aleph, \tag{1.42}$$

and

$$_tD_\infty^\alpha \psi(t) = \frac{(-1)^m}{\Gamma(m-\alpha)} \frac{d^m}{dt^m} \int_{t}^{\infty} (\xi-t)^{m-\alpha-1} \psi(\xi)d\xi, \quad m-1 < \alpha < m, m \in \aleph. \tag{1.43}$$

Proof
The proof is very similar to the proof of Definition 1.18. The only difference is that in Definition 1.18, the integral operator was there, but the differential operator is present here.

Note 1.10   It may be noted that the derivative of a constant (c) is not zero in the Riemann–Liouville sense, and mathematically it is written as:

$$D_t^\alpha c = \frac{ct^{-\alpha}}{\Gamma(1-\alpha)}, \quad 0 < \alpha < 1. \tag{1.44}$$

Proof
If $0 < \alpha < 1$ and $\psi(t) = c$ then Eq. (1.39) reduces to

$$D_t^\alpha c = \frac{1}{\Gamma(1-\alpha)} \frac{d}{dt} \int_0^t (t-\xi)^{1-\alpha-1} c \ d\xi. \tag{1.45}$$

Here, we have taken $m = 1$ since from Eq. (1.39) $m-1 < \alpha < m$. If we compare $m-1 < \alpha < m$ with $0 < \alpha < 1$, then we obtain $m = 1$. Accordingly, we may get Eq. (1.45). Now, simplifying Eq. (1.45), we have

$$D_t^\alpha c = \frac{c}{\Gamma(1-\alpha)} \frac{d}{dt} \int_0^t (t-\xi)^{-\alpha} d\xi = \frac{c}{\Gamma(1-\alpha)} \frac{d}{dt} \left[\frac{-\left((t-\xi)^{-\alpha+1}\right)}{(-\alpha+1)} \right]_0^t = \frac{c}{\Gamma(1-\alpha)} \frac{d}{dt} \left[\frac{t^{-\alpha+1}}{-\alpha+1} \right].$$

$$D_t^\alpha c = \frac{c}{\Gamma(1-\alpha)} (-\alpha+1) \left(\frac{t^{-\alpha}}{-\alpha+1} \right) = \frac{ct^{-\alpha}}{\Gamma(1-\alpha)}.$$

1.5 Caputo Fractional Derivative

The derivative of a constant is not zero in the Riemann–Liouville sense. As a result, the Riemann–Liouville derivative has fewer physical representations. Hence, the Caputo fractional derivative has been utilized in certain circumstances.

Definition 1.21 (Caputo Fractional Derivative)
The fractional-order Caputo derivative of an order α of a function $\psi(t)$ is defined as (Podlubny 1999; Samko et al. 2002):

$$
\begin{aligned}
D_t^\alpha \psi(t) &= D_t^{\alpha-m} D_t^m \psi(t) = J_t^{m-\alpha} D_t^m \psi(t) \\
&= \begin{cases} \dfrac{1}{\Gamma(m-\alpha)} \displaystyle\int_0^t (t-\xi)^{m-\alpha-1} \dfrac{d^m \psi(\xi)}{d\xi^m} d\xi, & m-1 < \alpha < m, m \in \aleph, \\[2ex] \dfrac{d^m}{dt^m} \psi(t), & \alpha = m, m \in \aleph. \end{cases}
\end{aligned}
\tag{1.46}
$$

Proof

Hints: The proof is similar to the proof of Definition 1.19, but the only difference is that in R-L fractional derivative, integer-order derivative lies outside the integration, but in the Caputo sense, integer-order derivative lies inside the integration.

Note 1.11 First, Lacroix introduced the integer-order derivatives of a function which is as follows:

$$D_t^n t^m = \frac{m!}{(m-n)!} t^{m-n} = \frac{\Gamma(m+1)}{\Gamma(m-n+1)} t^{m-n}, \quad \text{for } n \le m. \tag{1.47}$$

Later, he extended this integer-order derivative to fractional-order derivative in the Caputo sense as:

$$D_t^\alpha t^\beta = \frac{\Gamma(\beta+1)}{\Gamma(\beta-\alpha+1)} t^{\beta-\alpha}, \quad \text{for } \beta > \alpha - 1, \beta > -1. \tag{1.48}$$

Also, the derivative of a constant in the Caputo sense is zero. Mathematically,

$$D_t^\alpha c = 0.$$

Hints

In the Caputo derivative, integer-order derivative lies inside the integration, which yields the derivative of a constant as zero that is,

$$D_t^\alpha c = \frac{1}{\Gamma(m-\alpha)} \int_0^t (t-\xi)^{m-\alpha-1} \frac{d^m c}{d\xi^m} d\xi = \frac{1}{\Gamma(m-\alpha)} \int_0^t (t-\xi)^{m-\alpha-1} 0 \, d\xi = 0.$$

Example 1.2

$$\frac{d^{0.5} t}{dt^{0.5}} = \frac{\Gamma(1+1)}{\Gamma(1-0.5+1)} t^{1-0.5} = \frac{1}{\Gamma\left(\dfrac{3}{2}\right)} t^{0.5} = \frac{2}{\sqrt{\pi}} t^{1/2}.$$

Note 1.12 (i) The main advantages of Caputo fractional derivative are that the initial conditions for the fractional differential equations are the same form as that of ordinary differential equations. Another advantage is that the Caputo fractional derivative of a constant is zero, while the R-L fractional derivative of a constant is not zero.

(ii) Several properties in classical derivatives and integrations accept constant roles. However, these properties may not always hold in the fractional sense. For example, (a)$D_t^\alpha = D_t^{\alpha-m} D_t^m = J_t^{m-\alpha} D_t^m$ and (b) $D_t^\alpha = D_t^m D_t^{\alpha-m} = D_t^m J_t^{m-\alpha}$ for $m-1 < \alpha < m$ both look equal, but mathematically it is different.

Suppose

$$D_t^{7/5} t = D_t^{(7/5)-2} D_t^2 t = J_t^{2-(7/5)} D_t^2 t = J_t^{3/5} D_t^2 t = J_t^{3/5} 0 = 0, \text{(using (a))} \tag{1.49}$$

but

$$D_t^{7/5} t = D_t^2 D_t^{(7/5)-2} t = D_t^2 J_t^{2-(7/5)} t = D_t^2 J_t^{3/5} t, \text{(using (b))}$$

that is,

$$D_t^{7/5} t = D_t^2 J_t^{3/5} t. \tag{1.50}$$

Using Eq. (1.36), Eq. (1.50) reduces to

$$D_t^{7/5} t = D_t^2 \frac{\Gamma(1+1)}{\Gamma(1+(3/5)+1)} t^{8/5} = \frac{1}{\Gamma(13/5)} D_t^2 t^{8/5} = \frac{24/25}{\Gamma(13/5)} t^{-2/5}. \tag{1.51}$$

On simplifying Eq. (1.51), we obtain

$$D_t^{7/5} t = \frac{1}{\Gamma(3/5)} t^{-2/5} \ne 0.$$

So, $J_t^{m-\alpha} D_t^m$ is not always equal to $D_t^m J_t^{m-\alpha}$ in a fractional sense.

In the aforementioned Note 1.12 (ii), $(m-1 =) 1 < \frac{7}{5} < 2 (= m)$, that is, $m = 2$.

In the following section, we incorporate a theorem with respect to the interchange of derivatives and integration behavior.

Theorem 1.1 Kilbas et al. (2006); Podlubny (1999); Samko et al. (2002)

i) ${}_aD_t^\alpha\left[{}_aJ_t^\beta\psi(t)\right] = {}_aD_t^{\alpha-\beta}\psi(t).$

ii) ${}_aJ_t^\alpha\left[{}_aD_t^\beta\psi(t)\right] = {}_aJ_t^{\alpha-\beta}\psi(t) - \sum_{k=1}^{m}\frac{(t-a)^{\alpha-k}}{\Gamma(\alpha+1-k)}\,{}_aD_t^{\beta-k}\psi(t)|_{t=a}.$

Here $m = \lceil\beta\rceil + 1$, where $\lceil\beta\rceil$ represents the Ceiling function which means the least integer greater than or equal to β.

Proof

(i) Hints:

$${}_aD_t^\alpha\left[{}_aJ_t^\beta\psi(t)\right] = \frac{d^n}{dt^n}\left[{}_aJ_t^{n-\alpha}\left[{}_aJ_t^\beta\psi(t)\right]\right] = \frac{d^n}{dt^n}\left[{}_aJ_t^{n-(\alpha-\beta)}\right] = {}_aD_t^{\alpha-\beta}\psi(t).$$

(ii) Hints:

$$I = {}_aJ_t^\alpha\left[{}_aD_t^\beta\psi(t)\right] = \frac{1}{\Gamma(\alpha)}\int_a^t (t-\xi)^{\alpha-1}\left[{}_aD_t^\beta\psi(\xi)\right]d\xi.$$

$$I = \frac{1}{\Gamma(\alpha+1)}\int_a^t \frac{d}{dt}(t-\xi)^\alpha\left[{}_aD_t^\beta\psi(\xi)\right]d\xi.$$

Applying integration by parts, we have

$$I = -\frac{(t-a)^\alpha}{\Gamma(\alpha+1)}\left[{}_aD_t^\beta\psi(\xi)\Big|_{t=a}\right] + \frac{1}{\Gamma(\alpha+2)}\int_a^t \frac{d}{dt}(t-\xi)^{\alpha+1}\left[{}_aD_t^{\beta-1}\psi(\xi)\right]d\xi.$$

Again, successively applying integration by part, we may get the desired result.

1.6 Grünwald–Letnikov Fractional Derivative and Integral

Anton Karl Grünwald (1838–1920) and Aleksey Vasilievich Letnikov (1837–1888) proposed the Grünwald–Letnikov fractional derivative in 1867 and 1868, respectively. Finite differences are used to define the Grünwald–Letnikov fractional derivative, which is equivalent to the Riemann–Liouville definition.

Definition 1.22 (Grünwald–Letnikov Fractional Derivative)
The differential operator D_t^α of order α in the Grünwald–Letnikov sense is defined as (Podlubny 1999; Samko et al. 2002):

$${}_aD_t^\alpha\psi(t) = \lim_{h\to 0}\frac{1}{h^\alpha}\sum_{r=0}^{[(t-a)/h]}(-1)^r\frac{\Gamma(\alpha+1)}{r!\Gamma(\alpha-r+1)}\psi(t-rh). \tag{1.52}$$

Proof
Let $\psi(x) \in [a, b]$. Then the first-order derivative of the function $\psi(x)$ is defined as:

$$\psi^{(1)}(t) = \frac{d\psi}{dt} = \lim_{h\to 0}\frac{\psi(t)-\psi(t-h)}{h}.$$

Again applying a derivative operator on the aforementioned equation, we obtain

$$\psi^2(t) = \frac{d^2\psi}{dt^2} = \lim_{h\to 0}\frac{\psi(t)-2\psi(t-h)+\psi(t-2h)}{h^2}.$$

With the help of the method of induction, we may write n^{th}-order derivative as:

$$\psi^n(t) = \frac{d^n \psi}{dt^n} = \lim_{h \to 0} \frac{1}{h^n} \sum_{r=0}^{n} (-1)^r \binom{n}{r} \psi(t-rh). \tag{1.53}$$

where $a \le t \le b, h = \dfrac{t-a}{n}$, and $\dbinom{n}{r} = \dfrac{n(n-1)(n-2)(n-r+1)}{r!}$.

Now, we extend the nth-order derivative to fractional-order derivate as follows:

$$\psi^\alpha(t) = \frac{d^\alpha \psi}{dt^\alpha} = \lim_{h \to 0} \frac{1}{h^\alpha} \sum_{r=0}^{\alpha} (-1)^r \binom{\alpha}{r} \psi(t-rh), \tag{1.54}$$

where $\dbinom{\alpha}{r} = \dfrac{\alpha!}{r!(\alpha-r)!} = \dfrac{\Gamma(1+\alpha)}{r!\Gamma(\alpha-r+1)}$ and $h = \dfrac{t-a}{n} \Rightarrow n = \dfrac{t-a}{h}$. Since n is replaced by α and α is a non-integer. So, we can write $\alpha = \left[\dfrac{t-a}{h}\right]$.

Definition 1.23 (Grünwald–Letnikov Fractional Integral)
The fractional Grünwald–Letnikov integral operator is defined as (Podlubny 1999; Samko et al. 2002):

$$_aD_t^{-\alpha}\psi(t) = \lim_{h \to 0} h^\alpha \sum_{r=0}^{[(t-a)/h]} \frac{\Gamma(\alpha+r)}{r!\Gamma(\alpha)} \psi(t-rh). \tag{1.55}$$

Proof
Hints: Replacing α by $-\alpha$ of the Eq. (1.52), we get

$$_aD_t^{-\alpha}\psi(t) = \lim_{h \to 0} h^\alpha \sum_{r=0}^{[(t-a)/h]} (-1)^r \frac{\Gamma(-\alpha+1)}{r!\Gamma(-\alpha-r+1)} \psi(t-rh). \tag{1.56}$$

We know that

$$\begin{aligned}
\frac{\Gamma(-\alpha+1)}{r!\Gamma(-\alpha-r+1)} &= \binom{-\alpha}{r} \\
&= \frac{-\alpha(-\alpha-1)(-\alpha-2)...(-\alpha-r+1)}{r!} \\
&= \frac{(-1)^r \alpha(\alpha+1)(\alpha+2)...(\alpha+r-1)(\alpha-1)!}{r!(\alpha-1)!} \\
&= \frac{(-1)^r \Gamma(\alpha+r)}{r!\Gamma(\alpha)}
\end{aligned} \tag{1.57}$$

Substituting Eq. (1.57) into Eq. (1.56), we get Eq. (1.55).

1.7 Riesz Fractional Derivative and Integral

In this section, the definition of Riesz fractional integral and derivative has been presented.

Definition 1.24 (Riesz Fractional Integral)
The Riesz fractional integral of a function $\psi(t) \in C_\gamma, (\gamma \ge 1)$ of order α is defined as (Herrmann 2011; Podlubny 1999; Samko et al. 2002):

$$_0^R J_t^\alpha \psi(t) = c_\alpha \left(_{-\infty}J_t^\alpha + {_tJ_\infty^\alpha} \right) \psi(t) = \frac{c_\alpha}{\Gamma(\alpha)} \int_{-\infty}^{\infty} |t-\xi|^{\alpha-1} \psi(\xi) d\xi, \tag{1.58}$$

where $c_\alpha = \frac{1}{2\cos(\alpha\pi/2)}, \alpha \ne 1$ and $m-1 < \alpha \le m$,

where $_{-\infty}J_t^\alpha$ and $_tJ_\infty^\alpha$ are the left-hand and right-hand sides, respectively, of Riemann–Liouville fractional integral operators defined in Definition 1.18.

Definition 1.25 (Riesz Fractional Derivative)

The Riesz fractional differentiation of a function $\psi(t) \in C_\gamma$, $(\gamma \geq 1)$ of order α on the infinite domain $-\infty < t < \infty$ is defined as (Herrmann 2011; Podlubny 1999; Samko et al. 2002):

$$\frac{d^\alpha \psi(t)}{d|t|^\alpha} = -c_\alpha \left(_{-\infty}D_t^\alpha + {_t}D_\infty^\alpha \right) \psi(t), \tag{1.59}$$

where $c_\alpha = \dfrac{1}{2\cos(\alpha\pi/2)}$, $\alpha \neq 1$ and $m - 1 < \alpha \leq m$, where $_{-\infty}D_t^\alpha$ and $_tD_\infty^\alpha$ are the left-hand and right-hand sides, respectively, Riemann–Liouville differential operator defined in Definition 1.20.

In case of $a \leq t \leq b$ (i.e. t is defined in the finite interval), the Riesz fractional derivative of order α may be written as (Herrmann 2011; Podlubny 1999; Samko et al. 2002):

$$\frac{d^\alpha \psi(t)}{d|t|^\alpha} = -\frac{1}{2\cos(\alpha\pi/2)} \left(_aD_t^\alpha + {_t}D_b^\alpha \right) \psi(t), \quad m - 1 < \alpha \leq m \text{ and } \alpha \neq 1, \tag{1.60}$$

where

$$_aD_t^\alpha \psi(t) = \frac{1}{\Gamma(m-\alpha)} \frac{d^m}{dt^m} \int_a^t (t-\xi)^{m-\alpha-1} \psi(\xi) \; d\xi,$$

$$_tD_b^\alpha \psi(t) = \frac{(-1)^n}{\Gamma(m-\alpha)} \frac{d^m}{dt^m} \int_t^b (t-\xi)^{m-\alpha-1} \psi(\xi) \; d\xi.$$

Note 1.13 Let $\alpha > 0$ and $\beta > 0$ be such that $m - 1 < \alpha, \beta \leq m$, and $\alpha + \beta \leq m$, then we have the following index rule:

$$_0^RD_t^\alpha \left(_0^RD_t^\beta \psi(t) \right) = {_0^R}D_t^{\alpha+\beta}\psi(t).$$

Note 1.14 The Riesz fractional operator $_0^RD_t^{\alpha-1}\psi(t)$ of the order $0 < \alpha < 1$ can be expressed as Riesz fractional integral operator $_0^RJ_t^{1-\alpha}\psi(t)$ by the following identity (Herrmann 2011; Podlubny 1999; Samko et al. 2002):

$$_0^RD_t^{\alpha-1}\psi(t) = {_0^R}J_t^{1-\alpha}\psi(t), t \in T.$$

1.8 Modified Riemann–Liouville Derivative

This section presents another form of fractional derivative, namely the modified Riemann–Liouville derivative (Jumarie 2005, 2006) of order α, defined by the formula:

$$\psi^{(\alpha)}(t) = \lim_{h \to 0} \left(\frac{\sum_{k=0}^\infty (-1)^k \binom{\alpha}{k} \psi[t + (\alpha-k)h]}{h^\alpha} \right), \quad \alpha \in R, 0 < \alpha \leq 1. \tag{1.61}$$

Equation (1.61) can be written as:

$$D_t^\alpha \psi(t) = \begin{cases} \dfrac{1}{\Gamma(1-\alpha)} \dfrac{d}{dt} \displaystyle\int_0^t (t-\xi)^{-\alpha}[\psi(\xi) - \psi(0)]d\xi, & \text{if } 0 < \alpha \leq 1, \\[2ex] \left(\psi^{(\alpha-m)}(t) \right)^m, & \text{if } m \leq \alpha \leq m+1, m \geq 1. \end{cases} \tag{1.62}$$

1.9 Local Fractional Derivative

In this section, the theory and definitions of local fractional derivatives have been discussed.

1.9.1 Local Fractional Continuity of a Function

Definition 1.26 Let $\psi(x)$ is defined throughout some interval containing x_0 and all points near x_0, then $\psi(x)$ is said to be local fractional continuous at $x = x_0$, denoted by $\lim_{x \to x_0} \psi(x) = \psi(x_0)$, if to each positive ε and some positive constant k corresponds some positive δ such that (Yang 2012a, 2012b)

$$|\psi(x) - \psi(x_0)| < k\varepsilon^\alpha, 0 < \alpha \leq 1, \tag{1.63}$$

whenever $|x - x_0| < \delta$, $\varepsilon, \delta > 0$ and $\varepsilon, \delta \in R$. The function $\psi(x)$ is said to be local fractional continuous on the interval (a, b), represented by:

$$\psi(x) \in C_\alpha(a, \ b), \tag{1.64}$$

where α is the fractal dimension with $0 < \alpha \leq 1$.

Definition 1.27 A function $\psi(x) : R \to R, X \mapsto \psi(X)$ is called a non-differentiable function of exponent $\alpha, 0 < \alpha \leq 1$, which satisfies the Hölder function of exponent α, then for $x, y \in X$, we have (Yang 2012a, 2012b)

$$|\psi(x) - \psi(y)| < C|x - y|^\alpha. \tag{1.65}$$

Definition 1.28 A function $\psi(x) : R \to R, X \mapsto \psi(X)$ is said to be local fractional continuous of order $\alpha, 0 < \alpha \leq 1$, if we have (Yang 2012a, 2012b)

$$|\psi(x) - \psi(x_0)| < O((x - x_0)^\alpha). \tag{1.66}$$

Note 1.15 A function $\psi(x)$ is said to be in the space $C_\alpha(a, b)$ if and only if $|\psi(x) - \psi(x_0)| < O((x - x_0)^\alpha)$, with any $x_0 \in [a, b]$ and $0 < \alpha \leq 1$.

Theorem 1.2 **(Generalized Hadamard's Theorem)(Jumarie 2009)**
A function $\psi(x) \in C^\alpha(I)$ in a neighborhood of a point x_0 can be decomposed in the form:

$$\psi(x) = \psi(x_0) + \frac{(x - x_0)^\alpha}{\Gamma(1 + \alpha)} r(x),$$

where $r(x) \in C^{m\alpha}(I)$ (where m times αth differentiable on $I \subset R$).

1.9.2 Local Fractional Derivative

At $x = x_0$, if a function is not differentiable but has a fractional derivative of order α, then it is locally equivalent to the function:

$$\psi(x) = \psi(x_0) + \frac{(x - x_0)^\alpha}{\Gamma(1 + \alpha)} \psi^{(\alpha)}(x_0) + O((x - x_0)^{2\alpha}). \tag{1.67}$$

Definition 1.29 Given the aforementioned Eq. (1.67), the local fractional derivative of $\psi(x) \in C_\alpha[a, b]$ of order α at $x = x_0$ is defined as (Yang 2012a, 2012b):

$$\psi^{(\alpha)}(x_0) = \frac{d^\alpha \psi(x)}{dx^\alpha}\bigg|_{x = x_0} = \lim_{x \to x_0} \frac{\Delta^\alpha(\psi(x) - \psi(x_0))}{(x - x_0)^\alpha}, \tag{1.68}$$

where $\Delta^\alpha(\psi(x) - \psi(x_0)) \cong \Gamma(1 + \alpha)(\psi(x) - \psi(x_0))$ and $0 < \alpha \leq 1$.

Kolwankar and Gangal (1996) presented a different concept of local fractional derivative based on Cantor space theory, which is as follows.

Definition 1.30 The local fractional derivative of an order α $(0 < \alpha < 1)$ of a function $\psi \in C^0 : R \to R$ is defined as:

$$D^\alpha \psi(x) = \lim_{\xi \to x} D_x^\alpha (\psi(\xi) - \psi(x)), \tag{1.69}$$

if the limit exists in $R \cup \infty$.

If $\psi(x)$ is differentiable at the point other than $x = x_0$ with the nonzero value of the derivative, then it can be approximated locally as:

$$\psi(x) = \psi(x_0) + \psi'(x_0)(x - x_0) + O(x - x_0). \tag{1.70}$$

So, the local fractional derivative of $\psi(x)$ at $x = x_0$ becomes

$$
\begin{aligned}
D^\alpha \psi(x_0) &= \lim_{x \to x_0} \frac{d^\alpha(\psi(x) - \psi(x_0))}{d(x - x_0)^\alpha} \\
&= \psi'(x_0) \lim_{x \to x_0} \frac{d^\alpha(x - x_0)}{d(x - x_0)^\alpha}.
\end{aligned}
\tag{1.71}
$$

Note 1.16 Some properties of the local fractional derivative are given as follows (Hu et al., 2013):

i) $\dfrac{d^\alpha x^{k\alpha}}{dx^\alpha} = \dfrac{\Gamma(1 + k\alpha)}{\Gamma(1 + (k - 1)\alpha)} x^{(k-1)\alpha},$

ii) $\dfrac{d^\alpha E_\alpha(kx^\alpha)}{dx^\alpha} = kE_\alpha(kx^\alpha), k$ is a constant.

Note 1.17 (Hu et al. 2013)

(i) If $y(x) = (f \circ u)(x)$ where $u(x) = g(x)$, then we have

$$\frac{d^\alpha y(x)}{dx^\alpha} = f^{(\alpha)}(g(x)) \left(g^{(1)}(x) \right)^\alpha, \tag{1.72}$$

when $f^{(\alpha)}(g(x))$ and $g^{(1)}(x)$ exist.

(ii) If $y(x) = (f \circ u)(x)$ where $u(x) = g(x)$, then we have

$$\frac{d^\alpha y(x)}{dx^\alpha} = f^{(1)}(g(x)) g^{(\alpha)}(x), \tag{1.73}$$

when $f^{(1)}(g(x))$ and $g^{(\alpha)}(x)$ exist.

References

Baleanu, D., Diethelm, K., Scalas, E., and Trujillo, J.J. (2012). *Fractional Calculus: Models and Numerical Methods*. World Scientific Publishing Company.

Chakraverty, S., Jena, R.M., and Jena, S.K. (2020). *Time-Fractional Order Biological Systems with Uncertain Parameters. Synthesis Lectures on Mathematics and Statistics*, vol. 12(1), 1–160. Morgan & Claypool Publishers.

Das, S. (2011). *Functional Fractional Calculus*. Springer Science & Business Media.

Fox, C. (1962). The G and H functions as symmetrical fourier kernels. *Transactions of the American Mathematical Society* 98 (3): 395–429.

Haubold, H.J., Mathai, A.M., and Saxena, R.K. (2011). Mittag-Leffler functions and their applications. *Journal of Applied Mathematics* 2011 (298628): 1–51.

Herrmann, R. (2011). *Fractional Calculus: An Introduction for Physicists*. Singapore: World Scientific.

https://en.wikipedia.org/wiki/Cauchy_formula_for_repeated_integration.

Hu, M.S., Baleanu, D., and Yang, X.J. (2013). One-phase problems for discontinuous heat transfer in fractal media. *Mathematical Problems in Engineering* 2013 (358473): 1–3.

Jumarie, G. (2005). On the representation of fractional Brownian motion as an integral with respect to (dt)α. *Applied Mathematics Letters* 18: 739–748.

Jumarie, G. (2006). Modified Riemann-Liouville derivative and fractional Taylor series of nondifferentiable functions further results. *Computers & Mathematics with Applications* 51 (9–10): 1367–1376.

Jumarie, G. (2009). Table of some basic fractional calculus formulae derived from a modified Riemann–Liouville derivative for non-differentiable functions. *Applied Mathematics Letters* 22 (3): 378–385.

Kilbas, A.A., Srivastava, H.M., and Trujillo, J.J. (2006). *Theory and Application of Fractional Differential Equations*. Amsterdam: Elsevier Science B.V.

Kiryakova, V.S. (1993). *Generalized Fractional Calculus and Applications*. England: Longman Scientific and Technical.

Kolwankar, K.M. and Gangal, A.D. (1996). Fractional differentiability of nowhere differentiable functions and dimensions. Chaos: an Interdisciplinary. *Journal of Nonlinear Science* 6 (4): 505–513.

Kurulay, M. and Bayram, M. (2012). Some properties of the Mittag-Leffler functions and their relation with the Wright functions. *Advances in Difference Equations* 2012: 1–8.

Mathai, A.M. and Haubold, H.J. (2008). *Special Functions for Applied Scientists*. New York: Springer Verlag.

Miller, K.S. and Ross, B. (1993). *An Introduction to the Fractional Calculus and Fractional Differential Equations*. New York: Wiley.

Oldham, K.B. and Spanier, J. (1974). *The Fractional Calculus*. New York: Academic Press.

Podlubny, I. (1999). *Fractional Differential Equations*. New York: Academic Press.

Samko, S.G., Kilbas, A.A., and Marichev, O.I. (2002). *Fractional Integrals and Derivatives: Theory and Applications*. London: Taylor & Francis.

Wright, E.M. (1993). On the coefficients of power series having exponential singularities. *Journal of the London Mathematical Society* 8 (1): 71–79.

Yang, X.J. (2012a). *Advanced Local Fractional Calculus and Its Applications*. New York: World Science Publisher.

Yang, X.J. (2012b). A short note on local fractional calculus of function of one variable. *Journal of Applied Library and Information Science* 1 (1): 1–13.

2

Recent Trends in Fractional Dynamical Models and Mathematical Methods

2.1 Introduction

Mathematicians like Leibniz, L'Hôpital, Abel, Liouville, Riemann, and others conceptualized fractional calculus as the theory of integrals and derivatives of arbitrary real (and complex) order. Nonlocality, an intrinsic property of many complex systems, makes fractional derivatives suitable for modeling phenomena in various sciences and engineering disciplines. Fractional derivatives deal with the global evolution of the system rather than just focusing on the local dynamics. Therefore, fractional derivatives provide better representations of real-world behavior than ordinary derivatives. Although fractional calculus is three centuries old, now it is very popular with scientists and engineers. The beauty of this subject is that fractional derivatives (and integrals) are not a local (or point) property. This subject, therefore, considers the history and nonlocal distribution effects.

2.2 Fractional Calculus: A Generalization of Integer-Order Calculus

Consider an integer n, and when we say x^n we immediately visualize multiplying x by itself n times to give the result. We still obtain the result if n is not an integer, but it will not be easy to visualize how. Further, 2^e and 2^π are hard to visualize, but they exists. Similarly, although the fractional derivative $\frac{d^\pi}{dt^\pi} f(t)$ is hard to visualize, it exists. As real numbers exist between the integers, fractional derivatives and integrals exist between conventional integer-order derivatives and n-fold integrations. It is relatively easy to understand how integer-order derivatives are interpreted. In general, a first-order derivative measures the rate of change of a function or the slope of a tangent line at a particular point. Because fractional derivatives are inherently nonlocal, there is no clear geometric interpretation of fractional differentiation. As a result, the fractional derivatives of a function cannot be calculated simply from its behavior around a specific point. Indeed, the behavior of a function far away from the evaluation point can significantly impact the value of a fractional derivative. Because of this, providing a straightforward, understandable explanation of the geometrical meaning of fractional derivatives is challenging. In Shukla and Sapra (2019), Podlubny has shown that the geometric meaning of fractional integration is "Shadows on the walls," while the physical interpretation is "Shadows of the past."

In order to illustrate the usefulness of the fractional derivatives, let us take an example from reference (Diethelm 2010). It is well known that the relationship between stress $\sigma(t)$ and strain $\varepsilon(t)$ in material under the influence of external forces is

$$\sigma(t) = \eta \frac{d}{dt} \varepsilon(t). \tag{2.1}$$

Equation (2.1) is called Newton's law for a viscous liquid, with η representing the viscosity of the material, and

$$\sigma(t) = E\varepsilon(t), \tag{2.2}$$

is Hooke's law for an elastic solid, with E denoting the modulus of elasticity. We can rewrite Eqs. (2.1) and (2.2) as:

$$\sigma(t) = v \frac{d^\alpha}{dt^\alpha} \varepsilon(t), \tag{2.3}$$

with $\alpha = 0$ for elastic solids and $\alpha = 1$ for a viscous liquid. However, in the actual sense, viscoelastic materials have a behavior midway between an elastic solid and a viscous liquid. Therefore, it may be advantageous to give sense to the operator $\frac{d^\alpha}{dt^\alpha}$ if $0 < \alpha < 1$. There are various ways of defining fractional derivatives and integrals. Various fractional derivatives, integrals, and useful functions associated with fractional calculus have been presented in Chapter 1.

Computational Fractional Dynamical Systems: Fractional Differential Equations and Applications, First Edition.
Snehashish Chakraverty, Rajarama Mohan Jena, and Subrat Kumar Jena.

Fractional calculus has numerous applications in engineering and science, including electromagnetics, viscoelasticity, fluid mechanics, electrochemistry, biological population models, optics, and signal processing. Fractional differential equations have been found to be the better description of physical and engineering phenomena. Fractional derivatives are used to precisely model damping in systems. The following section covers some contemporary applications of fractional calculus and models of fractional dynamical systems.

2.3 Fractional Derivatives of Some Functions and Their Graphical Illustrations

In this section, we present explicit formulas to calculate fractional derivatives and integralsof some special functions and then depict their graphs.

a) **Unit function** (Dalir and Bashour 2010) If $f(x) = 1$ then $\dfrac{d^{\alpha}1}{dx^{\alpha}} = \dfrac{x^{-\alpha}}{\Gamma(1-\alpha)}$, for all α in Riemann–Liouville sense. Solution plot at different values of α is given in Figure 2.1.

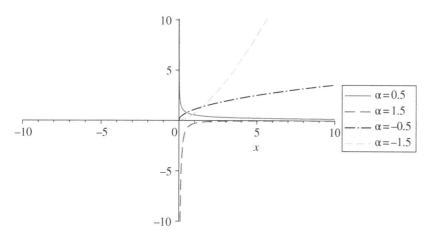

Figure 2.1 Solution plot of unit function.

b) **Identity function** (Dalir and Bashour 2010) If $f(x) = x$ then $\dfrac{d^{\alpha}x}{dx^{\alpha}} = \dfrac{x^{1-\alpha}}{\Gamma(2-\alpha)}$. Solution plot at different values of α is depicted in Figure 2.2.

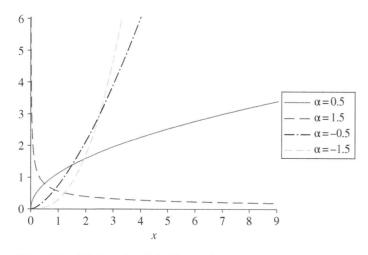

Figure 2.2 Solution plot of the identity function.

c) **Exponential function** (Dalir and Bashour 2010) Fractional derivative and integral of the function $f(x) = e^x$ is $\dfrac{d^{\alpha} e^x}{dx^{\alpha}} = \displaystyle\sum_{k}^{\infty} \dfrac{x^{k-\alpha}}{\Gamma(k-\alpha+1)}$. The solution plot at various values of α is illustrated in Figure 2.3.

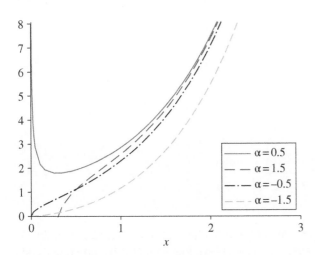

Figure 2.3   Solution plot of the exponential function.

d) **Sine function** (Dalir and Bashour 2010) Fractional derivative and integral of the sine function $f(x) = \sin(x)$ is written as $\dfrac{d^{\alpha} \sin(x)}{dx^{\alpha}} = \sin\left(x + \dfrac{\alpha\pi}{2}\right)$. The solution at various values of α is plotted in Figure 2.4.

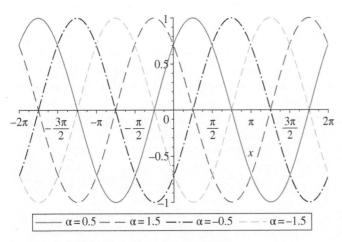

Figure 2.4   Solution plot of the sine function.

e) **Cosine function** (Dalir and Bashour 2010) Fractional derivative and integral of the cosine function $f(x) = \cos(x)$ is written as $\dfrac{d^{\alpha} \cos(x)}{dx^{\alpha}} = \cos\left(x + \dfrac{\alpha\pi}{2}\right)$. The solution is given in Figure 2.5 at various values of α.

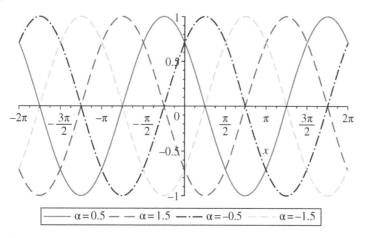

Figure 2.5 Solution plot of the cosine function.

2.4 Applications of Fractional Calculus

Although the fundamental mathematical ideas of fractional calculus (noninteger-order integral and differential operations) were developed long ago by mathematicians, such as Leibniz (1695), Liouville (1834), Riemann (1892), and others, and brought to the attention of the engineering world by Oliver Heaviside in the 1890s. The first book on this subject was not published until 1974 by Oldham and Spanier (1974). The use of fractional calculus in physics, continuum mechanics, signal processing, and electromagnetics has been highlighted in recent monographs and symposium proceedings. Some of the applications are listed in the following sections.

2.4.1 N.H. Abel and Tautochronous problem

Probably, the first application of fractional calculus was made by Abel (1823) during the year 1802–1829. He obtained the solution to an integral equation arising from the tautochronous problem or isochronic curve formulation. The purpose of this problem is to determine the geometry of a frictionless plane curve through the origin in a vertical plane along which a particle of mass m can descend in an independent time from its starting position. There are two ways of obtaining the solution: the first is using the standard approach and the second is through fractional calculus. N.H. Able proposed the solution based on the energy conservation principle, which states that the total amount of energy in an isolated system remains constant or, in other words, the sum between gravitational potential and kinetic energy remains constant. In this problem, he has considered no friction between particles. Therefore, the kinetic energy of the particle equals the difference between the potential energy and the initial point of the particle and the potential at the point where the particle is located.

If T is a constant for the sliding time, then the Abel integral equation (1823) is written as (Dalir and Bashour 2010):

$$\sqrt{2gT} = \int_0^{\eta} (\eta - y)^{-1/2} f'(y) dy, \tag{2.4}$$

where g is the acceleration due to gravity, (ξ, η) is the initial position, and $s = f(y)$ is the equation of the sliding curve. Equation (2.4) can be rewritten in fractional integral form as (Dalir and Bashour 2010):

$$T\sqrt{2g} = \Gamma\left(\frac{1}{2}\right) {}_0 D_{\eta}^{-1/2} f'(y). \tag{2.5}$$

2.4.2 Ultrasonic Wave Propagation in Human Cancellous Bone

In Sebaa et al. (2006), the viscous interactions between fluid and solid structures are described using fractional calculus. Blot's theory is used to calculate the reflection and transmission scattering operators for a slab of cancellous bone in an elastic frame. Experimental findings are compared to theoretical results for slow and fast waves transmitted through human cancellous bone samples.

2.4.3 Modeling of Speech Signals Using Fractional Calculus

This work (Assaleh and Ahmad 2007) presents a unique technique for speech signal modeling based on fractional calculus. The method used in Assaleh and Ahmad (2007) contrasts the well-known linear predictive coding (LPC) method based on integer-order models. Numerical simulations show that the speech signal may be appropriately described by utilizing a few integrals of fractional orders as basis functions.

2.4.4 Modeling the Cardiac Tissue Electrode Interface Using Fractional Calculus

In Magin (2008), all kinds of biopotential recording (e.g. electrocardiogram (ECG), electromyography (EMG), and electroencephalogram (EEG)) and functional electrical stimulation use the same tissue electrode interface (e.g. pacemaker, cochlear implant, and deep brain stimulation). The order of differentiation may be generalized by changing the defining current–voltage relationships in conventional lumped element circuit models of electrodes. Although fractional-order models enhance the explanation of known bioelectrode behavior, current experimental investigations of cardiac tissue show that different mathematical tools may be required to understand this complicated system fully.

2.4.5 Application of Fractional Calculus to the Sound Waves Propagation in Rigid Porous Materials

In Fellah and Depollier (2002), as the asymptotic expressions of stiffness and damping in porous materials are proportional to fractional powers of frequency, time derivatives of fractional order may be used to characterize the behavior of sound waves in these materials, including relaxation and frequency dependence.

2.4.6 Fractional Calculus for Lateral and Longitudinal Control of Autonomous Vehicles

In Súarez et al. (2004), the path-tracking problem of an autonomous electric vehicle is addressed here using fractional-order controllers (FOCs). To implement conventional controller and FOCs, the lateral dynamics of an industrial vehicle have been considered. Simulations and comparisons of several control schemes using these controllers have been made.

2.4.7 Application of Fractional Calculus in the Theory of Viscoelasticity

In Soczkiewicz (2002), using fractional derivatives in the viscoelasticity theory makes it possible to obtain constitutive equations for the complex elastic modulus of viscoelastic materials with only a few experimentally determined parameters. The fractional derivative has also been applied to studying complex moduli and impedances for various viscoelastic models.

2.4.8 Fractional Differentiation for Edge Detection

Generally, in image processing, integer-order differentiation operators are frequently utilized in edge identification, mainly order 1 for the gradient and order 2 for the Laplacian. The authors of Mathieu et al. (2003) demonstrated that noninteger (fractional) differentiation can be used to improve the criteria of thin detection or detection selectivity when dealing with parabolic luminance transitions, as well as the criterion of immunity to noise which can be interpreted in terms of robustness to noise in general.

2.4.9 Wave Propagation in Viscoelastic Horns Using a Fractional Calculus Rheology Model

In Timothy (2003), in order to describe the complex mechanical behavior of materials, fluid and solid models are used with fractional calculus. For molecular solutions, fractional derivatives have been demonstrated to represent the viscoelastic stress derived from polymer chain theory. This study investigates infinitesimal waves propagating along with one-dimensional horns with a small cross-sectional area change. This study examines explicitly linear, conical, exponential, and catenoidal shapes. The simulations use mathematical calculations and fractional rheology from Bagley data to analyze and predict the wave amplitudes versus frequency. It is possible to analyze design trade-offs of materials with real materials while employing fractional calculus representations to derive classic elastic and fluid "Webster equations" within the appropriate limits for engineering and scientific application.

2.4.10 Application of Fractional Calculus to Fluid Mechanics

In Kulish and Lage (2002), the authors have applied fractional calculus to obtain the solution to time-dependent, viscous–diffusion fluid mechanics problems. By combining fractional calculus and the Laplace transform method for the transient viscous – diffusion equation in a semi-infinite space, explicit analytical (fractional) solutions for the shear stress and fluid speed are obtained in the domain. Comparing the fractional results for boundary stress and fluid speed with the existing analytical solutions to the first and second Stokes problems shows that the fractional methodology is more straightforward and more powerful than the existing techniques.

2.4.11 Radioactivity, Exponential Decay, and Population Growth

The decomposition of radioactive substances is proportional to their quantity. Population growth follows a similar pattern. These models can be summarized by an ordinary differential equation of first order. The solution is exponential with a positive and negative argument for both models. In a radioactive model for decay, the proportionality constant depends on the material, while in a population growth model, it depends on the initial population. The signal of the proportionality constant makes the difference between these two models. To provide a more accurate description, the fractional models are considered, since the solution depends on the order of the fractional derivative. Let us consider the fractional differential equation as follows (Shukla and Sapra 2019):

$$\frac{d^\alpha m(t)}{dt^\alpha} = -k\,m(t), \quad 0 < \alpha \leq 1, \tag{2.6}$$

where k is the proportionality constant and $m(t)$ is the dependent variable which denotes the mass of the radioactive material and the population. With the help of the Laplace transform method, one can obtain the solution to the model mentioned earlier.

2.4.12 The Harmonic Oscillator

The equation of motion of the classical harmonic oscillator can be written as (Shukla and Sapra 2019):

$$m\frac{d^2 x(t)}{dt^2} = -kx(t), \tag{2.7}$$

where a particle of mass m constrained to move along the *x-axis* and bound to the equilibrium position $x = 0$ by a restoring force $-kx$. Using suitable initial condition, the solution of Eq. (2.7) will be harmonic oscillation as (Shukla and Sapra 2019):

$$x(t) = x_0 \cos\left(\sqrt{\frac{k}{m}}t\right). \tag{2.8}$$

Fractional derivative models are used to represent systems that need damping modeling accurately. Since all the natural systems have frictions, the solution provided by the damped harmonic oscillator is more accurate. The equation of simple harmonic oscillator in Caputo fractional sense may be written as (Shukla and Sapra 2019):

$$\frac{d^\alpha x(t)}{dt^\alpha} + \omega^\alpha x(t) = 0, \quad 1 < \alpha \leq 2. \tag{2.9}$$

Using the initial conditions $x(0) = x_0$ and $x'(0) = 0$, we find the solution of Eq. (2.9) as follows (Shukla and Sapra 2019):

$$x(t) = x_0 E_\alpha(-\omega^\alpha t^\alpha). \tag{2.10}$$

At $\alpha = 2$, Eq. (2.10) is written as

$$x(t) = x_0 E_2(-\omega^2 t^2) = x_0 \cos(\omega t). \tag{2.11}$$

This is the solution of an integer-order harmonic oscillator.

2.5 Overview of Some Analytical/Numerical Methods

Numerous physical and engineering problems are modeled by differential equations. In order to observe the qualitative characteristics and physical interpretation of a large number of fractional phenomena, the solutions of those models are required. It is sometimes challenging to obtain the solution (both analytical and numerical) of nonlinear partial differential equations of fractional order due to their nonlinear behavior. So for the last few decades, a great deal of attention has been directed toward the solution to these kinds of problems. Researchers throughout the globe are trying to develop various efficient methods to handle these problems. Although a few methods have been developed by other researchers to analyze the aforementioned problems, those are problem-dependent and sometimes are not efficient. So, the development of efficient computational methods for these problems is the recent challenge.

In the following section, the authors have presented some of the numerical and analytical methods used for obtaining the solution of fractional differential equations:

2.5.1 Fractional Adams–Bashforth/Moulton Methods

Diethelm et al. (2002, 2004) developed and studied an Adams–Bashforth type approach in their works. On the other hand, this methodology uses a two-step predictor–corrector method, which is notably different from the traditional method. Several authors have recently employed this approach to solve fractional differential equations. Zayernouri and Matzavinos (2016) developed an implicit–explicit (IMEX) splitting technique for linear and nonlinear fractional partial differential equations using the modified Adams–Bashforth and Adams–Moulton methods. They also investigated the Keller–Segel fractional chemotaxis system. Owolabi and Atangana (2017) suggested a new three-step fractional Adams–Bashforth approach using the Caputo–Fabrizio derivative for solving linear and nonlinear fractional differential equations. Owolabi et al. (2021) analyzed three nonlinear chaotic dynamical systems using a fractional version of the Adams–Bashforth approach described in the Liouville–Caputo derivative sense.

2.5.2 Fractional Euler Method

This approach is straightforward and provides solutions without linearization, perturbations, or other assumptions. The main characteristic of the approach is that the Euler method has an intuitive geometric meaning. Yu and Zhen (2020) adapted the implicit Euler method for nonlinear impulsive fractional differential equations. They also verified the convergence analysis of the method and concluded that the method is convergent of the first order. Improved Euler method and other relevant studies can be found in Odibat and Momani (2008) and Tong et al. (2013).

2.5.3 Finite Difference Method

Finite difference methods (FDMs) are well-known numerical methods to solve differential equations by approximating derivatives using different schemes (Hoffman and Frankel 2001; Bhat and Chakraverty 2004). The finite difference approximations are the simplest and oldest methods to solve differential equations. Vargas (2022) used the generalized finite difference approach to construct a straightforward discretization of space fractional derivatives. For fractional advection–dispersion equations with fractional derivative boundary conditions, Liu and Hou (2017) established an implicit finite difference approach. The approach has been shown to have first-order consistency, solvability, unconditional stability, and first-order convergence. Anley and Zheng (2020) proposed finite difference approximation for space fractional convection–diffusion model having space variable coefficients on the given bounded domain over time and space.

2.5.4 Finite Element Method

The finite element method (FEM) has wide applications in various science and engineering fields, viz. structural mechanics, biomechanics, and electromagnetic field problems, of which exact solutions may not be determined. The FEM serves as a numerical discretization approach that converts differential equations into algebraic equations. With the help of the space–time FEM, Lai et al. (2021) obtained the numerical solutions for linear Riesz space fractional partial differential equations with a second-order time derivative. Zheng et al. (2010) applied the FEM for the space fractional advection–diffusion equation with the nonhomogeneous initial–boundary condition defined in Caputo fractional derivative sense. Ford et al. (2011) consider the FEM for time fractional partial differential equations. The Lax–Milgram Lemma is also used to demonstrate the existence and uniqueness of the solutions. Recently, Li et al. (2021) combined the standard Galerkin FEM in the spatial direction, the fractional Crank–Nicolson method, and extrapolation methods in the temporal direction to handle the nonlinear time fractional parabolic problems with time delay.

2.5.5 Finite Volume Method

Another popular numerical methodology is the finite volume method (FVM) (Hejazi et al. 2013; Liu et al. 2014; Wang et al. 2015; Zhang et al. 2021). The fundamental conservation property of FVM makes it the preferred approach compared to other existing methods such as the FDM, and FEM. In this methodology, volumes (elements or cells) are evaluated at discrete points over a meshed geometry, similar to well-known numerical approaches such as FDM or FEM. Then, using the well-known divergence theorem (Godlewski and Raviart 2013), the associated volume integrals of the corresponding differential equation, including the divergence component, are transformed to surface integrals. In the last phase, the simulated differential equation is then converted over discrete volumes into a discrete system of algebraic equations. The system of algebraic equations is solved using conventional methods to find the dependent variables.

The aforementioned FVM provides many significant advantages when dealing with differential equations in science and engineering problems. A key feature of the FVM is the construction of physical space (domain of the differential equation) on unregulated polygonal meshes. Another advantage of the FVM is that it is effortless to apply multiple boundary conditions in a noninvasive manner. Because the unknown variables involved are computed at the centroids of the volume components rather than at their boundary faces (Liu 2018). These properties of the FVM have made it appropriate for numerical simulations in a wide range of applications.

2.5.6 Meshless Method

The meshless methods (Gu et al. 2011; Dehghan et al. 2015; Yang et al. 2015; Abbaszadeh and Dehghan 2017; Maalek et al. 2019) in the numerical analysis do not need the connection of nodes in the simulation domain; instead, they rely on the interaction of each node with all of its neighbors. As a result, initial comprehensive attributes like mass and kinetic energy are given to single nodes rather than mesh components. Numerical approaches, including the FDM, FVM, and FEM, were developed using data point meshes. Each point has a given number of predetermined neighbors, and this connection may be utilized to build mathematical operators such as the derivative. These operators are then used to create the simulation equations. However, in simulations where the material being simulated may move around (as in computational fluid dynamics) or when substantial material deformations can occur (as in plastic material simulations), maintaining mesh connectivity without adding error into the simulation may be problematic. The operators specified on the mesh may no longer yield proper values if tangled or degenerated during simulation. During simulation, the mesh can be regenerated (a process known as remeshing), but this might introduce errors since all previous data points must be mapped onto a new and different set of data points.

2.5.7 Reproducing Kernel Hilbert Space Method

The reproducing kernel Hilbert space (RKHS) technique has been widely utilized to regulate the solvability findings of numerous models, including different forms of differential–integral operators. Although the RKHS algorithm is still relatively new, it has several advantages. First, it has the ability to solve many fractional differential models with complex constraint conditions, which are challenging to solve. Second, the numerical solutions and their derivatives converge uniformly to the exact solutions and derivatives, respectively. And third, the algorithm is mesh-free, requires no time discretization, and is simple to implement. Many authors have applied this technique for handling a variety of complex phenomena. Djennadi et al. (2021) employed this method to solve the inverse source problem of the time–space fractional diffusion equation. Arqub et al. (2021) utilized this method to generate the pointwise numerical solution to the time fractional Burgers' model in the overdetermination Robin boundary condition. Solution of integro-differential equations of fractional order has been studied by Bushnaq et al. (2013) using the RKHS method. Some classes of time fractional partial differential equations subject to initial and Neumann boundary conditions have been solved by Omar Abu Arqub (2017) with the help of fitted RKHS method.

2.5.8 Wavelet Method

A wavelet approach is a useful tool for understanding dynamic systems and differential equations that arise in other fields of science and engineering. Wavelet is a wave-like oscillation with an amplitude that starts at zero and grows monotonically until decreasing back to zero. A wavelet series is frequently expressed as a collection of orthonormal basis functions or a complete square-integrable function. Wavelet classes are divided into three categories: discrete, continuous, and multire-solution-based wavelets. Discrete wavelets are often considered over a discrete subset of the upper half-plane, whereas continuous wavelets are projected over continuous function space. Only a finite number of wavelet coefficients are considered while evaluating discrete wavelets, which sometimes makes them mathematically difficult. Multiresolution-based wavelets are recommended in these situations.

Meyer (1989) investigated orthonormal wavelets over real line R in the mid-1980s. Ingrid Daubechies gradually constructed compactly supported orthogonal wavelets known as Daubechies wavelets (Daubechies 1988). Daubechies wavelets are sufficiently smooth, orthogonal, and compact. Nowadays, various orthogonal wavelets have been used in fractional real-life physical models. A thorough examination of wavelets and their applications may be found in Ur Rehman and Khan (2011), Hesameddini et al. (2012), Saeed (2014), and Ur Rehman et al. (2020).

2.5.9 The Sine-Gordon Expansion Method

This expansion method is based on the sine-Gordon equation to construct real- and complex-valued exact solutions. The governing equation is reduced to a classical ordinary differential equation via the suitable wave transform. The order of obtained polynomial-type solution is determined using the homogeneous balancing approach, which is based on the well-known sine-Gordon equation. A system of algebraic equations is formed by equating the coefficients of the solution powers. The resulting system provides the essential relationships between the parameters and coefficients to generate the solutions. For a specific set of parameters, several solutions are evaluated. Korkmaz et al. (2020) implemented the sine-Gordon expansion method to construct exact solutions for conformable time fractional equations in the regularized long wave (RLW) class. The Estevez–Mansfield–Clarkson (EMC) equation and the $(2 + 1)$-dimensional Riemann wave (RW) equation have remarkable applications in the field of plasma physics, fluid dynamics, optics, and image processing. Kundu et al. (2021) have solved these two equations using this technique. New traveling wave solutions of the time fractional Fitzhugh–Nagumo equation with the sine-Gordon expansion method have been obtained by Taşbozan and Ali (2020).

2.5.10 The Jacobi Elliptic Equation Method

A fractional partial differential equation is turned into an ordinary differential equation of integer order using this approach, which is based on a fractional complex transformation. The exact solutions are supposed to be represented as polynomials in the Jacobi elliptic functions, including the Jacobi sine, Jacobi cosine, and Jacobi elliptic function of the third type. The homogeneous balancing concept may be used to determine the polynomial degree. The obtained solutions include rational, trigonometric, and hyperbolic functions. Furthermore, complex-valued, periodic, and soliton solutions have been explored. Zheng (2014) investigated the exact solution of the space–time fractional Kortweg–de Vries (KdV) equation, the space–time fractional Benjamin–Bona–Mahony (BBM) equation, and the space–time fractional $(2 + 1)$-dimensional breaking soliton equations. Ünal et al. (2021) obtained the exact solutions of the space–time fractional symmetric regularized long wave (SRLW) equation using a direct method based on the Jacobi elliptic functions. Zheng and Feng (2014) implemented this method to solve the space fractional coupled Konopelchenko–Dubrovsky (KD) equations and the space–time fractional Fokas equation defined in the sense of the modified Riemann–Liouville derivative. Feng (2016) applied this method to seek Jacobi elliptic function solutions for the space–time fractional Kadomtsev–Petviashvili–Benjamin–Bona–Mahony (KPBBM) equation and the $(2 + 1)$-dimensional space–time fractional Nizhnik–Novikov–Veselov system.

2.5.11 The Generalized Kudryashov Method

This method follows the same procedure as the Jacobi elliptic and sine-Gordon expansion methods. In the first step, a fractional partial differential equation is transformed into an ordinary differential equation using a fractional complex transformation, and its solutions are represented as polynomials. Using the homogeneous balancing concept, the polynomial degree can then be determined. Next, a system of the algebraic equation can be generated. By solving this system of an algebraic equation and substituting these constants values into the original equation, one can determine the exact solution. The obtained solutions include symmetrical Fibonacci function solutions, hyperbolic function solutions, and rational solutions. This method is efficient and can establish new solutions for different types of fractional differential equations. Selvaraj et al. (2020) obtained the solution of the time fractional generalized Burgers–Fisher equation (TF-GBF) using the generalized Kudryashov method. Using the generalized Kudryashov method, the exact solutions of the nonlinear fractional double sinh-Poisson equation has been derived in Demiray and Bulut (2016). The modified Kudryashov method is applied to compute an approximation to the solutions of the space–time fractional modified BBM equation and the space–time fractional potential Kadomtsev–Petviashvili equation (Ege and Misirli 2014). Demiray et al. (2014) used this technique to find exact solutions of the time fractional Burgers equation, time fractional Cahn–Hilliard equation, and time fractional generalized third-order KdV equation.

Apart from the earlier-discussed methods, the interested author may follow different analytical/semi-analytical and expansion methods given in Chapters 3–25.

References

Abbaszadeh, M. and Dehghan, M. (2017). An improved meshless method for solving two-dimensional distributed order time-fractional diffusion-wave equation with error estimate. *Numerical Algorithms* 75 (1): 173–211.

Abel, N.H. (1823). Oplösning af et par opgaver ved hjelp af bestemte integraler. *Magazin for Naturvidenskaberne* I (2), Christiania.

Anley, E.F. and Zheng, Z. (2020). Finite difference approximation method for a space fractional convection-diffusion equation with variable coefficients. *Symmetry* 12 (3): 485(1–19).

Arqub, O.A. (2017). Fitted reproducing kernel Hilbert space method for the solutions of some certain classes of time-fractional partial differential equations subject to initial and Neumann boundary conditions. *Computers and Mathematics with Applications* 73 (6): 1243–1261.

Arqub, O.A., Al-Smadi, M., Gdairi, R.A. et al. (2021). Implementation of reproducing kernel Hilbert algorithm for pointwise numerical solvability of fractional Burgers' model in time-dependent variable domain regarding constraint boundary condition of Robin. *Results in Physics* 24: 104210.

Assaleh, K. and Ahmad, W.M. (2007). Modeling of speech signals using fractional calculus. *9th International Symposium on Signal Processing and Its Applications (ISSPA) 2007* (12–15 February 2007). Sharjah, United Arab Emirates: IEEE, pp. 1–4.

Bhat, R.B. and Chakraverty, S. (2004). *Numerical Analysis in Engineering*. London: Alpha Science International Ltd.

Bushnaq, S., Momani, S., and Zhou, Y. (2013). A reproducing kernel Hilbert space method for solving integro-differential equations of fractional order. *Journal of Optimization Theory and Applications* 156 (1): 96–105.

Dalir, M. and Bashour, M. (2010). Applications of fractional calculus. *Applied Mathematical Sciences* 4 (21): 1021–1032.

Daubechies, I. (1988). Orthonormal bases of compactly supported wavelets. *Communications on Pure and Applied Mathematics* 41 (7): 909–996.

Dehghan, M., Abbaszadeh, M., and Mohebbi, A. (2015). Meshless local Petrov-Galerkin and RBFs collocation methods for solving 2D fractional Klein-Kramers dynamics equation on irregular domains. *Computer Modeling in Engineering and Sciences* 107 (6): 481–516.

Demiray, S.T. and Bulut, H. (2016). Generalized Kudryashov method for nonlinear fractional double sinh-Poisson equation. *Journal of Nonlinear Sciences and Applications* 9: 1349–1355.

Demiray, S.T., Pandir, Y., and Bulut, H. (2014). Generalized Kudryashov method for time-fractional differential equations. *Abstract and Applied Analysis* 2014: 901540(1–13).

Diethelm, K. (2010). *The Analysis of Fractional Differential Equations*, 264p. Berlin: Springer Verlag ISBN: 9783642145735. DOI: 10.1007/ 978-3-642-14574-2.

Diethelm, K., Ford, N.J., and Freed, A.D. (2002). A predictor-corrector approach for the numerical solution of fractional differential equations. *Nonlinear Dynamics* 29 (1–4): 3–22.

Diethelm, K., Ford, N.J., and Freed, A.D. (2004). Detailed error analysis for a fractional Adams method. *Numerical Algorithms* 36 (1): 31–52.

Djennadi, S., Shawagfeh, N., and Arqub, O.A. (2021). A numerical algorithm in reproducing kernel-based approach for solving the inverse source problem of the time-space fractional diffusion equation. *Partial Differential Equations in Applied Mathematics* 4: 100164.

Ege, S.M. and Misirli, E. (2014). The modified Kudryashov method for solving some fractional-order nonlinear equations. *Advances in Difference Equations* 2014 (1): 1–13.

Fellah, Z.E.A. and Depollier, C. (2002). Application of fractional calculus to the sound waves propagation in rigid porous materials: validation via ultrasonic measurement. *Acta Acustica* 88 (2002): 34–39.

Feng, Q. (2016). Jacobi elliptic function solutions for fractional partial differential equations. *International Journal of Applied Mathematics* 46 (1): 1–9.

Ford, N.J., Xiao, J., and Yan, Y. (2011). A finite element method for time fractional partial differential equations. *Fractional Calculus and Applied Analysis* 14 (3): 454–474.

Godlewski, E. and Raviart, P.A. (2013). *Numerical Approximation of Hyperbolic Systems of Conservation Laws*, vol. 118. Springer Science & Business Media.

Gu, Y., Zhuang, P., and Liu, Q. (2011). An advanced meshless method for time fractional diffusion equation. *International Journal of Computational Methods* 8 (04): 653–665.

Hejazi, H., Moroney, T., and Liu, F. (2013). A finite volume method for solving the two-sided time-space fractional advection-dispersion equation. *Central European Journal of Physics* 11 (10): 1275–1283.

Hesameddini, E., Shekarpaz, S., and Latifizadeh, H. (2012). The Chebyshev wavelet method for numerical solutions of a fractional oscillator. *International Journal of Applied Mathematical Research* 1: 493–509.

Hoffman, J.D. and Frankel, S. (2001). *Numerical Methods for Engineers and Scientists*. New York: CRC Press.

Korkmaz, A., Hepson, O.E., Hosseini, K. et al. (2020). Sine-Gordon expansion method for exact solutions to conformable time fractional equations in RLW-class. *Journal of King Saud University-Science* 32 (1): 567–574.

Kulish, V.V. and Lage, J.L. (2002). Application of fractional calculus to fluid mechanics. *Journal of Fluids Engineering* 124 (3): 803.

Kundu, P.R., Fahim, M.R.A., Islam, M.E., and Akbar, M.A. (2021). The sine-Gordon expansion method for higher-dimensional NLEEs and parametric analysis. *Heliyon* 7 (3): e06459.

Lai, J., Liu, F., Anh, V.V., and Liu, Q. (2021). A space-time finite element method for solving linear Riesz space fractional partial differential equations. *Numerical Algorithms* 88: 499–520.

Li, L., She, M., and Niu, Y. (2021). Fractional Crank-Nicolson-Galerkin finite element methods for nonlinear time fractional parabolic problems with time delay. *Journal of Function Spaces* 2021: 1–10.

Liu, Z.L. (2018). Finite volume method. In: *Multiphysics in Porous Materials* (ed. Z.L. Liu), 385–395. Cham: Springer.

Liu, T. and Hou, M. (2017). A fast implicit finite difference method for fractional advection-dispersion equations with fractional derivative boundary conditions. *Advances in Mathematical Physics* 2017: 1–9.

Liu, F., Zhuang, P., Turner, I. et al. (2014). A new fractional finite volume method for solving the fractional diffusion equation. *Applied Mathematical Modelling* 38 (15–16): 3871–3878.

Maalek, F., Gharian, D., and Heydari, M.H. (2019). A meshless method for the variable-order time fractional telegraph equation. *Journal of Mathematical Extension* 13: 35–56.

Magin, R.L. (2008). Modeling the cardiac tissue electrode interface using fractional calculus. *Journal of Vibration and Control* 14 (9–10): 1431–1442.

Mathieu, B., Melchior, P., Oustaloup, A., and Ceyral, C. (2003). Fractional differentiation for edge detection. *Fractional Signal Processing and Applications* 83 (11): 2285–2480.

Meyer, Y. (1989). Orthonormal wavelets. In: *Wavelets* (ed. J.-M. Combes, A. Grossmann and P. Tchamitchian), 21–37. Berlin, Heidelberg: Springer.

Odibat, Z.M. and Momani, S. (2008). An algorithm for the numerical solution of differential equations of fractional order. *Journal of Applied Mathematics and Informatics* 26 (1–2): 15–27.

Oldham, K.B. and Spanier, J. (1974). *The Fractional Calculus*. Academic Press, Inc.

Owolabi, K.M. and Atangana, A. (2017). Analysis and application of new fractional Adams–Bashforth scheme with Caputo–Fabrizio derivative. *Chaos, Solitons and Fractals* 105: 111–119.

Owolabi, K.M., Atangana, A., and Gómez-Aguilar, J.F. (2021). Fractional Adams-Bashforth scheme with the Liouville-Caputo derivative and application to chaotic systems. *Discrete & Continuous Dynamical Systems – S* 14 (7): 2455.

Saeed, U. (2014). Hermite wavelet method for fractional delay differential equations. *Journal of Difference Equations* 2014: 1–8.

Sebaa, N., Fellah, Z.E.A., Lauriks, W., and Depollier, C. (2006). Application of fractional calculus to ultrasonic wave propagation in human cancellous bone. *Signal Processing Archive* 86 (10): 2668–2677.

Selvaraj, R., Venkatraman, S., Ashok, D.D., and Krishnaraja, K. (2020). Exact solutions of time-fractional generalised Burgers–Fisher equation using generalised Kudryashov method. *Pramana* 94 (1): 1–8.

Shukla, R.K. and Sapra, P. (2019). Fractional calculus and its applications for scientific professionals: a literature review. *International Journal of Modern Mathematical Sciences* 17 (2): 111–137.

Soczkiewicz, E. (2002). Application of fractional calculus in the theory of viscoelasticity. *Molecular and Quantum Acoustics* 23: 397–404.

Súarez, J.I., Vinagre, B.M., Calderon, A.J. et al. (2004). *Using Fractional Calculus for Lateral and Longitudinal Control of Autonomous Vehicles Lecture Notes in Computer Science*, 2809. Springer Berlin/Heidelberg.

Taşbozan, O. and Ali, K. (2020). The new travelling wave solutions of time fractional Fitzhugh-Nagumo equation with Sine-Gordon expansion method. *Adıyaman University Journal of Science* 10 (1): 256–263.

Timothy, M. (2003). Wave propagation in viscoelastic horns using a fractional calculus rheology model. *Acoustical Society of America Journal* 114 (4): 2442–2442.

Tong, P., Feng, Y.Q., and Lv, H.J. (2013). Euler's method for fractional differential equations. *WSEAS Transactions on Mathematics* 12: 1146–1153.

Ünal, S.Ç., Daşcioğlu, A., and Bayram, D.V. (2021). Jacobi elliptic function solutions of space-time fractional symmetric regularized long wave equation. *Mathematical Sciences and Applications E-Notes* 9 (2): 53–63.

Ur Rehman, M. and Khan, R.A. (2011). The Legendre wavelet method for solving fractional differential equations. *Communications in Nonlinear Science and Numerical Simulation* 16 (11): 4163–4173.

Ur Rehman, M., Baleanu, D., Alzabut, J. et al. (2020). Green–Haar wavelets method for generalized fractional differential equations. *Advances in Difference Equations* 2020 (1): 1–25.

Vargas, A.M. (2022). Finite difference method for solving fractional differential equations at irregular meshes. *Mathematics and Computers in Simulation* 193: 204–216.

Wang, H., Cheng, A., and Wang, K. (2015). Fast finite volume methods for space-fractional diffusion equations. *Discrete & Continuous Dynamical Systems-B* 20 (5): 1427.

Yang, J.Y., Zhao, Y.M., Liu, N. et al. (2015). An implicit MLS meshless method for 2-D time-dependent fractional diffusion–wave equation. *Applied Mathematical Modelling* 39 (3-4): 1229–1240.

Yu, Y. and Zhen, S. (2020). Convergence analysis of implicit Euler method for a class of nonlinear impulsive fractional differential equations. *Mathematical Problems in Engineering* 2020: 1–8.

Zayernouri, M. and Matzavinos, A. (2016). Fractional Adams–Bashforth/Moulton methods: an application to the fractional Keller–Segel chemotaxis system. *Journal of Computational Physics* 317: 1–14.

Zhang, M., Liu, F., Turner, I.W. et al. (2021). A finite volume method for the two-dimensional time and space variable-order fractional Bloch-Torrey equation with variable coefficients on irregular domains. *Computers & Mathematics with Applications* 98: 81–98.

Zheng, B. (2014). A new fractional Jacobi elliptic equation method for solving fractional partial differential equations. *Advances in Difference Equations* 2014 (1): 1–11.

Zheng, B. and Feng, Q. (2014). The Jacobi elliptic equation method for solving fractional partial differential equations. *Abstract and Applied Analysis* 2014: 1–9.

Zheng, Y., Li, C., and Zhao, Z. (2010). A note on the finite element method for the space-fractional advection diffusion equation. *Computers & Mathematics with Applications* 59 (5): 1718–1726.

3

Adomian Decomposition Method

3.1 Introduction

In Chapter 1, we have already discussed preliminaries and notations of fractional calculus. The development of recent and robust methods for solving linear and nonlinear ordinary/partial/fractional differential equations has been demonstrated in Chapter 2. In this chapter, we will discuss about Adomian decomposition method (ADM). The ADM was first introduced by Adomian in the early 1980s (Adomian 1990; Wazwaz 1998). It is a semi-analytical approach for solving linear and nonlinear ordinary/partial/fractional differential equations. It allows us to handle both nonlinear initial and boundary values problems. The method of solution of this method (Evans and Raslan 2005; Momani and Odibat 2006) is based primarily on decomposing the nonlinear operator equation to a set of functions. Each series term is constructed from a polynomial generated by expanding an analytic function into a power series. The theoretical formulation of this technique is usually quite simple, but the actual difficulty arises when calculating the polynomials involved or when proving the convergence of the series of functions. In this chapter, we present the ADM procedure to solve linear and nonlinear fractional partial differential equations (PDEs) along with examples.

In the subsequent sections, firstly, the theories behind the method with respect to fractional order are given. Then the systematic study of the technique, as mentioned earlier, and two problems are addressed. It is worth noting that the two simple example problems of fractional differential equations are investigated to have an easy understanding of the method.

3.2 Basic Idea of ADM

Let us consider a general nonlinear nonhomogeneous PDE in the following form (Bildik and Konuralp 2006):

$$Lu(x,t) + Ru(x,t) + Nu(x,t) = f(x,t), \tag{3.1}$$

where L is the highest order differential operator and easily invertible, R is the linear differential operator of the order less than L, $f(x,t)$ is the source term, and $Nu(x,t)$ represents the nonlinear term. The solution function $u(x,t)$ is assumed to be bounded, and the nonlinear term Nu satisfies the Lipschitz condition, that is $|Nu - Nv| \le c|u - v|$, where c is a positive constant. Applying the inverse operator L^{-1} on both sides of Eq. (3.1), we obtain

$$u(x,t) = -L^{-1}Ru(x,t) - L^{-1}Nu(x,t) + L^{-1}f(x,t) + \phi, \tag{3.2}$$

where ϕ satisfies $L\phi = 0$ and the initial conditions. If $L = D_t^\alpha$ in the Caputo sense, then $L^{-1} = J_t^\alpha$ whose expression is given in Chapter 1. Now, the solution may be defined by the method of decomposition provided in Eq. (3.2) by the following infinite series as:

$$u(x,t) = \sum_{n=0}^{\infty} u_n(x,t). \tag{3.3}$$

The nonlinear term Nu is then decomposed as:

$$Nu = \sum_{n=0}^{\infty} A_n, \tag{3.4}$$

Computational Fractional Dynamical Systems: Fractional Differential Equations and Applications, First Edition.
Snehashish Chakraverty, Rajarama Mohan Jena, and Subrat Kumar Jena.
© 2023 John Wiley & Sons, Inc. Published 2023 by John Wiley & Sons, Inc.

where A_n's are the Adomian polynomials and A_n depends on $u_0, u_1, u_2, ..., u_n$. The Adomian polynomials can be defined as (Adomian 1990; Wazwaz 1998):

$$A_n = \frac{1}{n!} \frac{d^n}{d\lambda^n} \left[N\left(\sum_{n=0}^{\infty} \lambda^n u_n \right) \right]_{\lambda=0}, \quad n = 0, 1, 2, ... \tag{3.5}$$

where λ is a grouping parameter of convenience.

For clarity, a few Adomian polynomials are listed as follows (Wazwaz 1999; Jafari and Daftardar-Gejji 2006; Ghorbani 2009):

$$\begin{cases} A_0 = N(u_0), \\ A_1 = \dfrac{d}{d\lambda} N(u_0 + u_1\lambda)\Big|_{\lambda=0} = u_1 N^{(1)}(u_0), \\ A_2 = \dfrac{1}{2!}\dfrac{d^2}{d\lambda^2} N(u_0 + u_1\lambda + u_2\lambda^2)\Big|_{\lambda=0} = u_2 N^{(1)}(u_0) + \dfrac{1}{2!}u_1^2 N^{(2)}(u_0), \\ A_3 = \dfrac{1}{3!}\dfrac{d^3}{d\lambda^3} N(u_0 + u_1\lambda + u_2\lambda^2 + u_3\lambda^3)\Big|_{\lambda=0} = u_3 N^{(1)}(u_0) + u_1 u_2 N^{(2)}(u_0) + \dfrac{1}{3!}u_1^3 N^{(3)}(u_0), \\ A_4 = \dfrac{1}{4!}\dfrac{d^4}{d\lambda^4} N(u_0 + u_1\lambda + u_2\lambda^2 + u_3\lambda^3 + u_4\lambda^4)\Big|_{\lambda=0} = u_4 N^{(1)}(u_0) + \left[\dfrac{1}{2!}u_2^2 + u_1 u_3\right] N^{(2)}(u_0) + \\ \qquad\qquad \dfrac{1}{2!}u_1^2 u_2 N^{(3)}(u_0) + \dfrac{1}{4!}u_1^4 N^{(4)}(u_0), \end{cases} \tag{3.6}$$

Then, substituting Eqs. (3.3) and (3.4) into Eq. (3.2), we have

$$u(x,t) = \sum_{n=0}^{\infty} u_n(x,t) = -L^{-1}R\left(\sum_{n=0}^{\infty} u_n(x,t)\right) - L^{-1}\left(\sum_{n=0}^{\infty} A_n\right) + L^{-1}f(x,t) + \phi. \tag{3.7}$$

The following expressions can be obtained from Eq. (3.7):

$$\left.\begin{aligned} u_0 &= L^{-1}f + \phi \\ u_1 &= -L^{-1}R(u_0) - L^{-1}A_0 \\ u_2 &= -L^{-1}R(u_1) - L^{-1}A_1 \\ &\vdots \\ u_{n+1} &= -L^{-1}R(u_n) - L^{-1}A_n \\ &\vdots \end{aligned}\right\} \tag{3.8}$$

If we calculate all the terms u_n's, then we get the exact solution. But, in actual practice, this process may take a longer time. To achieve an acceptable solution, it may be approximated by the truncated series $\sum_{n=0}^{N} u_n$ (by using the convergence of the series).

3.3 Numerical Examples

Here, we solve a linear fractional one-dimensional heat-like problem in Example 3.1 and a nonlinear advection equation in Example 3.2 for a clear understanding of the ADM. It is worth mentioning that the ADM can also be used for handling linear and nonlinear fractional differential equations.

Example 3.1 Let us consider the one-dimensional heat-like model (Özis and Agırseven 2008; Sadighi et al. 2008) as:

$$D_t^\alpha u(x,t) = \frac{1}{2}x^2 u_{xx}(x,t), \quad 0 < x < 1, \ t > 0, \ 0 < \alpha \le 1, \tag{3.9}$$

with the boundary conditions (BCs):

$$u(0,t) = 0, \quad u(1,t) = e^t. \tag{3.10}$$

and initial condition (IC):

$$u(x,0) = x^2. \tag{3.11}$$

Solution

Comparing Eq. (3.9) with Eq. (3.1), we have

$L = D_t^\alpha, Ru(x,t) = -\frac{1}{2}x^2 u_{xx}, Nu(x,t) = 0$ and source term $f(x,t) = 0$. Here $L^{-1} = J_t^\alpha$, the inverse operator of D_t^α. Now, applying the operator J_t^α on both sides of Eq. (3.9), we have

$$u(x,t) = J_t^\alpha\left(\frac{1}{2}x^2 u_{xx}(x,t)\right) + \phi. \tag{3.12}$$

Substituting Eq. (3.3) into Eq. (3.12), we get

$$u(x,t) = \sum_{n=0}^{\infty} u_n(x,t) = J_t^\alpha\left[\frac{1}{2}x^2\left(\sum_{n=0}^{\infty} u_n(x,t)\right)_{xx}\right] + \phi, \tag{3.13}$$

where $\phi = u(x,0) = x^2$.

From Eq. (3.8), the following expressions are obtained successively:

$$u_0 = L^{-1}f + \phi = x^2.$$

$$u_1 = -L^{-1}R(u_0) - L^{-1}A_0 = -J_t^\alpha\left(-\frac{1}{2}x^2(u_0(x,t))_{xx}\right) - 0 = x^2\frac{t^\alpha}{\Gamma(1+\alpha)}.$$

$$u_2 = -L^{-1}R(u_1) - L^{-1}A_1 = -J_t^\alpha\left(-\frac{1}{2}x^2(u_1(x,t))_{xx}\right) - 0 = x^2\frac{t^{2\alpha}}{\Gamma(1+2\alpha)}.$$

$$u_3 = -L^{-1}R(u_2) - L^{-1}A_2 = -J_t^\alpha\left(-\frac{1}{2}x^2(u_2(x,t))_{xx}\right) - 0 = x^2\frac{t^{3\alpha}}{\Gamma(1+3\alpha)}.$$

$$\vdots$$

$$u_{n+1} = -L^{-1}R(u_n) - L^{-1}A_n = -J_t^\alpha\left(-\frac{1}{2}x^2(u_n(x,t))_{xx}\right) - 0 = x^2\frac{t^{(n+1)\alpha}}{\Gamma(1+(n+1)\alpha)}.$$

So, the solution of the fractional heat-like Eq. (3.9) may be obtained as:

$$u(x,t) = \lim_{n\to\infty} u_n(x,t) = u_0(x,t) + u_1(x,t) + u_2(x,t) + \cdots$$

$$= x^2 + x^2\frac{t^\alpha}{\Gamma(1+\alpha)} + x^2\frac{t^{2\alpha}}{\Gamma(1+2\alpha)} + x^2\frac{t^{3\alpha}}{\Gamma(1+3\alpha)} + \cdots, \tag{3.14}$$

$$= x^2\left(1 + \frac{t^\alpha}{\Gamma(1+\alpha)} + \frac{t^{2\alpha}}{\Gamma(1+2\alpha)} + \frac{t^{3\alpha}}{\Gamma(1+3\alpha)} + \cdots\right) = x^2 E(t^\alpha),$$

where $E(t^\alpha)$ is called the Mittag-Leffer function (one may refer to Chapter 1).

In particular, at $\alpha = 1$, Eq. (3.14) reduces to $u(x,t) = x^2 e^t$ which is same as the solution of Sadighi et al. (2008). From Figure 3.1, it is concluded that if we increase the number of terms of solution, we may achieve a more accurate result. Figures 3.2–3.5 illustrate the fourth-order solution plots of Example 3.1 at different values of α.

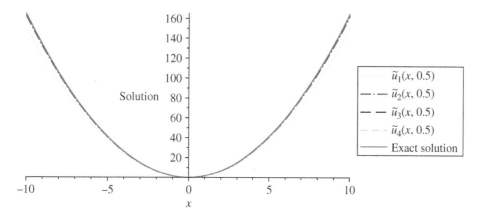

Figure 3.1 Comparison plot of the present solution with the exact solution of Example 3.1 taking a different number of terms of approximate solution.

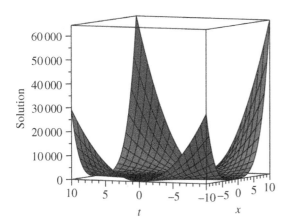

Figure 3.2 Fourth-order approximate solution plot of Example 3.1 at $\alpha = 1$.

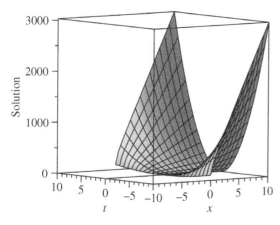

Figure 3.3 Fourth-order approximate solution plot of Example 3.1 at $\alpha = 0.3$.

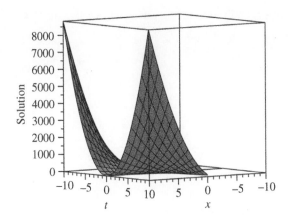

Figure 3.4 Fourth-order approximate solution plot of Example 3.1 at $\alpha = 0.5$.

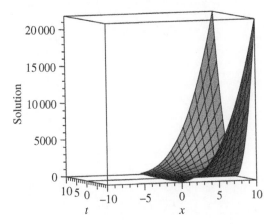

Figure 3.5 Fourth-order approximate solution plot of Example 3.1 at $\alpha = 0.7$.

Example 3.2 Consider the following nonlinear advection equation (Wazwaz 2007)

$$\frac{\partial^\alpha u}{\partial t^\alpha} + u\frac{\partial u}{\partial x} = 0, \quad 0 < \alpha \leq 1, \tag{3.15}$$

with the initial condition:

$$u(x, 0) = -x. \tag{3.16}$$

Solution

Equating Eq. (3.15) with Eq. (3.1), we have

$L = D_t^\alpha, Ru(x, t) = 0, Nu(x, t) = u\dfrac{\partial u}{\partial x}$ and source term $f(x, t) = 0$. On applying J_t^α on both sides of Eq. (3.15), we have

$$u(x, t) = -J_t^\alpha(u(x, t)u_x(x, t)) + \phi. \tag{3.17}$$

Then, in the decomposition method, each of the nonlinear terms uu_x is formally expanded in terms of power series as:

$$uu_x = \sum_{n=0}^{\infty} A_n, \tag{3.18}$$

where A_n are the Adomian polynomials. Substituting Eqs. (3.3) and (3.18), Eq. (3.17) reduces to

$$u(x, t) = \sum_{n=0}^{\infty} u_n(x, t) = -J_t^\alpha\left[\sum_{n=0}^{\infty} A_n\right] + \phi, \tag{3.19}$$

where $\phi = u(x, 0) = -x$.

In order to solve our problem, these Adomian polynomials have to be generalized in the following way (Adomian 1990; Wazwaz 1998):

$$A_n = \frac{1}{n!}\frac{d^n}{d\lambda^n}\left[N\left(\sum_{n=0}^{\infty}\lambda^n u_n\right)\left(\frac{\partial}{\partial x}\sum_{n=0}^{\infty}\lambda^n u_n\right)\right]_{\lambda=0}. \tag{3.20}$$

The first few terms of the Adomian polynomials are derived as follows:

$$A_0 = u_0\frac{\partial}{\partial x}u_0,$$

$$A_1 = \frac{d}{d\lambda}\left((u_0+u_1\lambda)\left(\frac{\partial u_0}{\partial x}+\lambda\frac{\partial u_1}{\partial x}\right)\right)\Bigg|_{\lambda=0} = u_0\frac{\partial u_1}{\partial x}+u_1\frac{\partial u_0}{\partial x},$$

$$A_2 = \frac{1}{2!}\frac{d^2}{d\lambda^2}\left((u_0+u_1\lambda+u_2\lambda^2)\left(\frac{\partial u_0}{\partial x}+\lambda\frac{\partial u_1}{\partial x}+\lambda^2\frac{\partial u_2}{\partial x}\right)\right)\Bigg|_{\lambda=0} = u_0\frac{\partial u_2}{\partial x}+u_1\frac{\partial u_1}{\partial x}+u_2\frac{\partial u_0}{\partial x},$$

similarly

$$A_3 = u_0\frac{\partial u_3}{\partial x}+u_1\frac{\partial u_2}{\partial x}+u_2\frac{\partial u_1}{\partial x}+u_3\frac{\partial u_0}{\partial x},$$

and so on.

Using Eq. (3.8), we may obtain

$$u_0 = L^{-1}f+\phi = -x.$$

$$u_1 = -L^{-1}R(u_0)-L^{-1}A_0 = -0-J_t^\alpha\left(u_0\frac{\partial}{\partial x}u_0\right) = -x\frac{t^\alpha}{\Gamma(1+\alpha)},$$

$$u_2 = -L^{-1}R(u_1)-L^{-1}A_1 = -0-J_t^\alpha\left(u_0\frac{\partial u_1}{\partial x}+u_1\frac{\partial u_0}{\partial x}\right) = -2x\frac{t^{2\alpha}}{\Gamma(1+2\alpha)},$$

$$u_3 = -L^{-1}R(u_2)-L^{-1}A_2 = -0-J_t^\alpha\left(u_0\frac{\partial u_2}{\partial x}+u_1\frac{\partial u_1}{\partial x}+u_2\frac{\partial u_0}{\partial x}\right) = -4x\frac{t^{3\alpha}}{\Gamma(1+3\alpha)}-\frac{x\Gamma(1+2\alpha)}{(\Gamma(1+\alpha))^2}\frac{t^{3\alpha}}{\Gamma(1+3\alpha)}.$$

$$\vdots$$

So, the solution of Eq. (3.15) may be obtained as:

$$u(x,t) = \lim_{n\to\infty}u_n(x,t) = u_0(x,t)+u_1(x,t)+u_2(x,t)+\cdots$$

$$= -x-x\frac{t^\alpha}{\Gamma(1+\alpha)}-2x\frac{t^{2\alpha}}{\Gamma(1+2\alpha)}-4x\frac{t^{3\alpha}}{\Gamma(1+3\alpha)}-\frac{x\Gamma(1+2\alpha)}{(\Gamma(1+\alpha))^2\Gamma(1+3\alpha)}-\cdots, \tag{3.21}$$

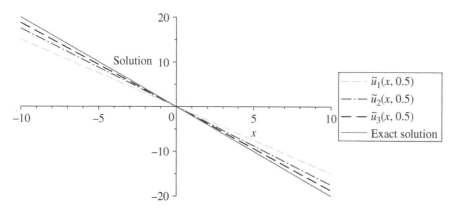

Figure 3.6 Comparison plot of the present solution with the exact solution of Example 3.2 taking a different number of terms of approximate solution.

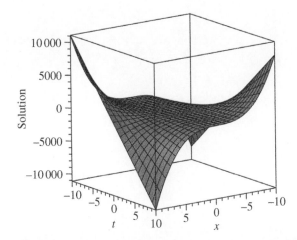

Figure 3.7 Third-order approximate solution plot of Example 3.2 at α = 1.

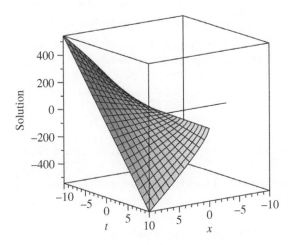

Figure 3.8 Third-order approximate solution plot of Example 3.2 at α = 0.3.

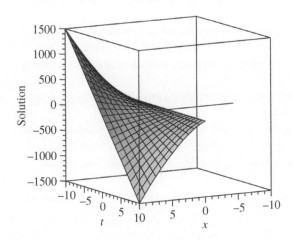

Figure 3.9 Third-order approximate solution plot of Example 3.2 at α = 0.5.

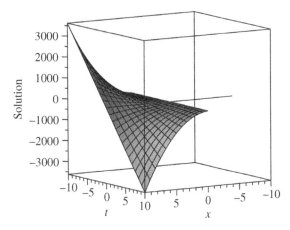

Figure 3.10 Third-order approximate solution plot of Example 3.2 at α = 0.7.

Particularly at $\alpha = 1$, Eq. (3.21) reduces to a closed-form solution $u(x, t) = \dfrac{x}{t-1}$, which is same as the solution of Wazwaz (2007). Figure 3.6 shows that increasing the number of terms in the solution may give a more accurate result. Figures 3.7–3.10 provide the third-order approximate solution plots of Example 3.2 for various α values.

References

Adomian, G. (1990). A review of the decomposition method and some recent results for nonlinear equations. *Mathematical and Computer Modelling* 13 (7): 17–43.

Bildik, N. and Konuralp, A. (2006). The use of variational iteration method, differential transform method and Adomian decomposition method for solving different types of nonlinear partial differential equations. *International Journal of Nonlinear Sciences and Numerical Simulation* 7 (1): 65–70.

Evans, D.J. and Raslan, K.R. (2005). The Adomian decomposition method for solving delay differential equation. *International Journal of Computer Mathematics* 82 (1): 49–54.

Ghorbani, A. (2009). Beyond Adomian polynomials: He polynomials. *Chaos, Solitons, and Fractals* 39 (3): 1486–1492.

Jafari, H. and Daftardar-Gejji, V. (2006). Revised Adomian decomposition method for solving systems of ordinary and fractional differential equations. *Applied Mathematics and Computation* 181 (1): 598–608.

Momani, S. and Odibat, Z. (2006). Analytical solution of a time-fractional Navier–Stokes equation by Adomian decomposition method. *Applied Mathematics and Computation* 177 (2): 488–494.

Özis, T. and Agırseven, D. (2008). He's homotopy perturbation method for solving heat-like and wave-like equations with variable coefficients. *Physics Letters A* 372: 5944–5950.

Sadighi, A., Ganji, D.D., Gorji, M., and Tolou, N. (2008). Numerical simulation of heat-like models with variable coefficients by the variational iteration method. *Journal of Physics: Conference Series* 96: 012083.

Wazwaz, A.M. (1998). A comparison between Adomian decomposition method and Taylor series method in the series solutions. *Applied Mathematics and Computation* 97 (1): 37–44.

Wazwaz, A.M. (1999). A reliable modification of Adomian decomposition method. *Applied Mathematics and Computation* 102 (1): 77–86.

Wazwaz, A.M. (2007). A comparison between the variational iteration method and adomian decomposition method. *Journal of Computational and Applied Mathematics* 207: 129–136.

4

Adomian Decomposition Transform Method

4.1 Introduction

In Chapter 3, we have already discussed the Adomian decomposition method (ADM), which is a semi-analytical approach for solving differential equations. In this chapter, we will discuss the hybrid methods, which are the coupling of ADM with various transform methods, viz. Laplace transform (LT), Sumudu transform (ST), Elzaki transform (ET), and Aboodh transform (AT). With the combination of these transform methods, ADM is called the Adomian decomposition transform method (ADTM) (Mohammed and Salim 2018; Ahmed et al. 2019; Thabet and Kendre 2019; Khalouta and Abdelouahab 2020). Although these four transform methods are helpful for solving fractional differential equations, these methods sometimes fail to handle nonlinear terms arising in the fractional differential equations. These difficulties may be overcome by coupling these transform methods with ADM. The theories behind the four transform methods with respect to fractional order are introduced in the subsequent sections. It is worth mentioning that two simple application example problems of fractional differential equations are investigated using all four methods.

4.2 Transform Methods for the Caputo Sense Derivatives

Definition 4.1   The LT of the Caputo fractional derivative is defined as (Baleanu and Jassim 2019; Faraz et al. 2010):

$$L\left[D_t^{n\alpha}u(x,t)\right] = s^{n\alpha}L[u(x,t)] - \sum_{k=0}^{n-1} s^{(n\alpha-k-1)}u^{(k)}(x,0), \quad n-1 < n\alpha \le 1, n \in N. \tag{4.1}$$

Definition 4.2   The ST of the Caputo fractional derivative is defined as (Jena and Chakraverty 2019):

$$S\left[D_t^{n\alpha}u(x,t)\right] = s^{-n\alpha}S[u(x,t)] - \sum_{k=0}^{n-1} s^{-n\alpha+k}u^{(k)}(x,0), \quad n-1 < n\alpha \le 1, n \in N. \tag{4.2}$$

Definition 4.3   The ET of the Caputo fractional derivative is defined as (Jena and Chakraverty 2019):

$$E\left[D_t^{n\alpha}u(x,t)\right] = \frac{E[u(x,t)]}{s^{n\alpha}} - \sum_{k=0}^{n-1} s^{k-n\alpha+2}u^{(k)}(x,0), \quad n-1 < n\alpha \le n, n \in N. \tag{4.3}$$

Definition 4.4   The AT of the Caputo fractional derivative is defined as (Aboodh et al. 2017; Jena and Chakraverty 2019):

$$A\left[D_t^{n\alpha}u(x,t)\right] = s^{n\alpha}A[u(x,t)] - \sum_{k=0}^{n-1} s^{-k+n\alpha-2}u^{(k)}(x,0), \quad n-1 < n\alpha \le n, n \in N. \tag{4.4}$$

Table 4.1 shows the transforms of some standard functions with respect to the aforementioned four transform methods and their definitions.

Following section deals with the systematic study of four hybrid methods, namely Adomian decomposition Laplace transform method (ADLTM), Adomian decomposition Sumudu transform method (ADSTM), Adomian decomposition Elzaki transform method (ADETM), and Adomian decomposition Aboodh transform method (ADATM), one after another.

Computational Fractional Dynamical Systems: Fractional Differential Equations and Applications, First Edition.
Snehashish Chakraverty, Rajarama Mohan Jena, and Subrat Kumar Jena.
© 2023 John Wiley & Sons, Inc. Published 2023 by John Wiley & Sons, Inc.

Table 4.1 Transforms of some essential functions.

Functions	Laplace transform	Sumudu transform	Elzaki transform	Aboodh transform
Definitions	$L[f(t)] = f(s)$	$S[g(t)] = g(s)$	$E[h(t)] = h(s)$	$A[p(t)] = p(s)$
	$= \int_0^\infty e^{-st} f(t) dt$	$= \int_0^\infty e^{-t} f(st) dt$	$= s \int_0^\infty e^{\frac{-t}{s}} f(t) dt$	$= \frac{1}{s} \int_0^\infty e^{-st} f(t) dt$
		$= \frac{1}{s} f\left(\frac{1}{s}\right)$	$= s f\left(\frac{1}{s}\right)$	$= \frac{1}{s} f(s)$
1	$\dfrac{1}{s}$	1	s^2	$\dfrac{1}{s^2}$
t^α	$\dfrac{\Gamma(1+\alpha)}{s^{\alpha+1}}$	$s^\alpha \Gamma(1+\alpha)$	$s^{\alpha+2} \Gamma(1+\alpha)$	$\dfrac{\Gamma(1+\alpha)}{s^{\alpha+2}}$
e^{at}	$\dfrac{1}{s-a}$	$\dfrac{1}{1-as}$	$\dfrac{s^2}{1-as}$	$\dfrac{1}{s(1-s)}$
$\sin(at)$	$\dfrac{a}{s^2+a^2}$	$\dfrac{as}{1+a^2 s^2}$	$\dfrac{as^3}{1+a^2 s^2}$	$\dfrac{a}{s(s^2+a^2)}$
$\cos(at)$	$\dfrac{s}{s^2+a^2}$	$\dfrac{1}{1+s^2 a^2}$	$\dfrac{s}{1+s^2 a^2}$	$\dfrac{1}{s^2+a^2}$

4.3 Adomian Decomposition Laplace Transform Method (ADLTM)

To clarify the fundamental idea of ADLTM, the fractional-order nonlinear nonhomogeneous partial differential equation (PDE) with initial conditions (ICs) are considered as:

$$D_t^{n\alpha} u(x,t) + Ru(x,t) + Nu(x,t) = f(x,t), \quad n-1 < n\alpha \le n. \tag{4.5}$$

subject to ICs

$$u^{(k)}(x,0) = g_k(x), \quad k = 0, 1, ..., n-1. \tag{4.6}$$

where $D_t^{n\alpha} = \dfrac{\partial^{n\alpha}}{\partial t^{n\alpha}}$ is the fractional differential operator in the Caputo sense, R, N are respectively linear and nonlinear differential operators, respectively, and $f(x,t)$ is the source term. The ADLTM approach involves mainly two stages. In the first stage, LT is taken on both sides of Eq. (4.5), and then in the second stage, ADM is applied where decomposition of the nonlinear term is done using Adomian polynomials. First, by operating LT on both sides of Eq. (4.5), we obtain

$$L[D_t^{n\alpha} u(x,t)] = L[f(x,t)] - L[Ru(x,t)] - L[Nu(x,t)]. \tag{4.7}$$

Using differentiation property (Eq. (4.1)) of LT, we obtain

$$s^{n\alpha} L[u(x,t)] - \sum_{k=0}^{n-1} s^{n\alpha-k-1} u^{(k)}(x,0) = L[f(x,t)] - L[Ru(x,t)] - L[Nu(x,t)]. \tag{4.8}$$

$$L[u(x,t)] = \frac{1}{s^{n\alpha}} \sum_{k=0}^{n-1} s^{n\alpha-k-1} u^{(k)}(x,0) + \frac{1}{s^{n\alpha}} L[f(x,t)] - \frac{1}{s^{n\alpha}} L[Ru(x,t)] - \frac{1}{s^{n\alpha}} L[Nu(x,t)]. \tag{4.9}$$

Applying inverse LT on both sides of Eq. (4.9), we find

$$u(x,t) = F(x,t) - L^{-1}\left(\frac{1}{s^{n\alpha}} L[Ru(x,t)]\right) - L^{-1}\left(\frac{1}{s^{n\alpha}} L[Nu(x,t)]\right). \tag{4.10}$$

Here, $F(x,t)$ represents the term coming from the IC and source term (first two terms on the right-hand side of Eq. (4.9)). Next, in order to implement ADM, first we need to consider the solution in series form as:

$$u(x,t) = \sum_{n=0}^{\infty} u_n(x,t). \tag{4.11}$$

and the nonlinear term may be decomposed by using Adomian polynomials (Wazwaz 1999) as:

$$Nu(x,t) = \sum_{n=0}^{\infty} A_n. \tag{4.12}$$

where A_n denotes the Adomian polynomials and which is defined as follows:

$$A_n(u_0, u_1, \ldots, u_n) = \frac{1}{n!} \frac{\partial^n}{\partial \lambda^n} \left[N\left(\sum_{i=0}^{\infty} \lambda^i u_i(x,t) \right) \right]_{\lambda=0}, \quad n = 0, 1, 2, \ldots \tag{4.13}$$

One may go through the reference (Wazwaz 1999) for a detailed derivation of Eq. (4.13), and the first few terms of A_n are included in Chapter 3. Substituting Eqs. (4.11) and (4.12) into Eq. (4.10), one may get the following expression:

$$\sum_{n=0}^{\infty} u_n(x,t) = F(x,t) - L^{-1}\left(\frac{1}{s^{n\alpha}} L\left[R \sum_{n=0}^{\infty} u_n(x,t) \right] \right) - L^{-1}\left(\frac{1}{s^{n\alpha}} L\left[\sum_{n=0}^{\infty} A_n \right] \right). \tag{4.14}$$

An iterative algorithm may be obtained by matching both sides of Eq. (4.14) as follows:

$$u_0(x,t) = F(x,t),$$

$$u_1(x,t) = -L^{-1}\left(\frac{1}{s^{n\alpha}} L[Ru_0(x,t)] \right) - L^{-1}\left(\frac{1}{s^{n\alpha}} L[A_0] \right),$$

$$u_2(x,t) = -L^{-1}\left(\frac{1}{s^{n\alpha}} L[Ru_1(x,t)] \right) - L^{-1}\left(\frac{1}{s^{n\alpha}} L[A_1] \right),$$

$$u_3(x,t) = -L^{-1}\left(\frac{1}{s^{n\alpha}} L[Ru_2(x,t)] \right) - L^{-1}\left(\frac{1}{s^{n\alpha}} L[A_2] \right),$$

$$\vdots$$

$$u_n(x,t) = -L^{-1}\left(\frac{1}{s^{n\alpha}} L[Ru_{n-1}(x,t)] \right) - L^{-1}\left(\frac{1}{s^{n\alpha}} L[A_{n-1}] \right)$$

$$\vdots$$

So, the solution of Eq. (4.5) may be obtained as:

$$u(x,t) = \lim_{n \to \infty} u_n(x,t) = u_0(x,t) + u_1(x,t) + u_2(x,t) + \cdots, \tag{4.15}$$

4.4 Adomian Decomposition Sumudu Transform Method (ADSTM)

Similar to the aforementioned procedure, applying ST on both sides of Eq. (4.5), we have

$$S[D_t^{n\alpha} u(x,t)] = S[f(x,t)] - S[Ru(x,t)] - S[Nu(x,t)]. \tag{4.16}$$

Using differentiation property of (Eq. (4.2)) for ST in Eq. (4.16), we obtain

$$s^{-n\alpha} S[u(x,t)] - \sum_{k=0}^{n-1} s^{-n\alpha+k} u^{(k)}(x,0) = S[f(x,t)] - S[Ru(x,t)] - S[Nu(x,t)]. \tag{4.17}$$

$$S[u(x,t)] = s^{n\alpha} \sum_{k=0}^{n-1} s^{-n\alpha+k} u^{(k)}(x,0) + s^{n\alpha} S[f(x,t)] - s^{n\alpha} S[Ru(x,t)] - s^{n\alpha} S[Nu(x,t)]. \tag{4.18}$$

Applying inverse ST on both sides of Eq. (4.18), we find

$$u(x,t) = F(x,t) - S^{-1}(s^{n\alpha} S[Ru(x,t)]) - S^{-1}(s^{n\alpha} S[Nu(x,t)]). \tag{4.19}$$

One may see reference (Jena and Chakraverty 2019) for a detailed description of this method. Substituting Eqs. (4.11) and (4.12) into Eq. (4.19), we may have the following expression:

$$\sum_{n=0}^{\infty} u_n(x,t) = F(x,t) - \left(S^{-1}\left(s^{n\alpha} S\left[R \sum_{n=0}^{\infty} u_n(x,t) \right] \right) + S^{-1}\left(s^{n\alpha} S\left[\sum_{n=0}^{\infty} A_n \right] \right) \right). \tag{4.20}$$

By comparing both sides of Eq. (4.20), we may have the following expressions:

$u_0(x,t) = F(x,t),$

$u_1(x,t) = -S^{-1}(s^{n\alpha}S[Ru_0(x,t)]) - S^{-1}(s^{n\alpha}S[A_0]),$

$u_2(x,t) = -S^{-1}(s^{n\alpha}S[Ru_1(x,t)]) - S^{-1}(s^{n\alpha}S[A_1]),$

$u_3(x,t) = -S^{-1}(s^{n\alpha}S[Ru_2(x,t)]) - S^{-1}(s^{n\alpha}S[A_2]),$

$\vdots$

$u_n(x,t) = -S^{-1}(s^{n\alpha}S[Ru_{n-1}(x,t)]) - S^{-1}(s^{n\alpha}S[A_{n-1}]),$

$\vdots$

So, the solution of Eq. (4.5) may be obtained as:

$$u(x,t) = \lim_{n \to \infty} u_n(x,t) = u_0(x,t) + u_1(x,t) + u_2(x,t) + \cdots, \tag{4.21}$$

4.5 Adomian Decomposition Elzaki Transform Method (ADETM)

Applying ET on both sides of Eq. (4.5), we obtain

$$E\left[D_t^{n\alpha}u(x,t)\right] = E[f(x,t)] - E[Ru(x,t)] - E[Nu(x,t)]. \tag{4.22}$$

Using differentiation property (Eq. (4.3)) of ET, we have

$$s^{-n\alpha}E[u(x,t)] - \sum_{k=0}^{n-1} s^{k-n\alpha+2}u^{(k)}(x,0) = E[f(x,t)] - E[Ru(x,t)] - E[Nu(x,t)]. \tag{4.23}$$

$$E[u(x,t)] = s^{n\alpha}\sum_{k=0}^{n-1} s^{k-n\alpha+2}u^{(k)}(x,0) + s^{n\alpha}E[f(x,t)] - s^{n\alpha}E[Ru(x,t)] - s^{n\alpha}E[Nu(x,t)]. \tag{4.24}$$

Inverse ET on both sides of Eq. (4.24) reduces to the following equation:

$$u(x,t) = F(x,t) - E^{-1}(s^{n\alpha}E[Ru(x,t)]) - E^{-1}(s^{n\alpha}E[Nu(x,t)]). \tag{4.25}$$

By plugging Eqs. (4.11) and (4.12) into Eq. (4.25), we have the expression as follows:

$$\sum_{n=0}^{\infty} u_n(x,t) = F(x,t) - \left(E^{-1}\left(s^{n\alpha} E\left[R \sum_{n=0}^{\infty} u_n(x,t) \right] \right) + E^{-1}\left(s^{n\alpha} E\left[\sum_{n=0}^{\infty} A_n \right] \right) \right). \tag{4.26}$$

Comparing both sides of Eq. (4.26), we have the following approximations successively:

$u_0(x,t) = F(x,t),$

$u_1(x,t) = -E^{-1}(s^{n\alpha}E[Ru_0(x,t)]) - E^{-1}(s^{n\alpha}E[A_0]),$

$u_2(x,t) = -E^{-1}(s^{n\alpha}E[Ru_1(x,t)]) - E^{-1}(s^{n\alpha}E[A_1]),$

$u_3(x,t) = -E^{-1}(s^{n\alpha}E[Ru_2(x,t)]) - E^{-1}(s^{n\alpha}E[A_2]),$

$\vdots$

$u_n(x,t) = -E^{-1}(s^{n\alpha}E[Ru_{n-1}(x,t)]) - E^{-1}(s^{n\alpha}E[A_{n-1}]),$

$\vdots$

So, the solution of Eq. (4.5) may be written as:

$$u(x, t) = \lim_{n \to \infty} u_n(x, t) = u_0(x, t) + u_1(x, t) + u_2(x, t) + \cdots, \tag{4.27}$$

4.6 Adomian Decomposition Aboodh Transform Method (ADATM)

Applying AT on both sides of Eq. (4.5) reduces to

$$A\left[D_t^{n\alpha} u(x, t)\right] = A[f(x, t)] - A[Ru(x, t)] - A[Nu(x, t)]. \tag{4.28}$$

Further, using Eq. (4.4), Eq. (4.28) reduces to the following expression:

$$s^{n\alpha} A[u(x, t)] - \sum_{k=0}^{n-1} s^{-k+n\alpha-2} u^{(k)}(x, 0) = A[f(x, t)] - A[Ru(x, t)] - A[Nu(x, t)]. \tag{4.29}$$

$$A[u(x, t)] = s^{-n\alpha} \sum_{k=0}^{n-1} s^{-k+n\alpha-2} u^{(k)}(x, 0) + s^{-n\alpha} A[f(x, t)] - s^{-n\alpha} A[Ru(x, t)] - s^{-n\alpha} A[Nu(x, t)]. \tag{4.30}$$

Applying inverse AT on both sides of Eq. (4.30), we get

$$u(x, t) = F(x, t) - A^{-1}(s^{-n\alpha} A[Ru(x, t)]) - A^{-1}(s^{-n\alpha} A[Nu(x, t)]). \tag{4.31}$$

A detailed description of this transform may be found in the reference (Aboodh et al. 2017; Jena and Chakraverty 2019). Putting Eqs. (4.11) and (4.12) into Eq. (4.31), we get

$$\sum_{n=0}^{\infty} u_n(x, t) = F(x, t) - \left(A^{-1}\left(s^{-n\alpha} A\left[R \sum_{n=0}^{\infty} u_n(x, t)\right]\right) + A^{-1}\left(s^{-n\alpha} A\left[\sum_{n=0}^{\infty} A_n\right]\right)\right). \tag{4.32}$$

Comparing both sides of Eq. (4.32), we have the successive approximations as:

$$u_0(x, t) = F(x, t),$$
$$u_1(x, t) = -A^{-1}(s^{-n\alpha} A[Ru_0(x, t)]) - A^{-1}(s^{-n\alpha} A[A_0]),$$
$$u_2(x, t) = -A^{-1}(s^{-n\alpha} A[Ru_1(x, t)]) - A^{-1}(s^{-n\alpha} A[A_1]),$$
$$u_3(x, t) = -A^{-1}(s^{-n\alpha} A[Ru_2(x, t)]) - A^{-1}(s^{-n\alpha} A[A_2]),$$
$$\vdots$$
$$u_n(x, t) = -A^{-1}(s^{-n\alpha} A[Ru_{n-1}(x, t)]) - A^{-1}(s^{-n\alpha} A[A_{n-1}]),$$
$$\vdots$$

So, the solution of Eq. (4.5) may be written as:

$$u(x, t) = \lim_{n \to \infty} u_n(x, t) = u_0(x, t) + u_1(x, t) + u_2(x, t) + \cdots, \tag{4.33}$$

Next, we solve two test problems in each of the aforementioned methods to demonstrate the present techniques.

4.7 Numerical Examples

4.7.1 Implementation of ADLTM

Example 4.1 Let us consider the one-dimensional heat-like model (Özis and Agırseven 2008; Sadighi et al. 2008) as:

$$D_t^\alpha u(x, t) = \frac{1}{2} x^2 u_{xx}(x, t), \quad 0 < x < 1, t > 0, \ 0 < \alpha \le 1, \tag{4.34}$$

subject to the boundary conditions (BCs):

$$u(0,t) = 0, u(1,t) = e^t, \tag{4.35}$$

and IC:

$$u(x,\ 0) = x^2. \tag{4.36}$$

Solution

Applying LT on both sides of Eq. (4.34) with IC and BCs, we have

$$s^\alpha L[u(x,t)] - s^{\alpha-1}u(x,0) = \frac{1}{2}L\left[x^2 u_{xx}\right]. \tag{4.37}$$

$$L[u(x,t)] = \frac{s^{-\alpha}}{2}L\left[x^2 u_{xx}\right] + \frac{u(x,0)}{s}. \tag{4.38}$$

$$L[u(x,t)] = \frac{s^{-\alpha}}{2}L\left[x^2 u_{xx}\right] + \frac{x^2}{s}. \tag{4.39}$$

Taking inverse LT on both sides of Eq. (4.39) gives

$$u(x,t) = L^{-1}\left(\frac{s^{-\alpha}}{2}L\left[x^2 u_{xx}\right]\right) + x^2. \tag{4.40}$$

According to the ADLTM, assuming a series solution for the PDE as Eq. (4.11) and by replacing the nonlinear term by Eq. (4.12) (in this problem, there has no nonlinear term), we obtain

$$\sum_{n=0}^{\infty} u_n(x,t) = x^2 + L^{-1}\left(\frac{s^{-\alpha}}{2}L\left[x^2 \sum_{n=0}^{\infty} (u_n(x,t))_{xx}\right]\right). \tag{4.41}$$

By comparing both sides of Eq. (4.41), we obtain

$$u_0(x,t) = x^2,$$

$$u_1(x,t) = L^{-1}\left(\frac{s^{-\alpha}}{2}L\left[x^2 u_{0xx}\right]\right)$$

$$= L^{-1}\left(\frac{s^{-\alpha}}{2}L\left[x^2 \frac{\partial^2}{\partial x^2}u_0\right]\right)$$

$$= L^{-1}\left(\frac{s^{-\alpha}}{2}L\left[2x^2\right]\right)$$

$$= L^{-1}\left(\frac{x^2}{s^{\alpha+1}}\right) = x^2\frac{t^\alpha}{\Gamma(1+\alpha)},$$

$$u_2(x,t) = L^{-1}\left(\frac{s^{-\alpha}}{2}L\left[x^2 u_{1xx}\right]\right)$$

$$= L^{-1}\left(\frac{s^{-\alpha}}{2}L\left[x^2 \frac{\partial^2}{\partial x^2}u_1\right]\right)$$

$$= L^{-1}\left(\frac{s^{-\alpha}}{2}L\left[\frac{2t^\alpha}{\Gamma(1+\alpha)}x^2\right]\right)$$

$$= L^{-1}\left(\frac{x^2}{s^{2\alpha+1}}\right) = x^2\frac{t^{2\alpha}}{\Gamma(1+2\alpha)},$$

$$u_3(x,t) = L^{-1}\left(\frac{s^{-\alpha}}{2}L\left[x^2 u_{2xx}\right]\right)$$

$$= x^2\frac{t^{3\alpha}}{\Gamma(1+3\alpha)},$$

$$\vdots$$

So, the solution of the fractional heat-like Eq. (4.34) may be obtained as:

$$u(x,t) = \lim_{n \to \infty} u_n(x,t) = u_0(x,t) + u_1(x,t) + u_2(x,t) + \cdots$$

$$= x^2 + x^2 \frac{t^\alpha}{\Gamma(1+\alpha)} + x^2 \frac{t^{2\alpha}}{\Gamma(1+2\alpha)} + x^2 \frac{t^{3\alpha}}{\Gamma(1+3\alpha)} + \cdots, \tag{4.42}$$

$$= x^2 \left(1 + \frac{t^\alpha}{\Gamma(1+\alpha)} + \frac{t^{2\alpha}}{\Gamma(1+2\alpha)} + \frac{t^{3\alpha}}{\Gamma(1+3\alpha)} + \cdots \right) = x^2 E(t^\alpha),$$

where $E(t^\alpha)$ is called the Mittag-Leffer function. In particular, at $\alpha = 1$, Eq. (4.42) reduces to $u(x,t) = x^2 e^t$, which is same as the solution of Sadighi et al. (2008).

Example 4.2 Consider the following nonlinear advection equation (Wazwaz 2007)

$$\frac{\partial^\alpha u}{\partial t^\alpha} + u \frac{\partial u}{\partial x} = 0, \quad 0 < \alpha \leq 1, \tag{4.43}$$

with IC:

$$u(x,0) = -x. \tag{4.44}$$

Solution

Applying LT on both sides of Eq. (4.43) with IC, we have

$$s^\alpha L[u(x,t)] - s^{\alpha-1} u(x,0) = -L[uu_x]. \tag{4.45}$$

$$L[u(x,t)] = -s^{-\alpha} L[uu_x] + \frac{u(x,0)}{s}. \tag{4.46}$$

$$L[u(x,t)] = -s^{-\alpha} L[uu_x] - \frac{x}{s}. \tag{4.47}$$

Operating inverse LT on both sides of Eq. (4.47) gives

$$u(x,t) = -L^{-1}(s^{-\alpha} L[uu_x]) - x. \tag{4.48}$$

As per the ADLTM, assuming a series solution for the PDE as Eq. (4.11) and by replacing the nonlinear term by Eq. (4.12), we get

$$\sum_{n=0}^{\infty} u_n(x,t) = -x - L^{-1}\left(s^{-\alpha} L\left[\sum_{n=0}^{\infty} A_n \right] \right). \tag{4.49}$$

Where the Adomian polynomials A_n for the nonlinear term uu_x are given by:

$$A_0 = u_0 \frac{\partial}{\partial x} u_0,$$

$$A_1 = u_0 \frac{\partial}{\partial x} u_1 + u_1 \frac{\partial}{\partial x} u_0, \tag{4.50}$$

$$A_2 = u_0 \frac{\partial}{\partial x} u_2 + u_1 \frac{\partial}{\partial x} u_1 + u_2 \frac{\partial}{\partial x} u_0,$$

$$\vdots$$

Next, the successive approximations to the solution of Eq. (4.43) may be found as:

$$u_0(x, t) = -x,$$

$$u_1(x, t) = -L^{-1}(s^{-\alpha}L[A_0])$$

$$= -L^{-1}\left(s^{-\alpha}L\left[u_0\frac{\partial}{\partial x}u_0\right]\right)$$

$$= -L^{-1}(s^{-\alpha}L[x])$$

$$= -L^{-1}\left(\frac{x}{s^{\alpha+1}}\right) = -x\frac{t^\alpha}{\Gamma(1+\alpha)},$$

$$u_2(x, t) = -L^{-1}(s^{-\alpha}L[A_1])$$

$$= -L^{-1}\left(s^{-\alpha}L\left[u_0\frac{\partial}{\partial x}u_1 + u_1\frac{\partial}{\partial x}u_0\right]\right)$$

$$= -L^{-1}\left(s^{-\alpha}L\left[2x\frac{t^\alpha}{\Gamma(1+\alpha)}\right]\right)$$

$$= -L^{-1}\left(\frac{2x}{s^{2\alpha+1}}\right) = -2x\frac{t^{2\alpha}}{\Gamma(1+2\alpha)},$$

$$u_3(x, t) = -L^{-1}(s^{-\alpha}L[A_2])$$

$$= -L^{-1}\left(s^{-\alpha}L\left[u_0\frac{\partial}{\partial x}u_2 + u_1\frac{\partial}{\partial x}u_1 + u_2\frac{\partial}{\partial x}u_0\right]\right)$$

$$= -L^{-1}\left(s^{-\alpha}L\left[4x\frac{t^{2\alpha}}{\Gamma(1+2\alpha)} + x\frac{t^{2\alpha}}{(\Gamma(1+\alpha))^2}\right]\right)$$

$$= -L^{-1}\left(\frac{4x}{s^{3\alpha+1}} + \frac{x\Gamma(1+2\alpha)}{(\Gamma(1+\alpha))^2 s^{3\alpha+1}}\right),$$

$$= -4x\frac{t^{3\alpha}}{\Gamma(1+3\alpha)} - \frac{x\Gamma(1+2\alpha)}{(\Gamma(1+\alpha))^2}\frac{t^{3\alpha}}{\Gamma(1+3\alpha)}$$

$$\vdots$$

So, the solution to Eq. (4.43) may be obtained as:

$$u(x, t) = \lim_{n\to\infty}u_n(x, t) = u_0(x, t) + u_1(x, t) + u_2(x, t) + \cdots$$

$$= -x - x\frac{t^\alpha}{\Gamma(1+\alpha)} - 2x\frac{t^{2\alpha}}{\Gamma(1+2\alpha)} - 4x\frac{t^{3\alpha}}{\Gamma(1+3\alpha)} - \frac{x\Gamma(1+2\alpha)}{(\Gamma(1+\alpha))^2}\frac{t^{3\alpha}}{\Gamma(1+3\alpha)} - \cdots, \tag{4.51}$$

Particularly at $\alpha = 1$, Eq. (4.51) reduces to closed-form solution $u(x, t) = \frac{x}{t-1}$, which is same as the solution of Waz-waz (2007).

4.7.2 Implementation of ADSTM

Applying ST on both sides of Eq. (4.34) with Eqs. (4.35) and (4.36), we obtain

$$s^{-\alpha}S[u(x, t)] - s^{-\alpha}u(x, 0) = \frac{1}{2}S[x^2 u_{xx}]. \tag{4.52}$$

$$S[u(x, t)] = \frac{s^\alpha}{2}S[x^2 u_{xx}] + u(x, 0). \tag{4.53}$$

$$S[u(x, t)] = \frac{s^\alpha}{2}S[x^2 u_{xx}] + x^2. \tag{4.54}$$

Taking inverse ST on both sides of Eq. (4.54) gives

$$u(x, t) = S^{-1}\left(\frac{s^\alpha}{2}S[x^2 u_{xx}]\right) + x^2. \tag{4.55}$$

Applying ADM, we have

$$\sum_{n=0}^{\infty} u_n(x,t) = x^2 + S^{-1}\left(\frac{s^\alpha}{2}S\left[x^2 \sum_{n=0}^{\infty} (u_n(x,t))_{xx}\right]\right). \tag{4.56}$$

Comparing both sides of Eq. (4.56), we get

$$u_0(x,t) = x^2,$$

$$u_1(x,t) = S^{-1}\left(\frac{s^\alpha}{2}S[x^2 u_{0xx}]\right)$$

$$= S^{-1}\left(\frac{s^\alpha}{2}S\left[x^2 \frac{\partial^2}{\partial x^2}u_0\right]\right)$$

$$= S^{-1}\left(\frac{s^\alpha}{2}S[2x^2]\right)$$

$$= S^{-1}(x^2 s^\alpha) = x^2 \frac{t^\alpha}{\Gamma(1+\alpha)},$$

$$u_2(x,t) = S^{-1}\left(\frac{s^\alpha}{2}S[x^2 u_{1xx}]\right)$$

$$= S^{-1}\left(\frac{s^\alpha}{2}S\left[x^2 \frac{\partial^2}{\partial x^2}u_1\right]\right)$$

$$= S^{-1}\left(\frac{s^\alpha}{2}S\left[\frac{2t^\alpha}{\Gamma(1+\alpha)}x^2\right]\right)$$

$$= S^{-1}(x^2 s^{2\alpha}) = x^2 \frac{t^{2\alpha}}{\Gamma(1+2\alpha)},$$

$$u_3(x,t) = S^{-1}\left(\frac{s^\alpha}{2}S[x^2 u_{2xx}]\right)$$

$$= x^2 \frac{t^{3\alpha}}{\Gamma(1+3\alpha)},$$

$$\vdots$$

So, the solution of the fractional heat-like Eq. (4.34) may be obtained as:

$$u(x,t) = x^2 + x^2\frac{t^\alpha}{\Gamma(1+\alpha)} + x^2\frac{t^{2\alpha}}{\Gamma(1+2\alpha)} + x^2\frac{t^{3\alpha}}{\Gamma(1+3\alpha)} + \cdots = x^2 E(t^\alpha), \tag{4.57}$$

In particular, at $\alpha = 1$, Eq. (4.57) reduces to $u(x,t) = x^2 e^t$, which is same as the solution of Sadighi et al. (2008). Further, operating ST on both sides of Eq. (4.43) with IC, we have

$$s^{-\alpha}S[u(x,t)] - s^{-\alpha}u(x,0) = -S[uu_x]. \tag{4.58}$$

$$S[u(x,t)] = -s^\alpha S[uu_x] + \frac{u(x,0)}{s^\alpha}. \tag{4.59}$$

$$S[u(x,t)] = -s^\alpha S[uu_x] - \frac{x}{s^\alpha}. \tag{4.60}$$

Applying inverse ST on both sides of Eq. (4.60) gives

$$u(x,t) = -S^{-1}(s^\alpha S[uu_x]) - x. \tag{4.61}$$

By ADM, we have

$$\sum_{n=0}^{\infty} u_n(x,t) = -x - S^{-1}\left(s^\alpha S\left[\sum_{n=0}^{\infty} A_n\right]\right). \tag{4.62}$$

Using Eq. (4.50) and comparing both sides of Eq. (4.62), we obtain

$$u_0(x, t) = -x,$$

$$
\begin{aligned}
u_1(x, t) &= -S^{-1}(s^\alpha S[A_0]) \\
&= -S^{-1}\left(s^\alpha S\left[u_0 \frac{\partial}{\partial x} u_0\right]\right) \\
&= -S^{-1}(s^\alpha S[x]) \\
&= -S^{-1}(xs^\alpha) = -x\frac{t^\alpha}{\Gamma(1+\alpha)},
\end{aligned}
$$

$$
\begin{aligned}
u_2(x, t) &= -S^{-1}(s^\alpha S[A_1]) \\
&= -S^{-1}\left(s^\alpha S\left[u_0 \frac{\partial}{\partial x} u_1 + u_1 \frac{\partial}{\partial x} u_0\right]\right) \\
&= -S^{-1}\left(s^\alpha S\left[2x\frac{t^\alpha}{\Gamma(1+\alpha)}\right]\right) \\
&= -S^{-1}(2xs^{2\alpha}) = -2x\frac{t^{2\alpha}}{\Gamma(1+2\alpha)},
\end{aligned}
$$

$$
\begin{aligned}
u_3(x, t) &= -S^{-1}(s^\alpha S[A_2]) \\
&= -S^{-1}\left(s^\alpha S\left[u_0 \frac{\partial}{\partial x} u_2 + u_1 \frac{\partial}{\partial x} u_1 + u_2 \frac{\partial}{\partial x} u_0\right]\right) \\
&= -S^{-1}\left(s^\alpha S\left[4x\frac{t^{2\alpha}}{\Gamma(1+2\alpha)} + x\frac{t^{2\alpha}}{(\Gamma(1+\alpha))^2}\right]\right) \\
&= -S^{-1}\left(4xs^{3\alpha} + \frac{x\Gamma(1+2\alpha)s^{3\alpha}}{(\Gamma(1+\alpha))^2}\right), \\
&= -4x\frac{t^{3\alpha}}{\Gamma(1+3\alpha)} - \frac{x\Gamma(1+2\alpha)}{(\Gamma(1+\alpha))^2}\frac{t^{3\alpha}}{\Gamma(1+3\alpha)}
\end{aligned}
$$

$$\vdots$$

So, the solution to Eq. (4.43) may be obtained as:

$$u(x, t) = = -x - x\frac{t^\alpha}{\Gamma(1+\alpha)} - 2x\frac{t^{2\alpha}}{\Gamma(1+2\alpha)} - 4x\frac{t^{3\alpha}}{\Gamma(1+3\alpha)} - \frac{x\Gamma(1+2\alpha)}{(\Gamma(1+\alpha))^2}\frac{t^{3\alpha}}{\Gamma(1+3\alpha)} - \ldots\ldots\ldots\ldots, \tag{4.63}$$

which is same as the solution of Wazwaz (Wazwaz, 2007) at $\alpha = 1$.

4.7.3 Implementation of ADETM

Now, taking ET on both sides of Eq. (4.34) with IC and BCs, we get

$$s^{-\alpha}E[u(x, t)] - s^{-\alpha+2}u(x, 0) = \frac{1}{2}E[x^2 u_{xx}]. \tag{4.64}$$

$$E[u(x, t)] = s^2 u(x, 0) + \frac{s^\alpha}{2}E[x^2 u_{xx}]. \tag{4.65}$$

$$E[u(x, t)] = s^2 x^2 + \frac{s^\alpha}{2}E[x^2 u_{xx}]. \tag{4.66}$$

Inverse ET on both sides of Eq. (4.66) gives

$$u(x, t) = x^2 + E^{-1}\left(\frac{s^\alpha}{2}E[x^2 u_{xx}]\right). \tag{4.67}$$

Through ADM, we have

$$\sum_{n=0}^{\infty} u_n(x,t) = x^2 + E^{-1}\left(\frac{s^\alpha}{2}E\left[x^2\sum_{n=0}^{\infty}(u_n(x,t))_{xx}\right]\right). \tag{4.68}$$

Comparing both sides of Eq. (4.68), we obtain

$$u_0(x,t) = x^2,$$

$$u_1(x,t) = E^{-1}\left(\frac{s^\alpha}{2}E\left[x^2 u_{0xx}\right]\right)$$

$$= E^{-1}\left(\frac{s^\alpha}{2}E\left[x^2\frac{\partial^2}{\partial x^2}u_0\right]\right)$$

$$= E^{-1}\left(\frac{s^\alpha}{2}E\left[2x^2\right]\right)$$

$$= E^{-1}(x^2 s^\alpha) = x^2\frac{t^\alpha}{\Gamma(1+\alpha)},$$

$$u_2(x,t) = E^{-1}\left(\frac{s^\alpha}{2}E\left[x^2 u_{1xx}\right]\right)$$

$$= E^{-1}\left(\frac{s^\alpha}{2}E\left[x^2\frac{\partial^2}{\partial x^2}u_1\right]\right)$$

$$= E^{-1}\left(\frac{s^\alpha}{2}E\left[\frac{2t^\alpha}{\Gamma(1+\alpha)}x^2\right]\right)$$

$$= E^{-1}(x^2 s^{2\alpha}) = x^2\frac{t^{2\alpha}}{\Gamma(1+2\alpha)},$$

$$u_3(x,t) = S^{-1}\left(\frac{s^\alpha}{2}E\left[x^2 u_{2xx}\right]\right)$$

$$= x^2\frac{t^{3\alpha}}{\Gamma(1+3\alpha)},$$

$$\vdots$$

So, the solution of Eq. (4.34) may be obtained as:

$$u(x,t) = = x^2 + x^2\frac{t^\alpha}{\Gamma(1+\alpha)} + x^2\frac{t^{2\alpha}}{\Gamma(1+2\alpha)} + x^2\frac{t^{3\alpha}}{\Gamma(1+3\alpha)} + \cdots \tag{4.69}$$

At $\alpha = 1$, Eq. (4.69) is same as the solution of Sadighi et al. (2008).
Again by operating ET on both sides of Eq. (4.43) with IC, we have

$$s^{-\alpha}E[u(x,t)] - s^{-\alpha+2}u(x,0) = -E[uu_x]. \tag{4.70}$$

$$E[u(x,t)] = -s^\alpha E[uu_x] + s^2 u(x,0). \tag{4.71}$$

$$E[u(x,t)] = -s^\alpha E[uu_x] - xs^2. \tag{4.72}$$

By inverse ET on both sides of Eq. (4.72), we have

$$u(x,t) = -E^{-1}(s^\alpha E[uu_x]) - x. \tag{4.73}$$

By ADM, we have

$$\sum_{n=0}^{\infty} u_n(x,t) = -x - E^{-1}\left(s^\alpha E\left[\sum_{n=0}^{\infty}A_n\right]\right). \tag{4.74}$$

Using Eq. (4.50) and comparing both sides of Eq. (4.74), we obtain

$$u_0(x,t) = -x,$$

$$u_1(x,t) = -E^{-1}(s^\alpha E[A_0])$$

$$= -E^{-1}\left(s^\alpha E\left[u_0\frac{\partial}{\partial x}u_0\right]\right)$$

$$= -E^{-1}(s^\alpha E[x])$$

$$= -E^{-1}(xs^{\alpha+2}) = -x\frac{t^\alpha}{\Gamma(1+\alpha)},$$

$$u_2(x,t) = -E^{-1}(s^\alpha E[A_1])$$

$$= -E^{-1}\left(s^\alpha E\left[u_0\frac{\partial}{\partial x}u_1 + u_1\frac{\partial}{\partial x}u_0\right]\right)$$

$$= -E^{-1}\left(s^\alpha E\left[2x\frac{t^\alpha}{\Gamma(1+\alpha)}\right]\right)$$

$$= -E^{-1}(2xs^{2\alpha+2}) = -2x\frac{t^{2\alpha}}{\Gamma(1+2\alpha)},$$

$$u_3(x,t) = -E^{-1}(s^\alpha E[A_2])$$

$$= -E^{-1}\left(s^\alpha E\left[u_0\frac{\partial}{\partial x}u_2 + u_1\frac{\partial}{\partial x}u_1 + u_2\frac{\partial}{\partial x}u_0\right]\right)$$

$$= -E^{-1}\left(s^\alpha E\left[4x\frac{t^{2\alpha}}{\Gamma(1+2\alpha)} + x\frac{t^{2\alpha}}{(\Gamma(1+\alpha))^2}\right]\right)$$

$$= -E^{-1}\left(4xs^{3\alpha+2} + \frac{x\Gamma(1+2\alpha)s^{3\alpha+2}}{(\Gamma(1+\alpha))^2}\right),$$

$$= -4x\frac{t^{3\alpha}}{\Gamma(1+3\alpha)} - \frac{x\Gamma(1+2\alpha)}{(\Gamma(1+\alpha))^2}\frac{t^{3\alpha}}{\Gamma(1+3\alpha)}$$

$$\vdots$$

So, the solution to Eq. (4.43) may be written as:

$$u(x,t) = -x - x\frac{t^\alpha}{\Gamma(1+\alpha)} - 2x\frac{t^{2\alpha}}{\Gamma(1+2\alpha)} - 4x\frac{t^{3\alpha}}{\Gamma(1+3\alpha)} - \frac{x\Gamma(1+2\alpha)}{(\Gamma(1+\alpha))^2}\frac{t^{3\alpha}}{\Gamma(1+3\alpha)} - \cdots, \tag{4.75}$$

which is same as the solution of Wazwaz (Wazwaz, 2007) at $\alpha = 1$.

4.7.4 Implementation of ADATM

Now, taking AT on both sides of Eq. (4.34) with IC and BCs, we have

$$s^\alpha A[u(x,t)] - s^{\alpha-2}u(x,0) = \frac{1}{2}A[x^2 u_{xx}]. \tag{4.76}$$

$$A[u(x,t)] = \frac{u(x,0)}{s^2} + \frac{s^{-\alpha}}{2}A[x^2 u_{xx}]. \tag{4.77}$$

$$A[u(x,t)] = \frac{x^2}{s^2} + \frac{s^{-\alpha}}{2}A[x^2 u_{xx}]. \tag{4.78}$$

Inverse AT on both sides of Eq. (4.78) reduces to

$$u(x,t) = x^2 + A^{-1}\left(\frac{s^{-\alpha}}{2}A\left[x^2 u_{xx}\right]\right). \tag{4.79}$$

Using ADM, we have

$$\sum_{n=0}^{\infty} u_n(x,t) = x^2 + A^{-1}\left(\frac{s^{-\alpha}}{2}A\left[x^2 \sum_{n=0}^{\infty} (u_n(x,t))_{xx}\right]\right). \tag{4.80}$$

Comparing both sides of Eq. (4.80), we obtain

$$u_0(x,t) = x^2,$$

$$u_1(x,t) = A^{-1}\left(\frac{s^{-\alpha}}{2}A\left[x^2 u_{0xx}\right]\right)$$

$$= A^{-1}\left(\frac{s^{-\alpha}}{2}A\left[x^2 \frac{\partial^2}{\partial x^2} u_0\right]\right)$$

$$= A^{-1}\left(\frac{s^{-\alpha}}{2}A\left[2x^2\right]\right)$$

$$= A^{-1}(x^2 s^{-\alpha-2}) = x^2 \frac{t^\alpha}{\Gamma(1+\alpha)},$$

$$u_2(x,t) = A^{-1}\left(\frac{s^{-\alpha}}{2}A\left[x^2 u_{1xx}\right]\right)$$

$$= A^{-1}\left(\frac{s^{-\alpha}}{2}A\left[x^2 \frac{\partial^2}{\partial x^2} u_1\right]\right)$$

$$= A^{-1}\left(\frac{s^{-\alpha}}{2}A\left[\frac{2t^\alpha}{\Gamma(1+\alpha)}x^2\right]\right)$$

$$= A^{-1}(x^2 s^{-2\alpha-2}) = x^2 \frac{t^{2\alpha}}{\Gamma(1+2\alpha)},$$

$$u_3(x,t) = A^{-1}\left(\frac{s^{-\alpha}}{2}A\left[x^2 u_{2xx}\right]\right)$$

$$= x^2 \frac{t^{3\alpha}}{\Gamma(1+3\alpha)},$$

$$\vdots$$

So, the solution of Eq. (4.34) may be obtained as:

$$u(x,t) = x^2 + x^2 \frac{t^\alpha}{\Gamma(1+\alpha)} + x^2 \frac{t^{2\alpha}}{\Gamma(1+2\alpha)} + x^2 \frac{t^{3\alpha}}{\Gamma(1+3\alpha)} + \cdots = x^2 E(t^\alpha), \tag{4.81}$$

In particular, at $\alpha = 1$, Eq. (4.81) reduces to $u(x, t) = x^2 e^t$, which is same as the solution of Sadighi et al. (2008). Again by operating AT on both sides of Eq. (4.43) with IC, we have

$$s^\alpha A[u(x,t)] - s^{\alpha-2}u(x,0) = -A[uu_x]. \tag{4.82}$$

$$A[u(x,t)] = -s^{-\alpha}A[uu_x] + \frac{u(x,0)}{s^2}. \tag{4.83}$$

$$A[u(x,t)] = -s^{-\alpha}A[uu_x] - \frac{x}{s^2}. \tag{4.84}$$

By inverse AT on both sides of Eq. (4.84), we have

$$u(x,t) = -A^{-1}(s^{-\alpha}A[uu_x]) - x. \tag{4.85}$$

By ADM, we have

$$\sum_{n=0}^{\infty} u_n(x,t) = -x - A^{-1}\left(s^{-\alpha}A\left[\sum_{n=0}^{\infty} A_n\right]\right). \tag{4.86}$$

Using Eq. (4.50) and comparing both sides of Eq. (4.86), we obtain

$$u_0(x,t) = -x,$$

$$u_1(x,t) = -A^{-1}(s^{-\alpha}A[A_0])$$

$$= -A^{-1}\left(s^{-\alpha}A\left[u_0\frac{\partial}{\partial x}u_0\right]\right)$$

$$= -A^{-1}(s^{-\alpha}A[x])$$

$$= -A^{-1}(xs^{-\alpha-2}) = -x\frac{t^{\alpha}}{\Gamma(1+\alpha)},$$

$$u_2(x,t) = -A^{-1}(s^{-\alpha}A[A_1])$$

$$= -A^{-1}\left(s^{-\alpha}A\left[u_0\frac{\partial}{\partial x}u_1 + u_1\frac{\partial}{\partial x}u_0\right]\right)$$

$$= -A^{-1}\left(s^{-\alpha}A\left[2x\frac{t^{\alpha}}{\Gamma(1+\alpha)}\right]\right)$$

$$= -A^{-1}(2xs^{-2\alpha-2}) = -2x\frac{t^{2\alpha}}{\Gamma(1+2\alpha)},$$

$$u_3(x,t) = -A^{-1}(s^{-\alpha}A[A_2])$$

$$= -A^{-1}\left(s^{-\alpha}A\left[u_0\frac{\partial}{\partial x}u_2 + u_1\frac{\partial}{\partial x}u_1 + u_2\frac{\partial}{\partial x}u_0\right]\right)$$

$$= -A^{-1}\left(s^{-\alpha}A\left[4x\frac{t^{2\alpha}}{\Gamma(1+2\alpha)} + x\frac{t^{2\alpha}}{(\Gamma(1+\alpha))^2}\right]\right)$$

$$= -A^{-1}\left(4xs^{-3\alpha-2} + \frac{x\Gamma(1+2\alpha)s^{-3\alpha-2}}{(\Gamma(1+\alpha))^2}\right),$$

$$= -4x\frac{t^{3\alpha}}{\Gamma(1+3\alpha)} - \frac{x\Gamma(1+2\alpha)}{(\Gamma(1+\alpha))^2}\frac{t^{3\alpha}}{\Gamma(1+3\alpha)}$$

$$\vdots$$

So, the solution to Eq. (4.43) may be written as:

$$u(x,t) = -x - x\frac{t^{\alpha}}{\Gamma(1+\alpha)} - 2x\frac{t^{2\alpha}}{\Gamma(1+2\alpha)} - 4x\frac{t^{3\alpha}}{\Gamma(1+3\alpha)} - \frac{x\Gamma(1+2\alpha)}{(\Gamma(1+\alpha))^2}\frac{t^{3\alpha}}{\Gamma(1+3\alpha)} - \cdots, \tag{4.87}$$

Particularly at $\alpha = 1$, Eq. (4.87) reduces to closed-form solution $u(x,t) = \frac{x}{t-1}$, which is same as the solution of Wazwaz (2007).

References

Aboodh, K.S., Idris, A., and Nuruddeen, R.I. (2017). On the Aboodh transform connections with some famous integral transforms. *International Journal of Engineering and Information Systems* 1 (9): 143–151.

Ahmed, S.S., Amin, M.B.M., and Hamasalih, S.A. (2019). Laplace Adomian decomposition and modify Laplace Adomian decomposition methods for solving linear Volterra integro-fractional differential equations with constant multi-time retarded delay. *Journal of University of Babylon for Pure and Applied Sciences* 27 (5): 35–53.

Baleanu, D. and Jassim, H.K. (2019). A modification fractional homotopy perturbation method for solving Helmholtz and coupled Helmholtz equations on cantor sets. *Fractal and Fractional* 3 (2): 30.

Faraz, N., Khan, Y., and Sankar, D.S. (2010). Decomposition-transform method for fractional differential equations. *International Journal of Nonlinear Sciences and Numerical Simulation* 11: 305–310.

Jena, R.M. and Chakraverty, S. (2019). Analytical solution of Bagley-Torvik equations using Sumudu transformation method. *SN Applied Sciences* 1 (3): 246.

Jena, R.M. and Chakraverty, S. (2019). Solving time-fractional Navier–Stokes equations using homotopy perturbation Elzaki transform. *SN Applied Sciences* 1 (1): 16 pages.

Jena, R.M. and Chakraverty, S. (2019). Q-Homotopy Analysis Aboodh Transform Method based solution of proportional delay time-fractional partial differential equations. *Journal of Interdisciplinary Mathematics* 22: 931–950.

Khalouta, A. and Abdelouahab, K. (2020). A comparative study of shehu variational iteration method and shehu decomposition method for solving nonlinear caputo time-fractional wave-like equations with variable coefficients. *Applications and Applied Mathematics: An International Journal* 15 (1): 430–445.

Mohammed, O.H. and Salim, H.A. (2018). Computational methods based laplace decomposition for solving nonlinear system of fractional order differential equations. *Alexandria Engineering Journal* 57: 3549–3557.

Özis, T. and Agırseven, D. (2008). He's homotopy perturbation method for solving heat-like and wave-like equations with variable coefficients. *Physics Letters A* 372: 5944–5950.

Sadighi, A., Ganji, D.D., Gorji, M., and Tolou, N. (2008). Numerical simulation of heat-like models with variable coefficients by the variational iteration method. *Journal of Physics: Conference Series* 96: 012083.

Thabet, H. and Kendre, S. (2019). New modification of Adomian decomposition method for solving a system of nonlinear fractional partial differential equations. *International Journal of Advances in Applied Mathematics and Mechanics* 6 (3): 1–13.

Wazwaz, A.M. (1999). A reliable modification of Adomian decomposition method. *Applied Mathematics and Computation* 102 (1): 77–86.

Wazwaz, A.M. (2007). A comparison between the variational iteration method and adomian decomposition method. *Journal of Computational and Applied Mathematics* 207: 129–136.

5

Homotopy Perturbation Method

5.1 Introduction

In this chapter, we will discuss about homotopy perturbation method (HPM), which is again a semi-analytical approach for solving linear and nonlinear ordinary/partial/fractional differential equations. HPM was first proposed by He (1999a). This approach has been established using artificial parameters (Liu 1997). Interested readers may visit references (He 2003, 2004) for more information. Almost all conventional perturbation methods are based on the assumption of small parameters. However, most nonlinear problems have no small parameters, and the determination of small parameters needs a unique art requiring special techniques. These small parameters are so sensitive that a slight change may influence the final result. The right choice of small parameters yields optimal performance. However, an inappropriate choice of small parameters leads to poor, even significant effects. Liu (1997) proposed the artificial parameter method and Liao (1995, 1997) contributed to the homotopy analysis method to eradicate the presumption of small parameters. He (1999a, 1999b) also established a technique called the variational iteration method (VIM), in which no small parameter assumptions are made, and is discussed in Chapter 9.

In the subsequent sections, firstly, the theories behind the method with respect to fractional order are addressed. Then the systematic step-by-step procedure of the technique along with two problems are introduced.

5.2 Procedure for HPM

In order to illustrate the fundamental idea of HPM, the fractional-order nonlinear nonhomogeneous partial differential equation with initial conditions (ICs) is considered as follows:

$$D_t^\alpha u(x,t) + Ru(x,t) + Nu(x,t) = f(x,t), \quad n-1 < \alpha \le n, \tag{5.1}$$

subject to ICs:

$$u^{(k)}(x,0) = g_k(x), \quad k = 0, 1, ..., n-1, \tag{5.2}$$

where $D_t^\alpha = \dfrac{\partial^\alpha}{\partial t^\alpha}$ is the differential operator, $D_t^\alpha u(x,t)$ is the derivative of $u(x,t)$ in the Caputo sense, R, N are the linear and nonlinear differential operators, and $f(x,t)$ is the source term. We shall next present the solution approach based on the standard HPM. Let us now construct the following homotopy of Eq. (5.1) as (He 1999a):

$$(1-p)D_t^\alpha u(x,t) + p\big(D_t^\alpha u(x,t) + Ru(x,t) + Nu(x,t) - f(x,t)\big) = 0, \tag{5.3}$$

or

$$D_t^\alpha u(x,t) + p(Ru(x,t) + Nu(x,t) - f(x,t)) = 0, \tag{5.4}$$

where $p \in [0, 1]$ is an embedding parameter. If $p = 0$, then Eqs. (5.3) and (5.4) become

$$D_t^\alpha u(x,t) = 0, \tag{5.5}$$

and when $p = 1$ Eqs. (5.3) and (5.4) turn out to be the original Eq. (5.1).

First, we need to consider the solution in series form containing the embedding parameter $p \in [0, 1]$ as:

$$u(x,t) = \sum_{n=0}^{\infty} p^n u_n(x,t), \tag{5.6}$$

and the nonlinear term may be decomposed by using He's polynomials (Ghorbani 2009) as:

$$Nu(x,t) = \sum_{n=0}^{\infty} p^n H_n(u). \tag{5.7}$$

where $H_n(u)$ denotes the He's polynomials and which is defined as follows:

$$H_n(u_0, \; u_1, ..., u_n) = \frac{1}{n!} \frac{\partial}{\partial p^n} \left[N \left(\sum_{n=0}^{\infty} p^n u_n(x,t) \right) \right]_{p=0}, \quad n = 0, 1, 2, ... \tag{5.8}$$

One may go through reference (Ghorbani 2009) for a detailed derivation of the Eq. (5.8). Substituting Eqs. (5.6) and (5.7) into Eq. (5.4), we get the following expression:

$$D_t^\alpha \left\{ \sum_{n=0}^{\infty} p^n u_n(x,t) \right\} = p \left(f(x,t) - R \sum_{n=0}^{\infty} p^n u_n(x,t) - N \sum_{n=0}^{\infty} p^n H_n(u) \right). \tag{5.9}$$

By comparing the coefficients of the same powers of "p" on both sides of Eq. (5.9), we may have the following approximations successively:

$$p^0 : D_t^\alpha \{u_0(x,t)\} = 0, \tag{5.10}$$

$$p^1 : D_t^\alpha \{u_1(x,t)\} = f(x,t) - Ru_0(x,t) - NH_0(u), \tag{5.11}$$

$$p^2 : D_t^\alpha \{u_2(x,t)\} = - Ru_1(x,t) - NH_1(u), \tag{5.12}$$

$$p^3 : D_t^\alpha \{u_3(x,t)\} = - Ru_2(x,t) - NH_2(u), \tag{5.13}$$

$$\vdots$$

$$p^n : D_t^\alpha \{u_n(x,t)\} = - Ru_{n-1}(x,t) - NH_{n-1}(u). \tag{5.14}$$

$$\vdots$$

Applying the operator J_t^α, the inverse operator of D_t^α which is given in the previous chapter on both sides of the aforementioned equations and using the ICs in Eq. (5.2), the first few terms of the HPM solution may be written as:

$$u_0(x,t) = \sum_{k=0}^{n-1} u^k(x,0) \frac{t^k}{k!} = \sum_{k=0}^{n-1} g_k(x) \frac{t^k}{k!}, \tag{5.15}$$

$$u_1(x,t) = J_t^\alpha[f(x,t)] - J_t^\alpha[Ru_0(x,t)] - J_t^\alpha[NH_0(u)], \tag{5.16}$$

$$u_2(x,t) = - J_t^\alpha[Ru_1(x,t)] - J_t^\alpha[NH_1(u)], \tag{5.17}$$

and so on. So, the solution of Eq. (5.1) may be obtained as:

$$u(x,t) = \lim_{n\to\infty} u_n(x,t) = u_0(x,t) + u_1(x,t) + u_2(x,t) + \cdots \tag{5.18}$$

5.3 Numerical Examples

Here, we use the present method to solve a linear fractional one-dimensional heat-like problem in Example 5.1 and a nonlinear advection equation in Example 5.2. It is worth mentioning that the HPM can also be used for handling linear and nonlinear fractional differential equations.

Example 5.1   Let us consider the one-dimensional heat-like model (Özis and Agırseven 2008; Sadighi et al. 2008) as:

$$D_t^\alpha u(x,t) = \frac{1}{2}x^2 u_{xx}(x,t), \quad 0 < x < 1, t > 0, 0 < \alpha \le 1, \tag{5.19}$$

with the boundary conditions (BCs):

$$u(0,t) = 0, \quad u(1,t) = e^t, \tag{5.20}$$

and IC:

$$u(x,\ 0) = x^2. \tag{5.21}$$

Solution

According to Eq. (5.4), let us construct the homotopy of Eq. (5.19) as follows:

$$D_t^\alpha u(x,t) - p\left(\frac{1}{2}x^2 u_{xx}(x,t)\right) = 0. \tag{5.22}$$

Substituting Eq. (5.6) into Eq. (5.19), we obtain

$$D_t^\alpha\left[\sum_{n=0}^\infty p^n u_n(x,t)\right] - p\left[\frac{1}{2}x^2\left(\sum_{n=0}^\infty p^n u_n(x,t)\right)_{xx}\right] = 0. \tag{5.23}$$

Collecting the like power of "p" yields the following expressions:

$$p^0 : D_t^\alpha\{u_0(x,t)\} = 0,$$

$$p^1 : D_t^\alpha\{u_1(x,t)\} = \frac{1}{2}x^2(u_0(x,t))_{xx},$$

$$p^2 : D_t^\alpha\{u_2(x,t)\} = \frac{1}{2}x^2(u_1(x,t))_{xx},$$

$$\vdots$$

Applying the operator J_t^α on both sides of the aforementioned expressions and using the initial as well as BCs Eqs. (5.20) and (5.21), we have

$$p^0 : u_0(x,t) = u(x,0) = x^2,$$

$$p^1 : u_1(x,t) = J_t^\alpha\left(\frac{1}{2}x^2(u_0(x,t))_{xx}\right) = x^2\frac{t^\alpha}{\Gamma(1+\alpha)},$$

$$p^2 : u_2(x,t) = J_t^\alpha\left(\frac{1}{2}x^2(u_1(x,t))_{xx}\right) = x^2\frac{t^{2\alpha}}{\Gamma(1+2\alpha)},$$

$$p^3 : u_3(x,t) = J_t^\alpha\left(\frac{1}{2}x^2(u_2(x,t))_{xx}\right) = x^2\frac{t^{3\alpha}}{\Gamma(1+3\alpha)},$$

$$\vdots$$

So, the solution of the fractional heat-like Eq. (5.19), as $p \to 1$, may be obtained as:

$$
\begin{aligned}
u(x,t) &= \lim_{n\to\infty} u_n(x,t) = u_0(x,t) + u_1(x,t) + u_2(x,t) + \cdots \\
&= x^2 + x^2\frac{t^\alpha}{\Gamma(1+\alpha)} + x^2\frac{t^{2\alpha}}{\Gamma(1+2\alpha)} + x^2\frac{t^{3\alpha}}{\Gamma(1+3\alpha)} + \cdots, \\
&= x^2\left(1 + \frac{t^\alpha}{\Gamma(1+\alpha)} + \frac{t^{2\alpha}}{\Gamma(1+2\alpha)} + \frac{t^{3\alpha}}{\Gamma(1+3\alpha)} + \cdots\right) = x^2 E(t^\alpha),
\end{aligned}
\tag{5.24}
$$

where $E(t^\alpha)$ is called the Mittag-Leffer function, which is given in Chapter 1.

In particular, at $\alpha = 1$, Eq. (5.24) reduces to $u(x,t) = x^2 e^t$, which is same as the solution of Sadighi et al. (2008). It is worth mentioning that by increasing the number of terms of solution, one may achieve a better approximate result, as shown in Figure 5.1. Figures 5.2–5.5 show the fourth-order approximate solution plots of Example 5.1 at different values of α.

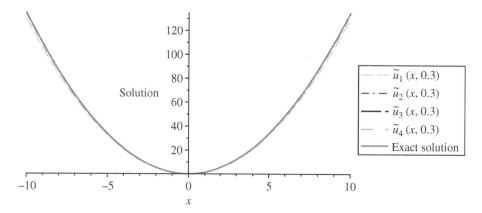

Figure 5.1 Comparison plot of the present solution with the exact solution of Example 5.1 taking a different number of terms of approximate solution.

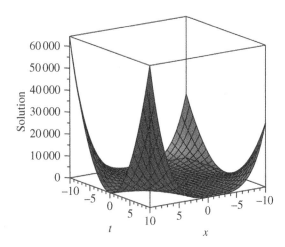

Figure 5.2 Fourth-order approximate solution plot of Example 5.1 at $\alpha = 1$.

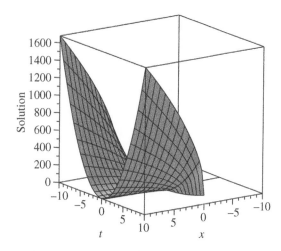

Figure 5.3 Fourth-order approximate solution plot of Example 5.1 at $\alpha = 0.2$.

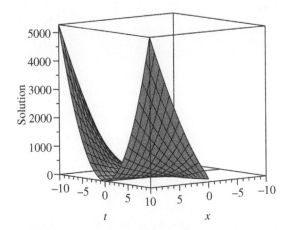

Figure 5.4 Fourth-order approximate solution plot of Example 5.1 at $\alpha = 0.4$.

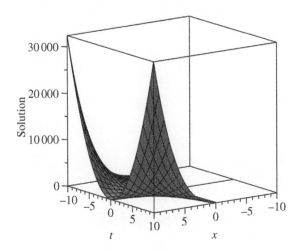

Figure 5.5 Fourth-order approximate solution plot of Example 5.1 at $\alpha = 0.8$.

Example 5.2 Consider the following nonlinear advection equation (Wazwaz 2007):

$$\frac{\partial^\alpha u}{\partial t^\alpha} + u\frac{\partial u}{\partial x} = 0, \quad 0 < \alpha \leq 1,$$ (5.25)

with IC:

$$u(x,0) = -x.$$ (5.26)

Solution

According to Eq. (5.4), we can construct the following homotopy of Eq. (5.25):

$$D_t^\alpha u(x,t) + p(u(x,t)u_x(x,t)) = 0.$$ (5.27)

Now using HPM, we obtain

$$D_t^\alpha\left[\sum_{n=0}^{\infty} p^n u_n(x,t)\right] + p\left[\sum_{n=0}^{\infty} p^n H_n(u)\right] = 0.$$ (5.28)

Comparing the same power of "p", the following expressions are obtained:

$$p^0 : D_t^\alpha \{u_0(x,t)\} = 0,$$
$$p^1 : D_t^\alpha \{u_1(x,t)\} = -H_0(u),$$
$$p^2 : D_t^\alpha \{u_2(x,t)\} = -H_1(u),$$
$$p^3 : D_t^\alpha \{u_3(x,t)\} = -H_2(u),$$
$$\vdots$$

(5.29)

Some of He's polynomials $H_n(u)$ (He 2003, 2004) for the term uu_x are

$$H_0(u) = u_0 \frac{\partial}{\partial x} u_0,$$
$$H_1(u) = u_0 \frac{\partial}{\partial x} u_1 + u_1 \frac{\partial}{\partial x} u_0,$$
$$H_2(u) = u_0 \frac{\partial}{\partial x} u_2 + u_1 \frac{\partial}{\partial x} u_1 + u_2 \frac{\partial}{\partial x} u_0,$$
$$\vdots$$

(5.30)

Putting Eq. (5.30) into Eq. (5.29), we get

$$p^0 : D_t^\alpha \{u_0(x,t)\} = 0,$$
$$p^1 : D_t^\alpha \{u_1(x,t)\} = -u_0 \frac{\partial}{\partial x} u_0,$$
$$p^2 : D_t^\alpha \{u_2(x,t)\} = -u_0 \frac{\partial}{\partial x} u_1 - u_1 \frac{\partial}{\partial x} u_0,$$
$$p^3 : D_t^\alpha \{u_3(x,t)\} = -u_0 \frac{\partial}{\partial x} u_2 - u_1 \frac{\partial}{\partial x} u_1 - u_2 \frac{\partial}{\partial x} u_0,$$
$$\vdots$$

(5.31)

Applying the operator J_t^α on both sides of the aforementioned expressions and using the IC, we obtain

$$p^0 : u_0(x,t) = -x,$$
$$p^1 : u_1(x,t) = J_t^\alpha[-x] = -x \frac{t^\alpha}{\Gamma(1+\alpha)},$$
$$p^2 : u_2(x,t) = -J_t^\alpha \left[2x \frac{t^\alpha}{\Gamma(1+\alpha)} \right] = -2x \frac{t^{2\alpha}}{\Gamma(1+2\alpha)},$$
$$p^3 : u_3(x,t) = -J_t^\alpha \left[4x \frac{t^{2\alpha}}{\Gamma(1+2\alpha)} + x \frac{t^{2\alpha}}{(\Gamma(1+\alpha))^2} \right] = -4x \frac{t^{3\alpha}}{\Gamma(1+3\alpha)} - \frac{x \Gamma(1+2\alpha)}{(\Gamma(1+\alpha))^2} \frac{t^{3\alpha}}{\Gamma(1+3\alpha)},$$
$$\vdots$$

So, the solution to Eq. (5.25), as $p \to 1$, may be obtained as:

$$u(x,t) = \lim_{n \to \infty} u_n(x,t) = u_0(x,t) + u_1(x,t) + u_2(x,t) + \cdots$$
$$= -x - x \frac{t^\alpha}{\Gamma(1+\alpha)} - 2x \frac{t^{2\alpha}}{\Gamma(1+2\alpha)} - 4x \frac{t^{3\alpha}}{\Gamma(1+3\alpha)} - \frac{x \Gamma(1+2\alpha)}{(\Gamma(1+\alpha))^2 \Gamma(1+3\alpha)} t^{3\alpha} - \cdots,$$

(5.32)

Particularly at $\alpha = 1$, Eq. (5.32) reduces to a closed-form solution $u(x,t) = \dfrac{x}{t-1}$, which is same as the solution of Wazwaz (2007). Figure 5.6 shows the comparison plots of the exact solution with the present solution by considering the increasing number of terms in the solution. Figures 5.7–5.10 give the third-order approximate solution plots of Example 5.2 for various α values.

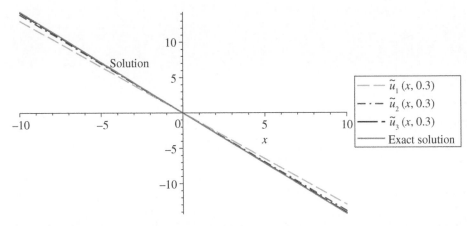

Figure 5.6 Comparison plot of the present solution with the exact solution of Example 5.2 taking a different number of terms of approximate solution.

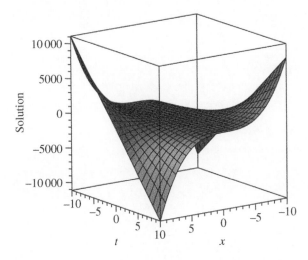

Figure 5.7 Third-order approximate solution plot of Example 5.2 at $\alpha = 1$.

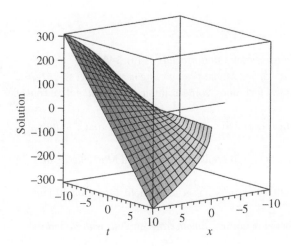

Figure 5.8 Third-order approximate solution plot of Example 5.2 at $\alpha = 0.2$.

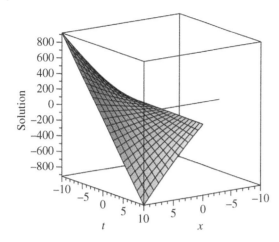

Figure 5.9 Third-order approximate solution plot of Example 5.2 at $\alpha = 0.4$.

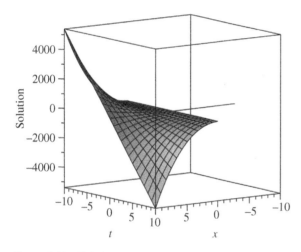

Figure 5.10 Third-order approximate solution plot of Example 5.2 at $\alpha = 0.8$.

References

Ghorbani, A. (2009). Beyond Adomian polynomials: He polynomials. *Chaos, Solitons, and Fractals* 39 (3): 1486–1492.

He, J.H. (1999a). Homotopy perturbation technique. *Computer Methods in Applied Mechanics and Engineering* 178 (3): 257–262.

He, J.-H. (1999b). Variational iteration method – a kind of nonlinear analytical technique: some examples. *International Journal of Non-Linear Mechanics* 34 (4): 699–708.

He, J.H. (2003). Homotopy perturbation method: a new nonlinear analytical technique. *Applied Mathematics and Computation* 135 (1): 73–79.

He, J.H. (2004). The homotopy perturbation method for nonlinear oscillators with discontinuities. *Applied Mathematics and Computation* 151 (1): 287–292.

Liao, S.J. (1995). An approximate solution technique not depending on small parameters: a special example. *International Journal of Non-Linear Mechanics* 30 (3): 371–380.

Liao, S.J. (1997). Boundary element method for general nonlinear differential operators. *Engineering Analysis with Boundary Elements* 20 (2): 91–99.

Liu, G.L. (1997). New research directions in singular perturbation theory: artificial parameter approach and inverse-perturbation technique. *Proceedings of the Seventh Conference of Modern Mathematics and Mechanics* (September 1997). Shanghai: Scientific Research Publishing, pp. 47–53.

Özis, T. and Agırseven, D. (2008). He's homotopy perturbation method for solving heat-like and wave-like equations with variable coefficients. *Physics Letters A* 372: 5944–5950.

Sadighi, A., Ganji, D.D., Gorji, M., and Tolou, N. (2008). Numerical simulation of heat-like models with variable coefficients by the variational iteration method. *Journal of Physics: Conference Series* 96: 012083.

Wazwaz, A.M. (2007). A comparison between the variational iteration method and adomian decomposition method. *Journal of Computational and Applied Mathematics* 207: 129–136.

6

Homotopy Perturbation Transform Method

6.1 Introduction

In Chapter 5, we have already discussed the homotopy perturbation method (HPM), which is a semi-analytical approach for solving linear and nonlinear ordinary/partial/fractional differential equations. In this chapter, we will discuss about the hybrid methods, which are the coupling of HPM with various transform methods, viz. Laplace transform (LT), Sumudu transform (ST), Elzaki transform (ET), and Aboodh transform (AT). As said earlier, HPM with the combination of these transform methods is called as homotopy perturbation transform method (HPTM) (Singh and Kumar 2011, 2012; Elzaki and Biazar 2013; Mahdy et al. 2015; Mohand and Mahgoub 2016; Sedeeg 2016; Olubanwo et al. 2019). Nowadays, these methods, namely homotopy perturbation Laplace transform method (HPLTM), homotopy perturbation Sumudu transform method (HPSTM), homotopy perturbation Elzaki transform method (HPETM), and homotopy perturbation Aboodh transform method (HPATM), are getting popular recently. Although these four transform methods are effective methods for solving fractional differential equations, but these methods sometimes fail to address nonlinear terms arising from the fractional differential equations. These difficulties may be overcome by coupling these transforms with that of HPM. In the subsequent sections, the theories behind the four transform methods with respect to fractional order are given. Then the systematic study of the earlier-mentioned four hybrid methods along with two problems for each of the methods is addressed.

6.2 Transform Methods for the Caputo Sense Derivatives

Definition 6.1 The LT of the Caputo fractional derivative is defined as (Baleanu and Jassim 2019):

$$L\left[D_t^{n\alpha}u(x,t)\right] = s^{n\alpha}L[u(x,t)] - \sum_{k=0}^{n-1} s^{(n\alpha-k-1)}u^{(k)}(x,\ 0), \quad n-1 < n\alpha \le 1, n \in N. \tag{6.1}$$

Definition 6.2 The ST of the Caputo fractional derivative is defined as (Jena and Chakraverty 2019):

$$S\left[D_t^{n\alpha}u(x,t)\right] = s^{-n\alpha}S[u(x,t)] - \sum_{k=0}^{n-1} s^{-n\alpha+k}u^{(k)}(x,0), \quad n-1 < n\alpha \le 1, n \in N. \tag{6.2}$$

Definition 6.3 The ET of the Caputo fractional derivative is defined as (Elzaki and Biazar 2013; Jena and Chakraverty 2019):

$$E\left[D_t^{n\alpha}u(x,t)\right] = \frac{E[u(x,t)]}{s^{n\alpha}} - \sum_{k=0}^{n-1} s^{k-n\alpha+2}u^{(k)}(x,0), \quad n-1 < n\alpha \le n, n \in N. \tag{6.3}$$

Definition 6.4 The AT of the Caputo fractional derivative is defined as (Aboodh 2013; Aboodh et al. 2017):

$$A\left[D_t^{n\alpha}u(x,t)\right] = s^{n\alpha}A[u(x,t)] - \sum_{k=0}^{n-1} s^{-k+n\alpha-2}u^{(k)}(x,0), \quad n-1 < n\alpha \le n, n \in N. \tag{6.4}$$

Computational Fractional Dynamical Systems: Fractional Differential Equations and Applications, First Edition.
Snehashish Chakraverty, Rajarama Mohan Jena, and Subrat Kumar Jena.
© 2023 John Wiley & Sons, Inc. Published 2023 by John Wiley & Sons, Inc.

Table 6.1 Transforms of some essential functions.

Functions	Laplace transform	Sumudu transform	Elzaki transform	Aboodh transform
Definitions	$L[f(t)] = f(s)$	$S[g(t)] = g(s)$	$E[h(t)] = h(s)$	$A[p(t)] = p(s)$
	$= \int\limits_0^\infty e^{-st} f(t) dt$	$= \int\limits_0^\infty e^{-t} f(st) dt$	$= s \int\limits_0^\infty e f(t) dt$	$= \dfrac{1}{s} \int\limits_0^\infty e^{-st} f(t) dt$
		$= \dfrac{1}{s}f\left(\dfrac{1}{s}\right)$	$= sf\left(\dfrac{1}{s}\right)$	$= \dfrac{1}{s}f(s)$
1	$\dfrac{1}{s}$	1	s^2	$\dfrac{1}{s^2}$
t^α	$\dfrac{\Gamma(1+\alpha)}{s^{\alpha+1}}$	$s^\alpha \Gamma(1+\alpha)$	$s^{\alpha+2}\Gamma(1+\alpha)$	$\dfrac{\Gamma(1+\alpha)}{s^{\alpha+2}}$
e^{at}	$\dfrac{1}{s-a}$	$\dfrac{1}{1-as}$	$\dfrac{s^2}{1-as}$	$\dfrac{1}{s(1-s)}$
$\sin(at)$	$\dfrac{a}{s^2+a^2}$	$\dfrac{as}{1+a^2s^2}$	$\dfrac{as^3}{1+a^2s^2}$	$\dfrac{a}{s(s^2+a^2)}$
$\cos(at)$	$\dfrac{s}{s^2+a^2}$	$\dfrac{1}{1+s^2a^2}$	$\dfrac{s}{1+s^2a^2}$	$\dfrac{1}{s^2+a^2}$

The aforementioned Table 6.1 shows the transforms of some standard functions with respect to the aforementioned four transform methods and their definitions.

Following section deals with the systematic study of four hybrid methods, namely HPLTM, HPSTM, HPETM, and HPATM, one after another.

6.3 Homotopy Perturbation Laplace Transform Method (HPLTM)

In order to clarify the basic idea of HPLTM, the fractional-order nonlinear nonhomogeneous partial differential equation with initial conditions (ICs) is considered as follows:

$$D_t^{n\alpha} u(x,t) + Ru(x,t) + Nu(x,t) = f(x,t), \quad n-1 < n\alpha \le n. \tag{6.5}$$

subject to ICs:

$$u^{(k)}(x,0) = g_k(x), \quad k = 0,1,...,n-1. \tag{6.6}$$

where $D_t^{n\alpha} = \dfrac{\partial^{n\alpha}}{\partial t^{n\alpha}}$ is the differential operator, $D_t^{n\alpha} u(x,t)$ is the derivative of $u(x, t)$ in the Caputo sense, R, N are the linear and nonlinear differential operators, and $f(x, t)$ is the source term. The HPLTM approach involves mainly two stages. In the first stage, LT is taken on both sides of Eq. (6.5), and then in the second stage, HPM is applied where decomposition of the nonlinear term is done using He's polynomials. First, by operating LT on both sides of Eq. (6.5), we obtain

$$L\left[D_t^{n\alpha} u(x,t)\right] = L[f(x,t)] - L[Ru(x,t)] - L[Nu(x,t)]. \tag{6.7}$$

Using differentiation property (Eq. (6.1)) of LT, we obtain

$$s^{n\alpha} L[u(x,t)] - \sum_{k=0}^{n-1} s^{n\alpha-k-1} u^{(k)}(x,0) = L[f(x,t)] - L[Ru(x,t)] - L[Nu(x,t)]. \tag{6.8}$$

$$L[u(x,t)] = \frac{1}{s^{n\alpha}} \sum_{k=0}^{n-1} s^{n\alpha-k-1} u^{(k)}(x,0) + \frac{1}{s^{n\alpha}} L[f(x,t)] - \frac{1}{s^{n\alpha}} L[Ru(x,t)] - \frac{1}{s^{n\alpha}} L[Nu(x,t)]. \tag{6.9}$$

Applying inverse LT on both sides of Eq. (6.9), we find

$$u(x,t) = F(x,t) - L^{-1}\left(\frac{1}{s^{n\alpha}}L[Ru(x,t)]\right) - L^{-1}\left(\frac{1}{s^{n\alpha}}L[Nu(x,t)]\right). \tag{6.10}$$

where $F(x, t)$ represents the term coming from the IC and source term (first two terms on the right-hand side of Eq. (6.9)).

Next, to implement HPM, first, we need to consider the solution as in series form containing the embedding parameters $p \in [0, 1]$ as:

$$u(x,t) = \sum_{n=0}^{\infty} p^n u_n(x,t). \tag{6.11}$$

and the nonlinear term may be decomposed by using He's polynomials (Ghorbani 2009) as:

$$Nu(x,t) = \sum_{n=0}^{\infty} p^n H_n(u). \tag{6.12}$$

where $H_n(u)$ denotes the He's polynomials and which is defined as follows:

$$H_n(u_0, \ u_1, ..., u_n) = \frac{1}{n!}\frac{\partial^n}{\partial p^n}\left[N\left(\sum_{n=0}^{\infty} p^n u_n(x,t)\right)\right]_{p=0}, \quad n = 0, 1, 2, ... \tag{6.13}$$

One may go through the reference (Ghorbani 2009) for a detailed derivation of the Eq. (6.13). Substituting Eqs. (6.11) and (6.12) into Eq. (6.10) and applying LT with HPM, one may get the following expression:

$$\sum_{n=0}^{\infty} p^n u_n(x,t) = F(x,t) - p\left(L^{-1}\left(\frac{1}{s^{n\alpha}}L\left[R\sum_{n=0}^{\infty} p^n u_n(x,t)\right]\right) + L^{-1}\left(\frac{1}{s^{n\alpha}}L\left[\sum_{n=0}^{\infty} p^n H_n(u)\right]\right)\right). \tag{6.14}$$

By comparing the coefficients of the same powers of "p" on both sides of Eq. (6.14), we may have the following approximations successively:

$$p^0 : u_0(x,t) = F(x,t),$$

$$p^1 : u_1(x,t) = -L^{-1}\left(\frac{1}{s^{n\alpha}}L[Ru_0(x,t)]\right) - L^{-1}\left(\frac{1}{s^{n\alpha}}L[H_0(u)]\right),$$

$$p^2 : u_2(x,t) = -L^{-1}\left(\frac{1}{s^{n\alpha}}L[Ru_1(x,t)]\right) - L^{-1}\left(\frac{1}{s^{n\alpha}}L[H_1(u)]\right),$$

$$p^3 : u_3(x,t) = -L^{-1}\left(\frac{1}{s^{n\alpha}}L[Ru_2(x,t)]\right) - L^{-1}\left(\frac{1}{s^{n\alpha}}L[H_2(u)]\right),$$

$$\vdots$$

$$p^n : u_n(x,t) = -L^{-1}\left(\frac{1}{s^{n\alpha}}L[Ru_{n-1}(x,t)]\right) - L^{-1}\left(\frac{1}{s^{n\alpha}}L[H_{n-1}(u)]\right)$$

$$\vdots$$

So, the solution of Eq. (6.5) may be obtained as:

$$u(x,t) = \lim_{n\to\infty} u_n(x,t) = u_0(x,t) + u_1(x,t) + u_2(x,t) + \cdots. \tag{6.15}$$

6.4 Homotopy Perturbation Sumudu Transform Method (HPSTM)

Similar to the aforementioned procedure, applying ST on both sides of the Eq. (6.5), we have

$$S[D_t^{n\alpha}u(x,t)] = S[f(x,t)] - S[Ru(x,t)] - S[Nu(x,t)]. \tag{6.16}$$

Using differentiation property of (Eq. 6.2) for ST in Eq. (6.16), we obtain

$$s^{-na}S[u(x,t)] - \sum_{k=0}^{n-1} s^{-na+k}u^{(k)}(x,0) = S[f(x,t)] - S[Ru(x,t)] - S[Nu(x,t)]. \tag{6.17}$$

$$S[u(x,t)] = s^{na}\sum_{k=0}^{n-1} s^{-na+k}u^{(k)}(x,0) + s^{na}S[f(x,t)] - s^{na}S[Ru(x,t)] - s^{na}S[Nu(x,t)]. \tag{6.18}$$

Applying inverse ST on both sides of Eq. (6.18), we find

$$u(x,t) = F(x,t) - S^{-1}(s^{na}S[Ru(x,t)]) - S^{-1}(s^{na}S[Nu(x,t)]). \tag{6.19}$$

One may see Jena and Chakraverty (2019) for a detailed description of this method. Substituting Eqs. (6.11) and (6.12) into Eq. (6.19) and applying ST with HPM, we may have the following expression:

$$\sum_{n=0}^{\infty} p^n u_n(x,t) = F(x,t) - p\left(S^{-1}\left(s^{na}S\left[R\sum_{n=0}^{\infty} p^n u_n(x,t)\right]\right) + S^{-1}\left(s^{na}S\left[\sum_{n=0}^{\infty} p^n H_n(u)\right]\right)\right). \tag{6.20}$$

By comparing the coefficients of like powers of "p" on both sides of Eq. (6.20), we may have the following approximations successively:

$$p^0 : u_0(x,t) = F(x,t),$$
$$p^1 : u_1(x,t) = -S^{-1}(s^{na}S[Ru_0(x,t)]) - S^{-1}(s^{na}S[H_0(u)]),$$
$$p^2 : u_2(x,t) = -S^{-1}(s^{na}S[Ru_1(x,t)]) - S^{-1}(s^{na}S[H_1(u)]),$$
$$p^3 : u_3(x,t) = -S^{-1}(s^{na}S[Ru_2(x,t)]) - S^{-1}(s^{na}S[H_2(u)]),$$
$$\vdots$$
$$p^n : u_n(x,t) = -S^{-1}(s^{na}S[Ru_{n-1}(x,t)]) - S^{-1}(s^{na}S[H_{n-1}(u)]),$$
$$\vdots$$

So, the solution of Eq. (6.5) may be obtained as:

$$u(x,t) = \lim_{n\to\infty} u_n(x,t) = u_0(x,t) + u_1(x,t) + u_2(x,t) + \cdots. \tag{6.21}$$

6.5 Homotopy Perturbation Elzaki Transform Method (HPETM)

Applying ET on both sides of the Eq. (6.5), we obtain

$$E[D_t^{na}u(x,t)] = E[f(x,t)] - E[Ru(x,t)] - E[Nu(x,t)]. \tag{6.22}$$

Using differentiation property (Eq. (6.3)) of ET, we have

$$s^{-na}E[u(x,t)] - \sum_{k=0}^{n-1} s^{k-na+2}u^{(k)}(x,0) = E[f(x,t)] - E[Ru(x,t)] - E[Nu(x,t)]. \tag{6.23}$$

$$E[u(x,t)] = s^{na}\sum_{k=0}^{n-1} s^{k-na+2}u^{(k)}(x,0) + s^{na}E[f(x,t)] - s^{na}E[Ru(x,t)] - s^{na}E[Nu(x,t)]. \tag{6.24}$$

Inverse ET on both sides of Eq. (6.24) reduces to the following equation:

$$u(x,t) = F(x,t) - E^{-1}(s^{na}E[Ru(x,t)]) - E^{-1}(s^{na}E[Nu(x,t)]). \tag{6.25}$$

Interested researchers may see Jena and Chakraverty (2019) for a detailed description and difference between LT and ET. By plugging Eqs. (6.11) and (6.12) into Eq. (6.25) and operating ET coupled with HPM, we have the expression as follows:

$$\sum_{n=0}^{\infty} p^n u_n(x,t) = F(x,t) - p\left(E^{-1}\left(s^{na}E\left[R\sum_{n=0}^{\infty} p^n u_n(x,t)\right]\right) + E^{-1}\left(s^{na}E\left[\sum_{n=0}^{\infty} p^n H_n(u)\right]\right)\right). \tag{6.26}$$

Comparing the coefficients of equal powers of "p" on both sides of Eq. (6.26), we have the following approximations successively:

$$p^0 : u_0(x,t) = F(x,t),$$
$$p^1 : u_1(x,t) = -E^{-1}(s^{n\alpha}E[Ru_0(x,t)]) - E^{-1}(s^{n\alpha}E[H_0(u)]),$$
$$p^2 : u_2(x,t) = -E^{-1}(s^{n\alpha}E[Ru_1(x,t)]) - E^{-1}(s^{n\alpha}E[H_1(u)]),$$
$$p^3 : u_3(x,t) = -E^{-1}(s^{n\alpha}E[Ru_2(x,t)]) - E^{-1}(s^{n\alpha}E[H_2(u)]),$$
$$\vdots$$
$$p^n : u_n(x,t) = -E^{-1}(s^{n\alpha}E[Ru_{n-1}(x,t)]) - E^{-1}(s^{n\alpha}E[H_{n-1}(u)]),$$
$$\vdots$$

So, the solution of Eq. (6.5) may be written as:

$$u(x,t) = \lim_{n\to\infty} u_n(x,t) = u_0(x,t) + u_1(x,t) + u_2(x,t) + \cdots. \tag{6.27}$$

6.6 Homotopy Perturbation Aboodh Transform Method (HPATM)

Applying AT on both sides of Eq. (6.5) reduces to

$$A[D_t^{n\alpha}u(x,t)] = A[f(x,t)] - A[Ru(x,t)] - A[Nu(x,t)]. \tag{6.28}$$

Further, using Eq. (6.4), Eq. (6.28) reduces to the following expression:

$$s^{n\alpha}A[u(x,t)] - \sum_{k=0}^{n-1} s^{-k+n\alpha-2}u^{(k)}(x,0) = A[f(x,t)] - A[Ru(x,t)] - A[Nu(x,t)]. \tag{6.29}$$

$$A[u(x,t)] = s^{-n\alpha}\sum_{k=0}^{n-1} s^{-k+n\alpha-2}u^{(k)}(x,0) + s^{-n\alpha}A[f(x,t)] - s^{-n\alpha}A[Ru(x,t)] - s^{-n\alpha}A[Nu(x,t)]. \tag{6.30}$$

Applying inverse AT on both sides of Eq. (6.30), we get

$$u(x,t) = F(x,t) - A^{-1}(s^{-n\alpha}A[Ru(x,t)]) - A^{-1}(s^{-n\alpha}A[Nu(x,t)]). \tag{6.31}$$

A detailed description of this transform may be found in Aboodh (2013), Aboodh et al. (2017), and Cherif and Ziane (2018). Putting Eqs. (6.11) and (6.12) into Eq. (6.31) and operating AT in addition to HPM, we get

$$\sum_{n=0}^{\infty} p^n u_n(x,t) = F(x,t) - p\left(A^{-1}\left(s^{-n\alpha}A\left[R\sum_{n=0}^{\infty} p^n u_n(x,t)\right]\right) + A^{-1}\left(s^{-n\alpha}A\left[\sum_{n=0}^{\infty} p^n H_n(u)\right]\right)\right). \tag{6.32}$$

Comparing the coefficients of the same powers of "p" on both sides of Eq. (6.32), we have the successive approximations as:

$$p^0 : u_0(x,t) = F(x,t),$$
$$p^1 : u_1(x,t) = -A^{-1}(s^{-n\alpha}A[Ru_0(x,t)]) - A^{-1}(s^{-n\alpha}A[H_0(u)]),$$
$$p^2 : u_2(x,t) = -A^{-1}(s^{-n\alpha}A[Ru_1(x,t)]) - A^{-1}(s^{-n\alpha}A[H_1(u)]),$$
$$p^3 : u_3(x,t) = -A^{-1}(s^{-n\alpha}A[Ru_2(x,t)]) - A^{-1}(s^{-n\alpha}A[H_2(u)]),$$
$$\vdots$$
$$p^n : u_n(x,t) = -A^{-1}(s^{-n\alpha}A[Ru_{n-1}(x,t)]) - A^{-1}(s^{-n\alpha}A[H_{n-1}(u)]),$$
$$\vdots$$

So, the solution of Eq. (6.5) may be written as:

$$u(x,t) = \lim_{n \to \infty} u_n(x,t) = u_0(x,t) + u_1(x,t) + u_2(x,t) + \cdots \tag{6.33}$$

Next, we solve two test problems to demonstrate the present methods.

6.7 Numerical Examples

6.7.1 Implementation of HPLTM

Example 6.1   Let us consider the one-dimensional heat-like model (Özis and Agırseven 2008; Sadighi et al. 2008) as:

$$D_t^\alpha u(x,t) = \frac{1}{2}x^2 u_{xx}(x,t), \quad 0 < x < 1, t > 0, 0 < \alpha \leq 1, \tag{6.34}$$

subject to the boundary conditions (BCs):

$$u(0,t) = 0, u(1,t) = e^t, \tag{6.35}$$

and IC:

$$u(x,0) = x^2. \tag{6.36}$$

Solution

Applying LT on both sides of Eq. (6.34) with IC and BCs, we have

$$s^\alpha L[u(x,t)] - s^{\alpha-1}u(x,0) = \frac{1}{2}L[x^2 u_{xx}]. \tag{6.37}$$

$$L[u(x,t)] = \frac{s^{-\alpha}}{2}L[x^2 u_{xx}] + \frac{u(x,0)}{s}. \tag{6.38}$$

$$L[u(x,t)] = \frac{s^{-\alpha}}{2}L[x^2 u_{xx}] + \frac{x^2}{s}. \tag{6.39}$$

Taking inverse LT on both sides of Eq. (6.39) gives

$$u(x,t) = L^{-1}\left(\frac{s^{-\alpha}}{2}L[x^2 u_{xx}]\right) + x^2. \tag{6.40}$$

Using HPM, we have

$$\sum_{n=0}^{\infty} p^n u_n(x,t) = x^2 + p\left(L^{-1}\left(\frac{s^{-\alpha}}{2}L\left[x^2 \sum_{n=0}^{\infty} (p^n u_n(x,t))_{xx}\right]\right)\right). \tag{6.41}$$

Comparing the coefficients of equal powers of p on both sides of Eq. (6.41), we obtain

$p^0 : u_0(x, t) = x^2,$

$p^1 : u_1(x, t) = L^{-1}\left(\dfrac{s^{-\alpha}}{2} L\left[x^2 u_{0xx}\right]\right)$

$\quad = L^{-1}\left(\dfrac{s^{-\alpha}}{2} L\left[x^2 \dfrac{\partial^2}{\partial x^2} u_0\right]\right)$

$\quad = L^{-1}\left(\dfrac{s^{-\alpha}}{2} L\left[2x^2\right]\right)$

$\quad = L^{-1}\left(\dfrac{x^2}{s^{\alpha+1}}\right) = x^2 \dfrac{t^\alpha}{\Gamma(1+\alpha)},$

$p^2 : u_2(x, t) = L^{-1}\left(\dfrac{s^{-\alpha}}{2} L\left[x^2 u_{1xx}\right]\right)$

$\quad = L^{-1}\left(\dfrac{s^{-\alpha}}{2} L\left[x^2 \dfrac{\partial^2}{\partial x^2} u_1\right]\right)$

$\quad = L^{-1}\left(\dfrac{s^{-\alpha}}{2} L\left[\dfrac{2t^\alpha}{\Gamma(1+\alpha)} x^2\right]\right)$

$\quad = L^{-1}\left(\dfrac{x^2}{s^{2\alpha+1}}\right) = x^2 \dfrac{t^{2\alpha}}{\Gamma(1+2\alpha)},$

$p^3 : u_3(x, t) = L^{-1}\left(\dfrac{s^{-\alpha}}{2} L\left[x^2 u_{2xx}\right]\right)$

$\quad = x^2 \dfrac{t^{3\alpha}}{\Gamma(1+3\alpha)},$

$\vdots$

So, the solution of the fractional heat-like Eq. (6.34), as $p \to 1$, may be obtained as:

$$u(x, t) = \lim_{n \to \infty} u_n(x, t) = u_0(x, t) + u_1(x, t) + u_2(x, t) + \cdots$$

$$= x^2 + x^2 \frac{t^\alpha}{\Gamma(1+\alpha)} + x^2 \frac{t^{2\alpha}}{\Gamma(1+2\alpha)} + x^2 \frac{t^{3\alpha}}{\Gamma(1+3\alpha)} + \cdots, \qquad (6.42)$$

$$= x^2\left(1 + \frac{t^\alpha}{\Gamma(1+\alpha)} + \frac{t^{2\alpha}}{\Gamma(1+2\alpha)} + \frac{t^{3\alpha}}{\Gamma(1+3\alpha)} + \cdots\right) = x^2 E(t^\alpha),$$

where $E(t^\alpha)$ is called the Mittag-Leffler function.

In particular, at $\alpha = 1$, Eq. (6.42) reduces to $u(x, t) = x^2 e^t$, which is same as the solution of Sadighi et al. (2008).

Example 6.2 Consider the following nonlinear advection equation (Wazwaz 2007):

$$\frac{\partial^\alpha u}{\partial t^\alpha} + u\frac{\partial u}{\partial x} = 0, \quad 0 < \alpha \leq 1, \qquad (6.43)$$

with IC

$$u(x, 0) = -x. \qquad (6.44)$$

Solution

Applying LT on both sides of Eq. (6.43) with IC, we have

$$s^\alpha L[u(x, t)] - s^{\alpha-1} u(x, 0) = -L[uu_x]. \qquad (6.45)$$

$$L[u(x, t)] = -s^{-\alpha} L[uu_x] + \frac{u(x, 0)}{s}. \qquad (6.46)$$

$$L[u(x, t)] = -s^{-\alpha} L[uu_x] - \frac{x}{s}. \qquad (6.47)$$

operating inverse LT on both sides of Eq. (6.47) gives

$$u(x, t) = -L^{-1}(s^{-\alpha}L[uu_x]) - x. \qquad (6.48)$$

By HPM, we have

$$\sum_{n=0}^{\infty} p^n u_n(x, t) = -x - p\left(L^{-1}\left(s^{-\alpha}L\left[\sum_{n=0}^{\infty} p^n H_n(u)\right]\right)\right). \qquad (6.49)$$

Some of He's polynomials $H_n(u)$ (Khan and Wu 2011) for the term uu_x are

$$
\begin{aligned}
H_0(u) &= u_0 \frac{\partial}{\partial x} u_0, \\
H_1(u) &= u_0 \frac{\partial}{\partial x} u_1 + u_1 \frac{\partial}{\partial x} u_0, \\
H_2(u) &= u_0 \frac{\partial}{\partial x} u_2 + u_1 \frac{\partial}{\partial x} u_1 + u_2 \frac{\partial}{\partial x} u_0, \\
&\vdots
\end{aligned}
\qquad (6.50)
$$

Comparing the coefficients of like powers of p on both sides of Eq. (6.49), we obtain

$$
\begin{aligned}
p^0 &: u_0(x, t) = -x, \\
p^1 &: u_1(x, t) = -L^{-1}(s^{-\alpha}L[H_0(u)]) \\
&= -L^{-1}\left(s^{-\alpha}L\left[u_0 \frac{\partial}{\partial x} u_0\right]\right) \\
&= -L^{-1}(s^{-\alpha}L[x]) \\
&= -L^{-1}\left(\frac{x}{s^{\alpha+1}}\right) = -x\frac{t^\alpha}{\Gamma(1+\alpha)}, \\
p^2 &: u_2(x, t) = -L^{-1}(s^{-\alpha}L[H_1(u)]) \\
&= -L^{-1}\left(s^{-\alpha}L\left[u_0 \frac{\partial}{\partial x} u_1 + u_1 \frac{\partial}{\partial x} u_0\right]\right) \\
&= -L^{-1}\left(s^{-\alpha}L\left[2x\frac{t^\alpha}{\Gamma(1+\alpha)}\right]\right) \\
&= -L^{-1}\left(\frac{2x}{s^{2\alpha+1}}\right) = -2x\frac{t^{2\alpha}}{\Gamma(1+2\alpha)}, \\
p^3 &: u_3(x, t) = -L^{-1}(s^{-\alpha}L[H_2(u)]) \\
&= -L^{-1}\left(s^{-\alpha}L\left[u_0 \frac{\partial}{\partial x} u_2 + u_1 \frac{\partial}{\partial x} u_1 + u_2 \frac{\partial}{\partial x} u_0\right]\right) \\
&= -L^{-1}\left(s^{-\alpha}L\left[4x\frac{t^{2\alpha}}{\Gamma(1+2\alpha)} + x\frac{t^{2\alpha}}{(\Gamma(1+\alpha))^2}\right]\right) \\
&= -L^{-1}\left(\frac{4x}{s^{3\alpha+1}} + \frac{x\Gamma(1+2\alpha)}{(\Gamma(1+\alpha))^2 s^{3\alpha+1}}\right), \\
&= -4x\frac{t^{3\alpha}}{\Gamma(1+3\alpha)} - \frac{x\Gamma(1+2\alpha)}{(\Gamma(1+\alpha))^2}\frac{t^{3\alpha}}{\Gamma(1+3\alpha)} \\
&\vdots
\end{aligned}
$$

So, the solution to Eq. (6.43), as $p \to 1$, may be obtained as:

$$
\begin{aligned}
u(x, t) &= \lim_{n \to \infty} u_n(x, t) = u_0(x, t) + u_1(x, t) + u_2(x, t) + \cdots \\
&= -x - x\frac{t^\alpha}{\Gamma(1+\alpha)} - 2x\frac{t^{2\alpha}}{\Gamma(1+2\alpha)} - 4x\frac{t^{3\alpha}}{\Gamma(1+3\alpha)} - \frac{x\Gamma(1+2\alpha)}{(\Gamma(1+\alpha))^2}\frac{t^{3\alpha}}{\Gamma(1+3\alpha)} - \cdots
\end{aligned}
\qquad (6.51)
$$

Particularly at $\alpha = 1$, Eq. (6.51) reduces to closed-form solution $u(x, t) = \dfrac{x}{t-1}$, which is same as the solution of Wazwaz (2007).

6.7.2   Implementation of HPSTM

Applying ST on both sides of Eq. (6.34) with Eqs. (6.35) and (6.36), we obtain

$$s^{-\alpha} S[u(x,t)] - s^{-\alpha} u(x,0) = \frac{1}{2} S[x^2 u_{xx}].$$ (6.52)

$$S[u(x,t)] = \frac{s^\alpha}{2} S[x^2 u_{xx}] + u(x,0).$$ (6.53)

$$S[u(x,t)] = \frac{s^\alpha}{2} S[x^2 u_{xx}] + x^2.$$ (6.54)

Taking inverse ST on both sides of Eq. (6.54) gives

$$u(x,t) = S^{-1}\left(\frac{s^\alpha}{2} S[x^2 u_{xx}]\right) + x^2.$$ (6.55)

Applying HPM, we have

$$\sum_{n=0}^{\infty} p^n u_n(x,t) = x^2 + p\left(S^{-1}\left(\frac{s^\alpha}{2} S\left[x^2 \sum_{n=0}^{\infty} (p^n u_n(x,t))_{xx}\right]\right)\right).$$ (6.56)

Comparing the coefficients of equal powers of p on both sides of Eq. (6.56), we get

$$p^0 : u_0(x,t) = x^2,$$

$$p^1 : u_1(x,t) = S^{-1}\left(\frac{s^\alpha}{2} S[x^2 u_{0xx}]\right)$$

$$= S^{-1}\left(\frac{s^\alpha}{2} S\left[x^2 \frac{\partial^2}{\partial x^2} u_0\right]\right)$$

$$= S^{-1}\left(\frac{s^\alpha}{2} S[2x^2]\right)$$

$$= S^{-1}(x^2 s^\alpha) = x^2 \frac{t^\alpha}{\Gamma(1+\alpha)},$$

$$p^2 : u_2(x,t) = S^{-1}\left(\frac{s^\alpha}{2} S[x^2 u_{1xx}]\right)$$

$$= S^{-1}\left(\frac{s^\alpha}{2} S\left[x^2 \frac{\partial^2}{\partial x^2} u_1\right]\right)$$

$$= S^{-1}\left(\frac{s^\alpha}{2} S\left[\frac{2t^\alpha}{\Gamma(1+\alpha)} x^2\right]\right)$$

$$= S^{-1}(x^2 s^{2\alpha}) = x^2 \frac{t^{2\alpha}}{\Gamma(1+2\alpha)},$$

$$p^3 : u_3(x,t) = S^{-1}\left(\frac{s^\alpha}{2} S[x^2 u_{2xx}]\right)$$

$$= x^2 \frac{t^{3\alpha}}{\Gamma(1+3\alpha)},$$

$$\vdots$$

So, the solution of the fractional heat-like Eq. (6.34), as $p \to 1$, may be obtained as:

$$u(x,t) = x^2 + x^2 \frac{t^\alpha}{\Gamma(1+\alpha)} + x^2 \frac{t^{2\alpha}}{\Gamma(1+2\alpha)} + x^2 \frac{t^{3\alpha}}{\Gamma(1+3\alpha)} + \cdots = x^2 E(t^\alpha).$$ (6.57)

In particular, at $\alpha = 1$, Eq. (6.57) reduces to $u(x,t) = x^2 e^t$, which is same as the solution of Sadighi et al. (2008). Further, operating ST on both sides of Eq. (6.43) with IC, we have

$$s^{-\alpha} S[u(x,t)] - s^{-\alpha} u(x,0) = -S[u u_x].$$ (6.58)

$$S[u(x,t)] = -s^\alpha S[uu_x] + \frac{u(x,0)}{s^\alpha}. \tag{6.59}$$

$$S[u(x,t)] = -s^\alpha S[uu_x] - \frac{x}{s^\alpha}. \tag{6.60}$$

Operating inverse ST on both sides of Eq. (6.60) gives

$$u(x,t) = -S^{-1}(s^\alpha S[uu_x]) - x. \tag{6.61}$$

By HPM, we have

$$\sum_{n=0}^{\infty} p^n u_n(x,t) = -x - p\left(S^{-1}\left(s^\alpha S\left[\sum_{n=0}^{\infty} p^n H_n(u)\right]\right)\right). \tag{6.62}$$

Using Eq. (6.50) and comparing the coefficients of like powers of p on both sides of Eq. (6.62), we obtain

$$p^0 : u_0(x,t) = -x,$$
$$p^1 : u_1(x,t) = -S^{-1}(s^\alpha S[H_0(u)])$$
$$= -S^{-1}\left(s^\alpha S\left[u_0 \frac{\partial}{\partial x} u_0\right]\right)$$
$$= -S^{-1}(s^\alpha S[x])$$
$$= -S^{-1}(xs^\alpha) = -x\frac{t^\alpha}{\Gamma(1+\alpha)},$$
$$p^2 : u_2(x,t) = -S^{-1}(s^\alpha S[H_1(u)])$$
$$= -S^{-1}\left(s^\alpha S\left[u_0 \frac{\partial}{\partial x} u_1 + u_1 \frac{\partial}{\partial x} u_0\right]\right)$$
$$= -S^{-1}\left(s^\alpha S\left[2x\frac{t^\alpha}{\Gamma(1+\alpha)}\right]\right)$$
$$= -S^{-1}(2xs^{2\alpha}) = -2x\frac{t^{2\alpha}}{\Gamma(1+2\alpha)},$$
$$p^3 : u_3(x,t) = -S^{-1}(s^\alpha S[H_2(u)])$$
$$= -S^{-1}\left(s^\alpha S\left[u_0 \frac{\partial}{\partial x} u_2 + u_1 \frac{\partial}{\partial x} u_1 + u_2 \frac{\partial}{\partial x} u_0\right]\right)$$
$$= -S^{-1}\left(s^\alpha S\left[4x\frac{t^{2\alpha}}{\Gamma(1+2\alpha)} + x\frac{t^{2\alpha}}{(\Gamma(1+\alpha))^2}\right]\right)$$
$$= -S^{-1}\left(4xs^{3\alpha} + \frac{x\Gamma(1+2\alpha)s^{3\alpha}}{(\Gamma(1+\alpha))^2}\right),$$
$$= -4x\frac{t^{3\alpha}}{\Gamma(1+3\alpha)} - \frac{x\Gamma(1+2\alpha)}{(\Gamma(1+\alpha))^2}\frac{t^{3\alpha}}{\Gamma(1+3\alpha)}$$
$$\vdots$$

So, the solution to Eq. (6.43) may be obtained as:

$$u(x,t) = = -x - x\frac{t^\alpha}{\Gamma(1+\alpha)} - 2x\frac{t^{2\alpha}}{\Gamma(1+2\alpha)} - 4x\frac{t^{3\alpha}}{\Gamma(1+3\alpha)} - \frac{x\Gamma(1+2\alpha)}{(\Gamma(1+\alpha))^2}\frac{t^{3\alpha}}{\Gamma(1+3\alpha)} - \cdots, \tag{6.63}$$

which is same as the solution of Wazwaz (2007) at $\alpha = 1$.

6.7.3 Implementation of HPETM

Now, taking ET on both sides of Eq. (6.34) with IC and BCs, we get

$$s^{-\alpha}E[u(x,t)] - s^{-\alpha+2}u(x,0) = \frac{1}{2}E\left[x^2 u_{xx}\right]. \tag{6.64}$$

$$E[u(x,t)] = s^2 u(x,0) + \frac{s^\alpha}{2}E\left[x^2 u_{xx}\right]. \tag{6.65}$$

$$E[u(x,t)] = s^2 x^2 + \frac{s^\alpha}{2}E\left[x^2 u_{xx}\right]. \tag{6.66}$$

Inverse ET on both sides of Eq. (6.66) gives

$$u(x,t) = x^2 + E^{-1}\left(\frac{s^\alpha}{2}E\left[x^2 u_{xx}\right]\right). \tag{6.67}$$

Through HPM, we have

$$\sum_{n=0}^{\infty} p^n u_n(x,t) = x^2 + p\left(E^{-1}\left(\frac{s^\alpha}{2}E\left[x^2 \sum_{n=0}^{\infty}(p^n u_n(x,t))_{xx}\right]\right)\right). \tag{6.68}$$

Comparing the coefficients of equal powers of p on both sides of Eq. (6.68), we obtain

$$p^0 : u_0(x,t) = x^2,$$

$$p^1 : u_1(x,t) = E^{-1}\left(\frac{s^\alpha}{2}E\left[x^2 u_{0xx}\right]\right)$$

$$= E^{-1}\left(\frac{s^\alpha}{2}E\left[x^2 \frac{\partial^2}{\partial x^2}u_0\right]\right)$$

$$= E^{-1}\left(\frac{s^\alpha}{2}E\left[2x^2\right]\right)$$

$$= E^{-1}(x^2 s^\alpha) = x^2 \frac{t^\alpha}{\Gamma(1+\alpha)},$$

$$p^2 : u_2(x,t) = E^{-1}\left(\frac{s^\alpha}{2}E\left[x^2 u_{1xx}\right]\right)$$

$$= E^{-1}\left(\frac{s^\alpha}{2}E\left[x^2 \frac{\partial^2}{\partial x^2}u_1\right]\right)$$

$$= E^{-1}\left(\frac{s^\alpha}{2}E\left[\frac{2t^\alpha}{\Gamma(1+\alpha)}x^2\right]\right)$$

$$= E^{-1}(x^2 s^{2\alpha}) = x^2 \frac{t^{2\alpha}}{\Gamma(1+2\alpha)},$$

$$p^3 : u_3(x,t) = S^{-1}\left(\frac{s^\alpha}{2}E\left[x^2 u_{2xx}\right]\right)$$

$$= x^2 \frac{t^{3\alpha}}{\Gamma(1+3\alpha)},$$

$$\vdots$$

So, the solution of the Eq. (6.34) may be obtained as:

$$u(x,t) = x^2 + x^2 \frac{t^\alpha}{\Gamma(1+\alpha)} + x^2 \frac{t^{2\alpha}}{\Gamma(1+2\alpha)} + x^2 \frac{t^{3\alpha}}{\Gamma(1+3\alpha)} + \cdots \tag{6.69}$$

At $\alpha = 1$, Eq. (6.69) is same as the solution of Sadighi et al. (2008).
Again by operating ET on both sides of Eq. (6.43) with IC, we have

$$s^{-\alpha}E[u(x,t)] - s^{-\alpha+2}u(x,0) = -E[uu_x]. \tag{6.70}$$

$$E[u(x,t)] = -s^\alpha E[uu_x] + s^2 u(x,0). \tag{6.71}$$

$$E[u(x,t)] = -s^\alpha E[uu_x] - xs^2. \tag{6.72}$$

By inverse ET on both sides of Eq. (6.72), we have

$$u(x, t) = -E^{-1}(s^\alpha E[uu_x]) - x. \tag{6.73}$$

By HPM, we have

$$\sum_{n=0}^{\infty} p^n u_n(x, t) = -x - p\left(E^{-1}\left(s^\alpha E\left[\sum_{n=0}^{\infty} p^n H_n(u)\right]\right)\right). \tag{6.74}$$

Using Eq. (6.50) and comparing the coefficients of like powers of p on both sides of Eq. (6.74), we obtain

$$p^0 : u_0(x, t) = -x,$$
$$p^1 : u_1(x, t) = -E^{-1}(s^\alpha E[H_0(u)])$$
$$= -E^{-1}\left(s^\alpha E\left[u_0 \frac{\partial}{\partial x} u_0\right]\right)$$
$$= -E^{-1}(s^\alpha E[x])$$
$$= -E^{-1}(xs^{\alpha+2}) = -x\frac{t^\alpha}{\Gamma(1 + \alpha)},$$
$$p^2 : u_2(x, t) = -E^{-1}(s^\alpha E[H_1(u)])$$
$$= -E^{-1}\left(s^\alpha E\left[u_0 \frac{\partial}{\partial x} u_1 + u_1 \frac{\partial}{\partial x} u_0\right]\right)$$
$$= -E^{-1}\left(s^\alpha E\left[2x\frac{t^\alpha}{\Gamma(1 + \alpha)}\right]\right)$$
$$= -E^{-1}(2xs^{2\alpha+2}) = -2x\frac{t^{2\alpha}}{\Gamma(1 + 2\alpha)},$$
$$p^3 : u_3(x, t) = -E^{-1}(s^\alpha E[H_2(u)])$$
$$= -E^{-1}\left(s^\alpha E\left[u_0 \frac{\partial}{\partial x} u_2 + u_1 \frac{\partial}{\partial x} u_1 + u_2 \frac{\partial}{\partial x} u_0\right]\right)$$
$$= -E^{-1}\left(s^\alpha E\left[4x\frac{t^{2\alpha}}{\Gamma(1 + 2\alpha)} + x\frac{t^{2\alpha}}{(\Gamma(1 + \alpha))^2}\right]\right)$$
$$= -E^{-1}\left(4xs^{3\alpha+2} + \frac{x\Gamma(1 + 2\alpha)s^{3\alpha+2}}{(\Gamma(1 + \alpha))^2}\right),$$
$$= -4x\frac{t^{3\alpha}}{\Gamma(1 + 3\alpha)} - \frac{x\Gamma(1 + 2\alpha)}{(\Gamma(1 + \alpha))^2}\frac{t^{3\alpha}}{\Gamma(1 + 3\alpha)}$$
$$\vdots$$

So, the solution to Eq. (6.43) may be written as:

$$u(x, t) = = -x - x\frac{t^\alpha}{\Gamma(1 + \alpha)} - 2x\frac{t^{2\alpha}}{\Gamma(1 + 2\alpha)} - 4x\frac{t^{3\alpha}}{\Gamma(1 + 3\alpha)} - \frac{x\Gamma(1 + 2\alpha)}{(\Gamma(1 + \alpha))^2}\frac{t^{3\alpha}}{\Gamma(1 + 3\alpha)} - \cdots, \tag{6.75}$$

which is same as the solution of Wazwaz (2007) at $\alpha = 1$.

6.7.4 Implementation of HPATM

Now, taking AT on both sides of Eq. (6.34) with IC and BCs, we have

$$s^\alpha A[u(x, t)] - s^{\alpha-2}u(x, 0) = \frac{1}{2}A[x^2 u_{xx}]. \tag{6.76}$$

$$A[u(x, t)] = \frac{u(x, 0)}{s^2} + \frac{s^{-\alpha}}{2}A[x^2 u_{xx}]. \tag{6.77}$$

$$A[u(x,t)] = \frac{x^2}{s^2} + \frac{s^{-\alpha}}{2} A\left[x^2 u_{xx}\right].$$

(6.78)

Inverse AT on both sides of Eq. (6.78) reduces to

$$u(x,t) = x^2 + A^{-1}\left(\frac{s^{-\alpha}}{2} A\left[x^2 u_{xx}\right]\right).$$

(6.79)

Using HPM, we have

$$\sum_{n=0}^{\infty} p^n u_n(x,t) = x^2 + p\left(A^{-1}\left(\frac{s^{-\alpha}}{2} A\left[x^2 \sum_{n=0}^{\infty} (p^n u_n(x,t))_{xx}\right]\right)\right).$$

(6.80)

Comparing the coefficients of equal powers of p on both sides of Eq. (6.80), we obtain

$$p^0 : u_0(x,t) = x^2,$$

$$\begin{aligned}
p^1 : u_1(x,t) &= A^{-1}\left(\frac{s^{-\alpha}}{2} A\left[x^2 u_{0xx}\right]\right) \\
&= A^{-1}\left(\frac{s^{-\alpha}}{2} A\left[x^2 \frac{\partial^2}{\partial x^2} u_0\right]\right) \\
&= A^{-1}\left(\frac{s^{-\alpha}}{2} A\left[2x^2\right]\right) \\
&= A^{-1}(x^2 s^{-\alpha-2}) = x^2 \frac{t^\alpha}{\Gamma(1+\alpha)},
\end{aligned}$$

$$\begin{aligned}
p^2 : u_2(x,t) &= A^{-1}\left(\frac{s^{-\alpha}}{2} A\left[x^2 u_{1xx}\right]\right) \\
&= A^{-1}\left(\frac{s^{-\alpha}}{2} A\left[x^2 \frac{\partial^2}{\partial x^2} u_1\right]\right) \\
&= A^{-1}\left(\frac{s^{-\alpha}}{2} A\left[\frac{2t^\alpha}{\Gamma(1+\alpha)} x^2\right]\right) \\
&= A^{-1}(x^2 s^{-2\alpha-2}) = x^2 \frac{t^{2\alpha}}{\Gamma(1+2\alpha)},
\end{aligned}$$

$$\begin{aligned}
p^3 : u_3(x,t) &= A^{-1}\left(\frac{s^{-\alpha}}{2} A\left[x^2 u_{2xx}\right]\right) \\
&= x^2 \frac{t^{3\alpha}}{\Gamma(1+3\alpha)},
\end{aligned}$$

$$\vdots$$

So, the solution of the Eq. (6.34) may be obtained as:

$$u(x,t) = x^2 + x^2 \frac{t^\alpha}{\Gamma(1+\alpha)} + x^2 \frac{t^{2\alpha}}{\Gamma(1+2\alpha)} + x^2 \frac{t^{3\alpha}}{\Gamma(1+3\alpha)} + \cdots = x^2 E(t^\alpha).$$

(6.81)

In particular, at $\alpha = 1$, Eq. (6.81) reduces to $u(x,t) = x^2 e^t$ which is same as the solution of Sadighi et al. (2008). Again by operating AT on both sides of Eq. (6.43) with IC, we have

$$s^\alpha A[u(x,t)] - s^{\alpha-2} u(x,0) = -A[uu_x].$$

(6.82)

$$A[u(x,t)] = -s^{-\alpha} A[uu_x] + \frac{u(x,0)}{s^2}.$$

(6.83)

$$A[u(x,t)] = -s^{-\alpha} A[uu_x] - \frac{x}{s^2}.$$

(6.84)

By inverse AT on both sides of Eq. (6.84), we have

$$u(x,t) = -A^{-1}(s^{-\alpha} A[uu_x]) - x.$$

(6.85)

By HPM, we have

$$\sum_{n=0}^{\infty} p^n u_n(x,t) = -x - p\left(A^{-1}\left(s^{-\alpha}A\left[\sum_{n=0}^{\infty} p^n H_n(u)\right]\right)\right).$$ (6.86)

Using Eq. (6.50) and comparing the coefficients of like powers of p on both sides of Eq. (6.86), we obtain

$$p^0 : u_0(x,t) = -x,$$
$$p^1 : u_1(x,t) = -A^{-1}(s^{-\alpha}A[H_0(u)])$$
$$= -A^{-1}\left(s^{-\alpha}A\left[u_0\frac{\partial}{\partial x}u_0\right]\right)$$
$$= -A^{-1}(s^{-\alpha}A[x])$$
$$= -A^{-1}(xs^{-\alpha-2}) = -x\frac{t^\alpha}{\Gamma(1+\alpha)},$$
$$p^2 : u_2(x,t) = -A^{-1}(s^{-\alpha}A[H_1(u)])$$
$$= -A^{-1}\left(s^{-\alpha}A\left[u_0\frac{\partial}{\partial x}u_1 + u_1\frac{\partial}{\partial x}u_0\right]\right)$$
$$= -A^{-1}\left(s^{-\alpha}A\left[2x\frac{t^\alpha}{\Gamma(1+\alpha)}\right]\right)$$
$$= -A^{-1}(2xs^{-2\alpha-2}) = -2x\frac{t^{2\alpha}}{\Gamma(1+2\alpha)},$$
$$p^3 : u_3(x,t) = -A^{-1}(s^{-\alpha}A[H_2(u)])$$
$$= -A^{-1}\left(s^{-\alpha}A\left[u_0\frac{\partial}{\partial x}u_2 + u_1\frac{\partial}{\partial x}u_1 + u_2\frac{\partial}{\partial x}u_0\right]\right)$$
$$= -A^{-1}\left(s^{-\alpha}A\left[4x\frac{t^{2\alpha}}{\Gamma(1+2\alpha)} + x\frac{t^{2\alpha}}{(\Gamma(1+\alpha))^2}\right]\right)$$
$$= -A^{-1}\left(4xs^{-3\alpha-2} + \frac{x\Gamma(1+2\alpha)s^{-3\alpha-2}}{(\Gamma(1+\alpha))^2}\right),$$
$$= -4x\frac{t^{3\alpha}}{\Gamma(1+3\alpha)} - \frac{x\Gamma(1+2\alpha)}{(\Gamma(1+\alpha))^2}\frac{t^{3\alpha}}{\Gamma(1+3\alpha)}$$
$$\vdots$$

So, the solution to Eq. (6.43) may be written as:

$$u(x,t) = = -x - x\frac{t^\alpha}{\Gamma(1+\alpha)} - 2x\frac{t^{2\alpha}}{\Gamma(1+2\alpha)} - 4x\frac{t^{3\alpha}}{\Gamma(1+3\alpha)} - \frac{x\Gamma(1+2\alpha)}{(\Gamma(1+\alpha))^2}\frac{t^{3\alpha}}{\Gamma(1+3\alpha)} - \cdots$$ (6.87)

Particularly at $\alpha = 1$, Eq. (6.87) reduces to closed-form solution $u(x,t) = \frac{x}{t-1}$, which is same as the solution of Waz-waz (2007).

References

Aboodh, K.S. (2013). The new integral transform "Aboodh transform". *Global Journal of Pure and Applied Mathematics* 9 (1): 35–43.

Aboodh, K.S., Idris, A., and Nuruddeen, R.I. (2017). On the Aboodh transform connections with some famous integral transforms. *The International Journal of Engineering Science* 1 (9): 143–151.

Baleanu, D. and Jassim, H.K. (2019). A modification fractional homotopy perturbation method for solving Helmholtz and coupled Helmholtz equations on cantor sets. *Fractal and Fractional* 3 (2): 30.

Cherif, M.H. and Ziane, D. (2018). Variational iteration method combined with new transform to solve fractional partial differential equations. *Universal Journal of Mathematics and Applications* 1 (2): 113–120.

Elzaki, T.M. and Biazar, J. (2013). Homotopy perturbation method and Elzaki transform for solving system of nonlinear partial differential equations. *World Applied Sciences Journal* 24 (7): 944–948.

Ghorbani, A. (2009). Beyond Adomian polynomials: He polynomials. *Chaos, Solitons, and Fractals* 39 (3): 1486–1492.

Jena, R.M. and Chakraverty, S. (2019). Analytical solution of Bagley-Torvik equations using Sumudu transformation method. *SN Applied Sciences* 1 (3): 246.

Jena, R.M. and Chakraverty, S. (2019). Solving time-fractional Navier–Stokes equations using homotopy perturbation Elzaki transform. *SN Applied Sciences* 1 (1): 16.

Khan, Y. and Wu, Q. (2011). Homotopy perturbation transform method for nonlinear equations using He's polynomials. *Computers and Mathematics with Applications* 61 (8): 1963–1967.

Mahdy, A., Mohamed, A.S., and Mtawa, A.A.H. (2015). Implementation of the homotopy perturbation Sumudu transform method for solving Klein-Gordon equation. *Applied Mathematics* 6: 136–148.

Mohand, M. and Mahgoub, A. (2016). A coupling method of homotopy perturbation and Aboodh transform for solving nonlinear fractional heat – like equations. *International Journal of Systems Science and Applied Mathematics* 1 (4): 63–68.

Olubanwo, O.O., Odetunde, O.S., and Talabi, A.T. (2019). Aboodh homotopy perturbation method of solving Burgers equation. *Asian Journal of Applied Sciences* 7 (2): 295–302.

Özis, T. and Agırseven, D. (2008). He's homotopy perturbation method for solving heat-like and wave-like equations with variable coefficients. *Physical Letters A* 372: 5944–5950.

Sadighi, A., Ganji, D.D., Gorji, M., and Tolou, N. (2008). Numerical simulation of heat-like models with variable coefficients by the variational iteration method. *Journal of Physics: Conference Series* 96: 012083.

Sedeeg, A.K.H. (2016). A coupling Elzaki transform and homotopy perturbation method for solving nonlinear fractional heat-like equations. *American Journal of Mathematical and Computer Modelling* 1 (1): 15–20.

Singh, J., Kumar, D., and Sushila (2011). Homotopy perturbation Sumudu transform method for nonlinear equations. *Advances in Applied Mathematics and Mechanics* 4 (4): 165–175.

Singh, J., Kumar, D., and Sushila (2012). Homotopy perturbation algorithm using Laplace transform for gas dynamics equation. *Journal of the Applied Mathematics Statistics and Informatics* 8 (1): 55–61.

Wazwaz, A.M. (2007). A comparison between the variational iteration method and adomian decomposition method. *Journal of Computational and Applied Mathematics*. 207: 129–136.

7

Fractional Differential Transform Method

7.1 Introduction

In Chapter 6, we have already discussed the hybrid method, which combines homotopy perturbation method (HPM) and various transform methods. DTM is a semi-analytical method based on the Taylor series expansion, which constructs an analytical solution in the form of a polynomial. Zhou (1986) first proposed the differential transform method (DTM) and initially applied it to initial value problems used in electrical circuits. Jang et al. (2001) claimed that DTM is an iterative process to obtain the solution of differential equations in the Taylor series form. Taylor's standard high-order series method involves symbolic computation. Although the Taylor series method needs more computational work for higher orders, this method reduces the computational domain size and applies to several problems (Fatma 2004; Hassan 2004; Erturk and Momani 2008). In this chapter, we use DTM with a combination of Caputo fractional derivatives, which is called as fractional differential transform method (FDTM). The present method is based on the combination of the classical one-dimensional FDTM and generalized Taylor's formula. The concerned authors may follow (Arikoglu and Ozkol 2006, 2007; Nazari and Shahmorad 2010; Methi 2016) for more details.

7.2 Fractional Differential Transform Method

There are various ways to generalize the notion of differentiation of fractional orders. The fractional differentiation in Riemann–Liouville sense is described as (Arikoglu and Ozkol 2007):

$$D_0^\beta f(x) = \frac{1}{\Gamma(n-\beta)} \frac{d^n}{dx^n} \left[\int_0^x (x-t)^{n-\beta-1} f(t) dt \right], \tag{7.1}$$

for $n-1 < \beta \le n$, $n \in Z^+$ and $x > 0$. Let us expand the analytical and continuous function $f(x)$ in terms of fractional power series as:

$$f(x) = \sum_{k=0}^{\infty} F(k)(x-x_0)^{k/\alpha}, \tag{7.2}$$

where α is the order of fraction and $F(k)$ is the fractional differential transform of $f(x)$. Regarding practical applications in different branches of science, fractional initial conditions (ICs) are often not available, and their physical meaning may not be obvious. Therefore, to deal with integer-ordered ICs in Caputo sense (Caputo 1967), the definition in Eq. (7.1) should be modified as follows (Arikoglu and Ozkol 2007):

$$D_0^\beta \left[f(x) - \sum_{k=0}^{n-1} \frac{1}{k!}(x-x_0)^k f^k(0) \right] = \frac{1}{\Gamma(n-\beta)} \frac{d^n}{dx^n} \left[\int_0^x \left\{ \frac{f(t) - \sum_{k=0}^{n-1} \frac{1}{k!}(t-x_0)^k f^k(0)}{(x-t)^{\beta-n+1}} \right\} dt \right]. \tag{7.3}$$

Computational Fractional Dynamical Systems: Fractional Differential Equations and Applications, First Edition.
Snehashish Chakraverty, Rajarama Mohan Jena, and Subrat Kumar Jena.
© 2023 John Wiley & Sons, Inc. Published 2023 by John Wiley & Sons, Inc.

Since the ICs are extended to integer-order derivatives, so the transformation of the ICs is defined as:

$$
F(k) = \begin{cases} \dfrac{1}{(k/\alpha)!} \left[\dfrac{d^{k/\alpha} f(x)}{dx^{k/\alpha}} \right]_{x = x_0}, & \text{If } \left(\dfrac{k}{\alpha} \right) \in Z^+, \\[3mm] 0, & \text{If } \left(\dfrac{k}{\alpha} \right) \notin Z^+, \end{cases} \quad \text{for } k = 0, 1, 2, \ldots, (n\alpha - 1) \tag{7.4}
$$

where n is the order of the fractional differential equation (FDE). The following theorems are obtained using Eqs. (7.1) and (7.2).

Theorem 7.1   (Arikoglu and Ozkol 2007)

a) If $f(x) = g(x) \pm h(x)$ then $F(k) = G(k) \pm H(k)$.

b) If $f(x) = g(x) h(x)$,   then $F(k) = \displaystyle\sum_{k=0}^{l} G(k) H(l - k)$

c) If $f(x) = g_1(x) g_2(x) g_3(x) \ldots g_{n-1}(x) g_n(x)$, then

$$
F(k) = \sum_{k_{n-1}=0}^{k} \sum_{k_{n-2}=0}^{k_{n-1}} \cdots \sum_{k_3=0}^{k_4} \sum_{k_2=0}^{k_3} \sum_{k_1=0}^{k_2} G(k_1) G(k_2 - k_1) G(k_3 - k_2) \ldots G(k_{n-1} - k_{n-2}) G(k - k_{n-1}).
$$

d) If $f(x) = (x - x_0)^n$,   then $F(k) = \delta(k - n\alpha)$ where $\delta(k) = \begin{cases} 1 & \text{if } k = 0 \\ 0 & \text{if } k \neq 0 \end{cases}$

Proof: One may see Arikoglu and Ozkol (2007) for the explanations of all the aforementioned results.

Theorem 7.2   (Arikoglu and Ozkol 2007)

If $f(x) = D_{x_0}^{\beta}[g(x)]$,   then $F(k) = \dfrac{\Gamma\left(1 + \beta + \frac{k}{\alpha}\right)}{\Gamma\left(1 + \dfrac{k}{\alpha}\right)} G(k + \alpha\beta)$,

where α is the order of the FDE.

Proof: One may refer to Arikoglu and Ozkol (2007) for the explanation.

Theorem 7.3   (Arikoglu and Ozkol 2007)

If $f(x) = \dfrac{d^{\beta_1}}{dx^{\beta_1}}[g_1(x)] \dfrac{d^{\beta_2}}{dx^{\beta_2}}[g_2(x)] \ldots \dfrac{d^{\beta_{n-1}}}{dx^{\beta_{n-1}}}[g_{n-1}(x)] \dfrac{d^{\beta_n}}{dx^{\beta_n}}[g_n(x)]$, then

$$
F(k) = \sum_{k_{n-1}=0}^{k} \sum_{k_{n-2}=0}^{k_{n-1}} \cdots \sum_{k_3=0}^{k_4} \sum_{k_2=0}^{k_3} \sum_{k_1=0}^{k_2} \frac{\Gamma\left(1 + \beta_1 + \dfrac{k_1}{\alpha}\right)}{\Gamma\left(1 + \dfrac{k_1}{\alpha}\right)} \frac{\Gamma\left(1 + \beta_2 + \dfrac{k_2 - k_1}{\alpha}\right)}{\Gamma\left(1 + \dfrac{k_2 - k_1}{\alpha}\right)} \cdots \frac{\Gamma\left(1 + \beta_{n-1} + \dfrac{k_{n-1} - k_{n-2}}{\alpha}\right)}{\Gamma\left(1 + \dfrac{k_{n-1} - k_{n-2}}{\alpha}\right)}
$$

$$
\frac{\Gamma\left(1 + \beta_n + \dfrac{k - k_{n-1}}{\alpha}\right)}{\Gamma\left(1 + \dfrac{k - k_{n-1}}{\alpha}\right)} G_1(k_1 + \alpha\beta_1) G_2(k_2 - k_1 + \alpha\beta_2) \ldots G_{n-1}(k_{n-1} - k_{n-2} + \alpha\beta_{n-1}) G_n(k - k_{n-1} + \alpha\beta_n),
$$

where $\alpha\beta_i \in Z^+$ for $i = 1, 2, 3, \ldots, n$.

Proof: The proof of this result can be seen in Arikoglu and Ozkol (2007).

Theorem 7.4   (Arikoglu and Ozkol 2007)

a) If $f(x) = \displaystyle\int_{x_0}^{x} g(t)\,dt$,   then $F(k) = \dfrac{\alpha G(k - \alpha)}{k}$,   for $k \geq \alpha$.

b) If $f(x) = g(x) \displaystyle\int_{x_0}^{x} h(t)\,dt$,   then $F(k) = \alpha \displaystyle\sum_{k_1=\alpha}^{k} \dfrac{H(k_1 - \alpha)}{k_1} G(k - k_1)$,   for $k \geq \alpha$.

From the aforementioned theorems, one may see that DTM is a subset of FDTM. Eq. (7.2) is the classical form of expanding an analytical function in the Taylor series in this particular case.

7.3 Illustrative Examples

In order to demonstrate the effectiveness of the FDTM, we use the present method to solve a linear fractional one-dimensional heat-like problem in Example 7.1 and a nonlinear advection equation in Example 7.2.

Example 7.1 Let us consider the one-dimensional heat-like model (Özis and Agırseven 2008; Sadighi et al. 2008) as:

$$D_t^\beta u(x,t) = \frac{1}{2} x^2 u_{xx}(x,t), \quad 0 < x < 1, t > 0, 0 < \beta \le 1, \tag{7.5}$$

with the boundary conditions (BCs):

$$u(0,t) = 0, \quad u(1,t) = e^t, \tag{7.6}$$

and IC:

$$u(x,0) = x^2. \tag{7.7}$$

Solution

Since the value of β $(0 < \beta \le 1)$ lies between 0 and 1. So, we have considered the value of $\alpha = 10$ in order to get integer order by multiplying 10 to β. The author may consider the order of the FDE as $\alpha = 20$ and the result of the problem will be the same, but it will take more computation as compared to the earlier one. Taking IC $u(x,0) = f(x) = x^2$ and using Theorem 7.2, Eq. (7.5) reduces to the following expression:

$$\frac{\Gamma(1 + \beta + (k/10))}{\Gamma(1 + (k/10))} U(k + 10\beta) = \frac{1}{2} x^2 \frac{\partial^2}{\partial x^2} U(k), \tag{7.8}$$

and using Eq. (7.4), IC and BC are transformed to

$$U(k) = \begin{cases} \dfrac{1}{(k/\beta)!} \left[\dfrac{d^{k/\beta} f(x)}{dx^{k/\beta}} \right]_{x=0}, & \dfrac{k}{\beta} \in Z^+, \\[4mm] 0, & \dfrac{k}{\beta} \notin Z^+, \end{cases} \quad \text{for } k = 0,1,2,\ldots,10\beta - 1. \tag{7.9}$$

From Eq. (7.9), we obtain

$$k = 0 : \frac{0}{\beta} = 0 \in Z^+, \quad U(0) = \frac{1}{0!}[u(x,0)] = x^2,$$

$$k = 1 : \frac{1}{\beta} \notin Z^+, \quad U(1) = 0,$$

$$\vdots$$

$$k = \beta : \frac{\beta}{\beta} = 1 \in Z^+, \quad U(\beta) = \frac{1}{1!} \left[\frac{d}{dx} f(x) \right]_{x=0} = 0,$$

$$\vdots$$

$$k = 9\beta : \frac{9\beta}{\beta} = 9 \in Z^+, \quad U(9\beta) = \frac{1}{9!} \left[\frac{d^9}{dx^9} f(x) \right]_{x=0} = 0.$$

Using Eq. (7.8), the following results are obtained:

$$k = 0 : U(10\beta) = \frac{1}{\Gamma(1+\beta)} \left[\frac{1}{2}x^2 \frac{\partial^2}{\partial x^2} U(0) \right] = \frac{x^2}{\Gamma(1+\beta)},$$

$$k = 1 : U(10\beta + 1) = \frac{\Gamma\left(1 + \frac{1}{10}\right)}{\Gamma\left(1 + \beta + \frac{1}{10}\right)} \left[\frac{1}{2}x^2 \frac{\partial^2}{\partial x^2} U(1) \right] = 0,$$

$$\vdots$$

$$k = \beta : U(11\beta) = \frac{\Gamma\left(1 + \frac{\beta}{10}\right)}{\Gamma\left(1 + \beta + \frac{\beta}{10}\right)} \left[\frac{1}{2}x^2 \frac{\partial^2}{\partial x^2} U(\beta) \right] = 0,$$

$$\vdots$$

$$k = 10\beta : U(20\beta) = \frac{\Gamma(1+\beta)}{\Gamma(1+2\beta)} \left[\frac{1}{2}x^2 \frac{\partial^2}{\partial x^2} U(10\beta) \right] = \frac{x^2}{\Gamma(1+2\beta)},$$

$$\vdots$$

$$k = 20\beta : U(30\beta) = \frac{\Gamma(1+2\beta)}{\Gamma(1+3\beta)} \left[\frac{1}{2}x^2 \frac{\partial^2}{\partial x^2} U(20\beta) \right] = \frac{x^2}{\Gamma(1+3\beta)},$$

$$\vdots$$

Now, applying inverse differential transform, we have

$$\begin{aligned} u(x,t) &= \sum_{k=0}^{\infty} U(k) t^{\frac{k}{10}} = x^2 + \frac{x^2 t^\beta}{\Gamma(1+\beta)} + \frac{x^2 t^{2\beta}}{\Gamma(1+2\beta)} + \frac{x^2 t^{3\beta}}{\Gamma(1+3\beta)} + \dots \\ &= x^2 \left(1 + \frac{t^\beta}{\Gamma(1+\beta)} + \frac{t^{2\beta}}{\Gamma(1+2\beta)} + \frac{t^{3\beta}}{\Gamma(1+3\beta)} + \dots \right) = x^2 E(t^\beta), \end{aligned}$$

(7.10)

where $E(t^\beta)$ is called the Mittag-Leffler function, which is given in Chapter 1.

In particular, at $\beta = 1$, Eq. (7.10) reduces to $u(x, t) = x^2 e^t$, which is same as the solution of Sadighi et al. (2008). It is worth mentioning that increasing the number of terms of the solution may lead to a better result, as shown in Figure 7.1. Figures 7.1–7.5 depict the fourth-order approximate solution plots of Example 7.1 with different β values.

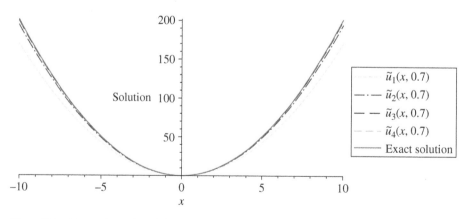

Figure 7.1 Comparison plot of the present solution with the exact solution of Example 7.1 taking a different number of terms of approximate solution.

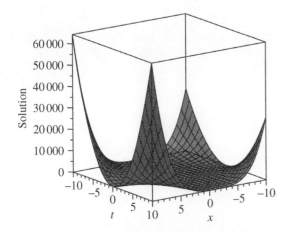

Figure 7.2 Fourth-order approximate solution plot of Example 7.1 at $\beta = 1$.

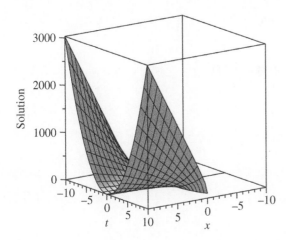

Figure 7.3 Fourth-order approximate solution plot of Example 7.1 at $\beta = 0.3$.

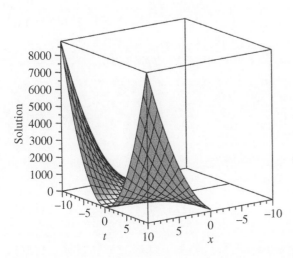

Figure 7.4 Fourth-order approximate solution plot of Example 7.1 at $\beta = 0.5$.

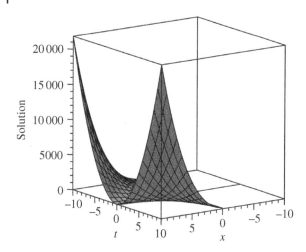

Figure 7.5 Fourth-order approximate solution plot of Example 7.1 at $\beta = 0.7$.

Example 7.2 Consider the following nonlinear advection equation (Wazwaz 2007):

$$\frac{\partial^\beta u}{\partial t^\beta} + u\frac{\partial u}{\partial x} = 0, \quad 0 < \beta \le 1, \tag{7.11}$$

with IC:

$$u(x, 0) = -x. \tag{7.12}$$

Solution

By choosing $\alpha = 10$, $f(x) = -x$ and using Theorems 7.1 (b) and 7.2, Eq. (7.11) is transformed to

$$\frac{\Gamma(1 + \beta + (k/10))}{\Gamma(1 + (k/10))} U(k + 10\beta) = -\left[\sum_{l=0}^{k} U(l)\frac{\partial}{\partial x}U(k - l)\right]. \tag{7.13}$$

As per Eq. (7.9), IC of Eq. (7.12) may be written as:

$$\begin{aligned} U(0) &= -x, \\ U(k) &= 0, \quad k = 1, 2, ..., \beta, ..., 2\beta, ..., (10\beta - 1). \end{aligned} \tag{7.14}$$

The following recurrence relations may be obtained from Eq. (7.13):

$$k = 0 : U(10\beta) = \frac{-1}{\Gamma(1 + \beta)}\left[U(0)\frac{\partial}{\partial x}U(0)\right] = \frac{-x}{\Gamma(1 + \beta)}, \tag{7.15}$$

$$U(10\beta + k) = 0, \quad \text{for } k = 1, ..., \beta, ..., 2\beta, ..., (10\beta - 1). \tag{7.16}$$

$$k = 10\beta : U(20\beta) = \frac{-\Gamma(1 + \beta)}{\Gamma(1 + 2\beta)}\left[\sum_{l=0}^{10\beta} U(l)\frac{\partial}{\partial x}U(10\beta - l)\right]. \tag{7.17}$$

Using Eqs. (7.14)–(7.16), Eq. (7.17) may be calculated as:

$$U(20\beta) = \frac{-\Gamma(1+\beta)}{\Gamma(1+2\beta)}\left[U(0)\frac{\partial}{\partial x}U(10\beta) + U(10\beta)\frac{\partial}{\partial x}U(0)\right] = \frac{-2x}{\Gamma(1+2\beta)}. \tag{7.18}$$

Now,

$$U(20\beta + k) = 0, \quad k = 1, \ldots, \beta, \ldots, 2\beta, \ldots, (10\beta - 1). \tag{7.19}$$

Substituting $k = 20\beta$ in Eq. (7.13), we have

$$U(30\beta) = \frac{-\Gamma(1+2\beta)}{\Gamma(1+3\beta)}\left[\sum_{l=0}^{20\beta}U(l)\frac{\partial}{\partial x}U(20\beta - l)\right]. \tag{7.20}$$

Using Eqs. (7.14)–(7.19), Eq. (7.20) may be computed as:

$$\begin{aligned} U(30\beta) &= \frac{-\Gamma(1+2\beta)}{\Gamma(1+3\beta)}\left[U(0)\frac{\partial}{\partial x}U(20\beta) + U(10\beta)\frac{\partial}{\partial x}U(10\beta) + U(20\beta)\frac{\partial}{\partial x}U(0)\right] \\ &= \frac{-4x}{\Gamma(1+3\beta)} - \frac{x\,\Gamma(1+2\beta)}{(\Gamma(1+\beta))^2\Gamma(1+3\beta)}. \end{aligned} \tag{7.21}$$

$$\vdots$$

Applying inverse differential transformation, $u(x, t)$ can be evaluated as:

$$\begin{aligned} u(x,t) &= \sum_{k=0}^{\infty}U(k)t^{k/10} = -x + \frac{-x}{\Gamma(1+\beta)}t^{\beta} + \frac{-2x}{\Gamma(1+2\beta)}t^{2\beta} \\ &\quad + \left(\frac{-4x}{\Gamma(1+3\beta)} - \frac{x\,\Gamma(1+2\beta)}{(\Gamma(1+\beta))^2\Gamma(1+3\beta)}\right)t^{3\beta} + \cdots \end{aligned} \tag{7.22}$$

Particularly at $\beta = 1$, Eq. (7.22) reduces to a closed-form solution $u(x, t) = \dfrac{x}{t-1}$, which is same as the solution of Wazwaz (2007). In Figure 7.6, the exact solution is compared with the present solution by considering the increasing number of terms. Plots of third-order approximate solution of Example 7.2 for various β values appear in Figures 7.7–7.10.

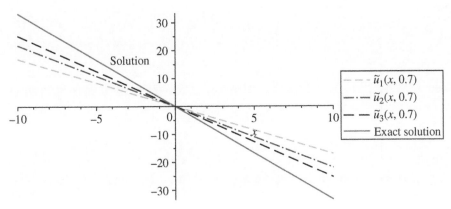

Figure 7.6 Comparison plot of the present solution with the exact solution of Example 7.2 taking a different number of terms of approximate solution.

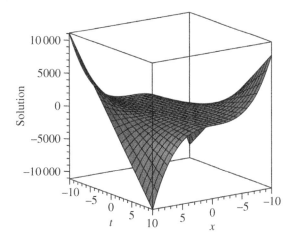

Figure 7.7 Third-order approximate solution plot of Example 7.2 at $\beta = 1$.

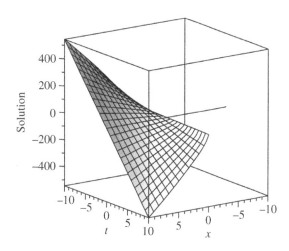

Figure 7.8 Third-order approximate solution plot of Example 7.2 at $\beta = 0.3$.

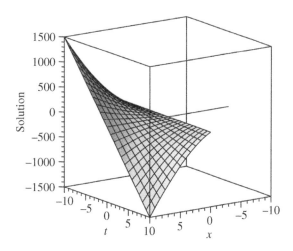

Figure 7.9 Third-order approximate solution plot of Example 7.2 at $\beta = 0.5$.

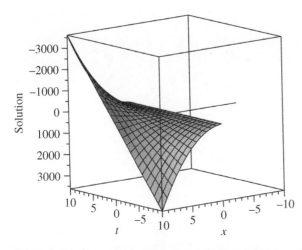

Figure 7.10 Third-order approximate solution plot of Example 7.2 at $\beta = 0.7$.

References

Arikoglu, A. and Ozkol, I. (2006). Solution of difference equations by using differential transform method. *Applied Mathematics and Computation* 174: 442–454.

Arikoglu, A. and Ozkol, I. (2007). Solution of fractional differential equations by using differential transform method. *Chaos, Solitons and Fractals* 34: 1473–1481.

Caputo, M. (1967). Linear models of dissipation whose Q is almost frequency independent. Part II. *Geophysical Journal of the Royal Astronomical Society* 13: 529–539.

Erturk, V.S. and Momani, S. (2008). Solving system of fractional differential equations using differential transform method. *Journal of Computational and Applied Mathematics* 214: 142–151.

Fatma, A. (2004). Solution of the system of differential equations by differential transform. *Applied Mathematics and Computation* 147: 547–567.

Hassan, I.H.A.H. (2004). Differential transformation technique for solving higher-order initial value problems. *Applied Mathematics and Computation* 154: 299–311.

Jang, M.J., Chen, C.L., and Liu, Y.C. (2001). Two-dimensional differential transform for partial differential equations. *Applied Mathematics and Computation* 121: 261–270.

Methi, G. (2016). Solution of differential equations using differential transform method. *Asian Journal of Mathematics and Statistics* 9: 1–5.

Nazari, D. and Shahmorad, S. (2010). Application of the fractional differential transform method to fractional-order integro-differential equations with nonlocal boundary conditions. *Journal of Computational and Applied Mathematics* 234: 883–891.

Özis, T. and Agırseven, D. (2008). He's a homotopy perturbation method for solving heat-like and wave-like equations with variable coefficients. *Physics Letters A* 372: 5944–5950.

Sadighi, A., Ganji, D.D., Gorji, M., and Tolou, N. (2008). Numerical simulation of heat-like models with variable coefficients by the variational iteration method. *Journal of Physics: Conference Series* 96: 012083.

Wazwaz, A.M. (2007). A comparison between the variational iteration method and adomian decomposition method. *Journal of Computational and Applied Mathematics* 207: 129–136.

Zhou, J.K. (1986). *Differential Transformation and its Application for Electrical Circuits*. Wuhan, China: Huarjung University Press.

8

Fractional Reduced Differential Transform Method

8.1 Introduction

In Chapter 7, we have discussed the differential transform method (DTM), which was first proposed by Zhou (1986), and it was used to solve linear and nonlinear initial value problems arising in electric circuit analysis. Later, DTM was used to solve partial differential equations. Another analytical version of DTM is the reduced differential transform method (RDTM). Recently, the RDTM has been found to be effective and reliable for handling linear and nonlinear partial differential equations and integral equations (Abazari and Kılıçman 2013; Abazari and Soltanalizadeh, 2013; Saravanan and Magesh 2013). Keskin and Oturanc (2010) applied this method to solve fractional differential equations with some modifications. Later this method was named as fractional reduced differential transform method (FRDTM). FRDTM has been successfully applied to solve various types of fractional partial differential equations (see Jena et al. (2019a, 2019b) and Saravanan and Magesh (2016)) and also higher-dimensional problems (Rawashdeh 2015). In Chapters 3 and 5, we have implemented HPM and ADM, respectively, to solve some fractional-order differential equations. The significant disadvantages of these approaches are that it requires complicated and plenty of calculations. In order to overcome such types of drawbacks, FRDTM has been introduced. The FRDTM is a computationally efficient and implementable analytical technique and provides an approximate analytical solution for both linear and nonlinear fractional differential equations. It does not use any discretization, transformation, perturbation, or restrictive conditions. This method needs a lesser size of computations than HPM and ADM.

First, we briefly describe the FRDTM in the following section.

8.2 Description of FRDTM

In order to understand the basic concept of FRDTM (Jena et al. 2019a, 2019b; Jafari et al. 2016), first, we recall and review the local fractional Taylor's theorems, and then, we extend it to FRDTM for fractional derivative.

Theorem 8.1 (Local fractional Taylor's theorem)

Suppose that $\zeta^{(k+1)\alpha} \in C_\alpha(a, b)$, for $k = 0, 1, ..., n$ and $0 < \alpha \le 1$, then we have

$$\zeta(x) = \sum_{k=0}^{\infty} \zeta^{(k\alpha)}(0) \frac{(x - x_0)^{k\alpha}}{\Gamma(1 + k\alpha)}, \qquad (8.1)$$

where

$$a < x_0 < x < b, \quad \forall x \in (a, b) \quad \text{and} \quad \zeta^{(k+1)\alpha}(x) = \underbrace{D_x^\alpha D_x^\alpha D_x^\alpha ... D_x^\alpha}_{k+1 \quad times} \zeta(x).$$

Computational Fractional Dynamical Systems: Fractional Differential Equations and Applications, First Edition.
Snehashish Chakraverty, Rajarama Mohan Jena, and Subrat Kumar Jena.

Definition 8.1   The fractional reduced differential transform $\zeta_k(x)$ of an analytic function $\zeta(x, t)$ is defined as the following formula:

$$\zeta_k(x) = \frac{1}{\Gamma(1 + \alpha k)} \left(\frac{\partial^{k\varepsilon} \zeta(x,t)}{\partial t^{k\alpha}} \right)_{t=0} \quad \text{where } k = 0, 1, ..., n, \tag{8.2}$$

Definition 8.2   The fractional inverse differential transform of $\zeta_k(x)$ is defined as follows:

$$\zeta(x, t) = \sum_{k=0}^{\infty} \zeta_k(x) t^{k\alpha}. \tag{8.3}$$

Using Eqs. (8.2) and (8.3), the following theorems on FRDTM are deduced.

Theorem 8.2   Let $\psi(x,\ t)$, $\xi(x,\ t)$ and $\zeta(x,\ t)$ are three analytical functions such that $\psi(x,t) = R_D^{-1}[\psi_k(x)]$, $\xi(x,t) = R_D^{-1}[\xi_k(x)]$ and $\zeta(x,t) = R_D^{-1}[\zeta_k(x)]$. Hence,

i) If $\psi(x,t) = c_1 \xi(x,t) \pm c_2 \zeta(x,t)$,   then $\psi_k(x) = c_1 \xi_k(x) \pm c_2 \zeta_k(x)$, where c_1 and c_2 are constants.

ii) If $\psi(x,t) = a\ \xi(x,t)$,   then $\psi_k(x) = a\ \xi_k(x)$.

iii) If $\psi(x,t) = \xi(x,t)\zeta(x,t)$,   then $\psi_k(x) = \sum_{i=0}^{j} \xi_i(x)\zeta_{j-i}(x) = \sum_{i=0}^{j} \zeta_i(x)\xi_{j-i}(x)$.

iv) If $\psi(x,t) = \frac{\partial^m}{\partial x^m}\xi(x,t)$,   then $\psi_k(x) = \frac{\partial^m}{\partial x^m}\xi_k(x)$.

v) If $\psi(x,t) = \frac{\partial^{n\alpha}}{\partial t^{n\alpha}}\xi(x,t)$,   then $\psi_k(x) = \frac{\Gamma(1 + (k+n)\alpha)}{(1 + k\alpha)}\xi_{k+n}(x)$.

Here, R_D^{-1} denotes the inverse reduced differential transform operator.

Proof

i) From Eq. (8.2), we have

$$\psi_k(x) = \frac{1}{\Gamma(1 + \alpha k)} \left(\frac{\partial^{k\alpha} \psi(x,t)}{\partial t^{k\alpha}} \right)_{t=0}$$

$$= \frac{1}{\Gamma(1 + \alpha k)} \left(\frac{\partial^{k\alpha}}{\partial t^{k\alpha}} (c_1 \xi(x,t) \pm c_2 \zeta(x,t)) \right)_{t=0}$$

$$= \frac{1}{\Gamma(1 + \alpha k)} \left(c_1 \frac{\partial^{k\alpha} \xi(x,t)}{\partial t^{k\alpha}} \pm c_2 \frac{\partial^{k\alpha} \zeta(x,t)}{\partial t^{k\alpha}} \right)_{t=0} \tag{8.4}$$

$$= c_1 \frac{1}{\Gamma(1 + \alpha k)} \left(\frac{\partial^{k\alpha} \xi(x,t)}{\partial t^{k\alpha}} \right)_{t=0} \pm c_2 \frac{1}{\Gamma(1 + \alpha k)} \left(\frac{\partial^{k\alpha} \zeta(x,t)}{\partial t^{k\alpha}} \right)_{t=0},$$

$$\psi_k(x) = c_1 \xi_k(x) \pm c_2 \zeta_k(x).$$

ii) From Eq. (8.2), we have

$$\psi_k(x) = \frac{1}{\Gamma(1 + \alpha k)} \left(\frac{\partial^{k\alpha} \psi(x,t)}{\partial t^{k\alpha}} \right)_{t=0}$$

$$= \frac{1}{\Gamma(1 + \alpha k)} \left(\frac{\partial^{k\alpha}}{\partial t^{k\alpha}} (a\xi(x,t)) \right)_{t=0} \tag{8.5}$$

$$= a \frac{1}{\Gamma(1 + \alpha k)} \left(\frac{\partial^{k\alpha} \xi(x,t)}{\partial t^{k\alpha}} \right)_{t=0} = a\xi_k(x).$$

iii) From Eq. (8.3), we obtain

$$\psi(x,t) = \left(\sum_{k=0}^{\infty} \xi_k(x) t^{k\alpha} \right) \left(\sum_{k=0}^{\infty} \zeta_k(x) t^{k\alpha} \right)$$

$$= \left(\xi_0(x) + \xi_1(x) t^{\alpha} + \xi_2(x) t^{2\alpha} + \cdots \right) \left(\zeta_0(x) + \zeta_1(x) t^{\alpha} + \zeta_2(x) t^{2\alpha} + \cdots \right)$$

$$= \begin{pmatrix} \xi_0(x)\zeta_0(x) + (\xi_0(x)\zeta_1(x) + \xi_1(x)\zeta_0(x))t^{\alpha} + (\xi_0(x)\zeta_2(x) + \xi_1(x)\zeta_1(x) + \xi_2(x)\zeta_0(x))t^{2\alpha} + \cdots \\ + (\xi_0(x)\zeta_k(x) + \xi_1(x)\zeta_{k-1}(x) + \cdots + \xi_{k-1}(x)\zeta_1(x) + \xi_k(x)\zeta_0(x))t^{2\alpha} \end{pmatrix} \quad (8.6)$$

$$\psi_k(x) = \sum_{i=0}^{j} \xi_i(x)\zeta_{j-i}(x) = \sum_{i=0}^{j} \zeta_i(x)\xi_{j-i}(x).$$

iv) From Eq. (8.2), we have

$$\psi_k(x) = \frac{1}{\Gamma(1+\alpha k)} \left(\frac{\partial^{k\alpha} \psi(x,t)}{\partial t^{k\alpha}} \right)_{t=0}$$

$$= \frac{1}{\Gamma(1+\alpha k)} \left(\frac{\partial^{k\alpha}}{\partial t^{k\alpha}} \left(\frac{\partial^m}{\partial x^m} \xi(x,t) \right) \right)_{t=0} \quad (8.7)$$

$$= \frac{1}{\Gamma(1+\alpha k)} \left(\frac{\partial^m}{\partial x^m} \left(\frac{\partial^{k\alpha} \xi(x,t)}{\partial t^{k\alpha}} \right) \right)_{t=t_0} = \frac{\partial^m \xi_k(x)}{\partial x^m}.$$

v) From Eq. (8.2), we have

$$\psi_k(x) = \frac{1}{\Gamma(1+\alpha k)} \left(\frac{\partial^{k\alpha} \psi(x,t)}{\partial t^{k\alpha}} \right)_{t=0}$$

$$= \frac{1}{\Gamma(1+\alpha k)} \left(\frac{\partial^{k\alpha}}{\partial t^{k\alpha}} \left(\frac{\partial^{n\alpha}}{\partial t^{n\alpha}} \xi(x,t) \right) \right)_{t=0} \quad (8.8)$$

$$= \frac{1}{\Gamma(1+\alpha k)} \left(\frac{\partial^{(k+n)\alpha}}{\partial t^{(k+n)\alpha}} (\xi(x,t)) \right)_{t=0}$$

$$\psi_k(x) = \frac{\Gamma(1+(k+n)\alpha)}{(1+k\alpha)} \xi_{k+n}(x).$$

Interested authors may see Jena et al. (2019a, 2019b) and Jafari et al. (2016) for further details about the FRDTM.

8.3 Numerical Examples

We apply the present method to solve the one-dimensional heat-like fractional model in Example 8.1 and the nonlinear fractional advection equation in Example 8.2. It is worth mentioning that the FRDTM can also be used for handling linear and nonlinear fractional ordinary and partial differential equations.

Example 8.1 Let us consider the one-dimensional heat-like model (Özis and Agırseven 2008; Sadighi et al. 2008) as:

$$D_t^{\alpha} u(x,t) = \frac{1}{2} x^2 u_{xx}(x,t), \quad 0 < x < 1, t > 0, 0 < \alpha \leq 1, \quad (8.9)$$

subject to the boundary conditions (BCs):

$$u(0,t) = 0, \quad u(1,t) = e^t, \tag{8.10}$$

and initial condition (IC):

$$u(x, \ 0) = x^2. \tag{8.11}$$

Solution

By applying FRDTM to Eq. (8.9) and using Theorems 8.1 (iv) and (v), the following recurrence relation is obtained:

$$\frac{\Gamma(1 + (k+1)\alpha)}{(1+k\alpha)} u_{k+1}(x) = \frac{1}{2} x^2 \frac{\partial^2}{\partial x^2} u_k(x), \tag{8.12}$$

By using FRDTM to the IC of Eq. (8.11), we obtain

$$u_0(x) = x^2. \tag{8.13}$$

Using Eq. (8.13) into Eq. (8.12), the following values of $u_k(x)$ for $k = 0, 1, 2, \dots$ are obtained successively:

$$k = 0 : u_1(x) = \frac{\Gamma(1)}{\Gamma(1+\alpha)} \left(\frac{1}{2} x^2 \frac{\partial^2}{\partial x^2} u_0(x) \right) = \frac{1}{\Gamma(1+\alpha)} \left(\frac{1}{2} 2x^2 \right) = \frac{x^2}{\Gamma(1+\alpha)}, \tag{8.14}$$

$$k = 1 : u_2(x) = \frac{\Gamma(1+\alpha)}{\Gamma(1+2\alpha)} \left(\frac{1}{2} x^2 \frac{\partial^2}{\partial x^2} u_1(x) \right) = \frac{\Gamma(1+\alpha)}{\Gamma(1+2\alpha)} \left(\frac{1}{2} x^2 \frac{2}{\Gamma(1+\alpha)} \right) = \frac{x^2}{\Gamma(1+2\alpha)}, \tag{8.15}$$

$$k = 2 : u_3(x) = \frac{\Gamma(1+2\alpha)}{\Gamma(1+3\alpha)} \left(\frac{1}{2} x^2 \frac{\partial^2}{\partial x^2} u_2(x) \right) = \frac{\Gamma(1+2\alpha)}{\Gamma(1+3\alpha)} \left(\frac{1}{2} x^2 \frac{2}{\Gamma(1+2\alpha)} \right) = \frac{x^2}{\Gamma(1+3\alpha)}, \tag{8.16}$$

and so on. Similarly, the rest of the components of the iteration formula Eq. (8.12) can be obtained.

So, the solution of Eq. (8.9) may be written as:

$$\begin{aligned} u(x,t) &= \sum_{k=0}^{\infty} u_k(x) t^{\alpha k}, \\ &= u_0(x) + u_1(x) t^{\alpha} + u_2(x) t^{2\alpha} + \cdots, \\ &= x^2 + \frac{x^2 t^{\alpha}}{\Gamma(1+\alpha)} + \frac{x^2 t^{2\alpha}}{\Gamma(1+2\alpha)} + \frac{x^2 t^{3\alpha}}{\Gamma(1+3\alpha)} + \cdots = x^2 E(t^{\alpha}), \end{aligned} \tag{8.17}$$

where $E(t^{\alpha})$ is called the Mittag-Leffler function. Equation (8.17) is same as the solution given in Özis and Agırseven (2008) and Sadighi et al. (2008) at $\alpha = 1$. The obtained solution is compared with the exact solution, which is shown in Figure 8.1. Figures 8.2–8.5 depict the fourth-order approximate solution plots of Example 8.1 at various values of α.

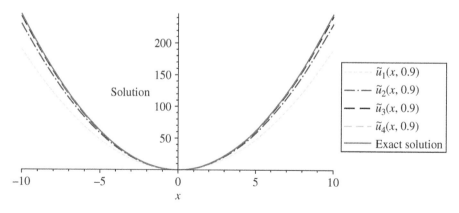

Figure 8.1 Comparison plot of the present solution with the exact solution of Example 8.1.

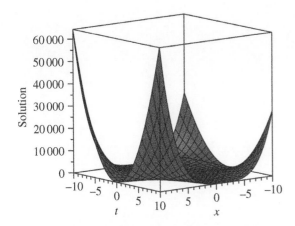

Figure 8.2 Fourth-order approximate solution plot of Example 8.1 at $\alpha = 1$.

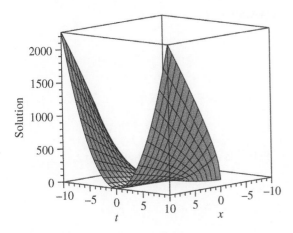

Figure 8.3 Fourth-order approximate solution plot of Example 8.1 at $\alpha = 0.25$.

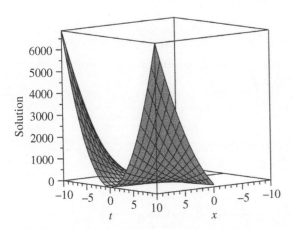

Figure 8.4 Fourth-order approximate solution plot of Example 8.1 at $\alpha = 0.45$.

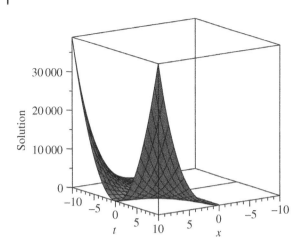

Figure 8.5 Fourth-order approximate solution plot of Example 8.1 at $\alpha = 0.85$.

Example 8.2 Consider the following nonlinear advection equation (Wazwaz 2007):

$$\frac{\partial^{\alpha} u}{\partial t^{\alpha}} + u\frac{\partial u}{\partial x} = 0, \quad 0 < \alpha \leq 1, \tag{8.18}$$

with an IC:

$$u(x,0) = -x. \tag{8.19}$$

Solution

Using FRDTM on the aforementioned two equations, we obtained the following recurrence relation:

$$\begin{cases} \dfrac{\Gamma(1 + (k+1)\alpha)}{(1 + k\alpha)} u_{k+1}(x) = -\left(\displaystyle\sum_{i=0}^{k} u_i(x)\frac{\partial}{\partial x}u_{k-i}(x)\right), \\ u_0(x) = -x. \end{cases} \tag{8.20}$$

On solving Eq. (8.20), we have

$$k = 0 : u_1(x) = \frac{\Gamma(1)}{\Gamma(1+\alpha)}\left(-u_0(x)\frac{\partial}{\partial x}u_0(x)\right) = \frac{-1}{\Gamma(1+\alpha)}(-x)(-1) = \frac{-x}{\Gamma(1+\alpha)}, \tag{8.21}$$

$$k = 1 : u_2(x) = \frac{-\Gamma(1+\alpha)}{\Gamma(1+2\alpha)}\left(u_0(x)\frac{\partial}{\partial x}u_1(x) + u_1(x)\frac{\partial}{\partial x}u_0(x)\right)$$

$$= \frac{-\Gamma(1+\alpha)}{\Gamma(1+2\alpha)}\left\{(-x)\left(\frac{-1}{\Gamma(1+\alpha)}\right) + \left(\frac{-x}{\Gamma(1+\alpha)}\right)(-1)\right\} = \frac{-2x}{\Gamma(1+2\alpha)}, \tag{8.22}$$

$$k = 2 : u_3(x) = \frac{-\Gamma(1+2\alpha)}{\Gamma(1+3\alpha)}\left\{u_0\frac{\partial}{\partial x}u_2 + u_1\frac{\partial}{\partial x}u_1 + u_2\frac{\partial}{\partial x}u_0\right\}$$

$$= \frac{-\Gamma(1+2\alpha)}{\Gamma(1+3\alpha)}\left\{(-x)\left(\frac{-2}{\Gamma(1+2\alpha)}\right) + \left(\frac{-x}{\Gamma(1+\alpha)}\right)\left(\frac{-1}{\Gamma(1+\alpha)}\right) + \left(\frac{-2x}{\Gamma(1+2\alpha)}\right)(-1)\right\} \tag{8.23}$$

$$= \frac{-\Gamma(1+2\alpha)}{\Gamma(1+3\alpha)}\left\{\frac{4x}{\Gamma(1+2\alpha)} + \frac{x}{(\Gamma(1+\alpha))^2}\right\} = \frac{-4x}{\Gamma(1+3\alpha)} + \frac{-\Gamma(1+2\alpha)}{\Gamma(1+3\alpha)}\frac{x}{(\Gamma(1+\alpha))^2}.$$

and so on. Similarly, the rest of the components $u_k(x)$ for $k = 3, 4, \ldots$ can be calculated.

So, the solution of Eq. (8.18) may be written as:

$$u(x,t) = \sum_{k=0}^{\infty} u_k(x) t^{\alpha k},$$

$$= u_0(x) + u_1(x) t^{\alpha} + u_2(x) t^{2\alpha} + \cdots,$$

$$= -x + \frac{-x}{\Gamma(1+\alpha)} t^{\alpha} + \frac{-2x}{\Gamma(1+2\alpha)} t^{2\alpha} + \frac{-4x}{\Gamma(1+3\alpha)} t^{3\alpha} + \frac{-\Gamma(1+2\alpha)}{\Gamma(1+3\alpha)} \frac{x}{(\Gamma(1+\alpha))^2} t^{3\alpha} + \cdots$$

(8.24)

In particular, at $\alpha = 1$, Eq. (8.24) reduces to

$$u(x,t) = -x - xt - xt^2 - xt^3 - \cdots = -x(1 + t + t^2 + t^3 + \cdots) = \frac{x}{t-1}.$$

(8.25)

which is same as the solution of Wazwaz (2007). The exact solution is compared with the present solution in Figure 8.6 by considering the increasing number of terms of solution. Figures 8.7–8.10 show the solution plots of the third-order approximate solutions to Example 8.2 for various values α.

The test problems confirm that the FRDTM is an efficient method for solving linear/nonlinear fractional partial differential equations. The series usually converges with an increase in the number of terms, but one may not always expect the compact form solution.

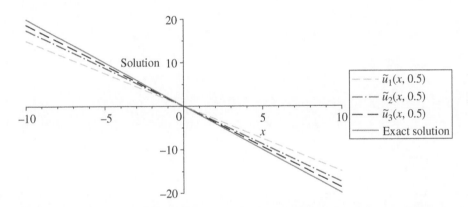

Figure 8.6 Comparison plot of the present solution with the exact solution of Example 8.2.

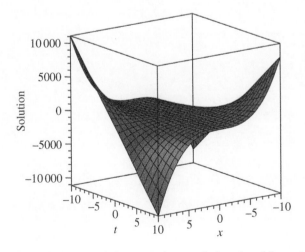

Figure 8.7 Third-order approximate solution plot of Example 8.2 at $\alpha = 1$.

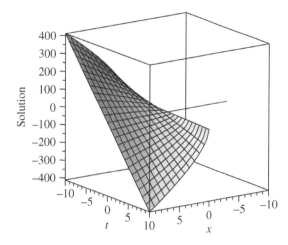

Figure 8.8 Third-order approximate solution plot of Example 8.2 at $\alpha = 0.25$.

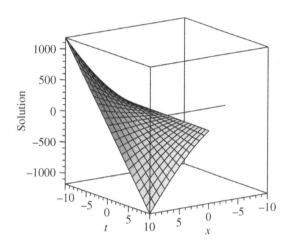

Figure 8.9 Third-order approximate solution plot of Example 8.2 at $\alpha = 0.45$.

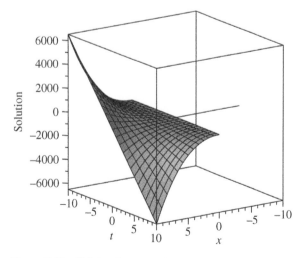

Figure 8.10 Third-order approximate solution plot of Example 8.2 at $\alpha = 0.85$.

References

Abazari, R. and Kılıçman, A. (2013). Numerical study of two-dimensional Volterra integral equations by RDTM and comparison with DTM. *Abstract and Applied Analysis* 10: 929478.

Abazari, R. and Soltanalizadeh, B. (2013). Reduced differential transform method and its application on Kawahara equations. *Thai Journal of Mathematics* 11: 199–216.

Jafari, H., Jassim, H.K., Moshokoa, S.P. et al. (2016). Reduced differential transform method for partial differential equations within local fractional derivative operators. *Advances in Mechanical Engineering* 8 (4): 1–6.

Jena, R.M., Chakraverty, S., and Baleanu, D. (2019a). On new solutions of time-fractional wave equations arising in Shallow water wave propagation. *Mathematics* 7: 722.

Jena, R.M., Chakraverty, S., and Baleanu, D. (2019b). On the solution of imprecisely defined nonlinear time-fractional dynamical model of marriage. *Mathematics* 7: 689.

Keskin, Y. and Oturanc, G. (2010). Reduced differential transform method: a new approach to fractional partial differential equations. *Nonlinear Science Letters A* 1: 207–218.

Özis, T. and Agırseven, D. (2008). He's homotopy perturbation method for solving heat-like and wave-like equations, with variable coefficients. *Physics Letters A* 372: 5944–5950.

Rawashdeh, M.S. (2015). An efficient approach for time–fractional damped burger and time–Sharma–Tasso–Olver equations using the FRDTM. *Applied Mathematics and Information Sciences* 9 (3): 1239–1246.

Sadighi, A., Ganji, D.D., Gorji, M., and Tolou, N. (2008). Numerical simulation of heat-like models with variable coefficients by the variational iteration method. *Journal of Physics: Conference Series* 96: 012083.

Saravanan, A. and Magesh, N. (2013). A comparison between the reduced differential transform method and the Adomian decomposition method for the Newell–Whitehead–Segel equation. *Journal of the Egyptian Mathematical Society* 21: 259–265.

Saravanan, A. and Magesh, N. (2016). An efficient computational technique for solving the Fokker–Planck equation with space and time-fractional derivatives. *Journal of King Saud University-Science* 28: 160–166.

Wazwaz, A.M. (2007). A comparison between the variational iteration method and adomian decomposition method. *Journal of Computational and Applied Mathematics* 207: 129–136.

Zhou, J. (1986). *Differential Transformation and Its Application for Electrical Circuits*. Wuhan, China: Huazhong University Press.

9

Variational Iterative Method

9.1 Introduction

In the previous chapters, we have discussed two semi-analytical approaches, such as Adomian decomposition method (ADM) and homotopy perturbation method (HPM), to solve linear and nonlinear ordinary/partial/fractional differential equations. In this chapter, we will discuss the variational iterative method (VIM). VIM was first proposed by Chinese mathematician He (1998) and He and Wu (2007). He successfully used this technique for solving ordinary and partial differential equations. Subsequently, several researchers used this method for solving linear, nonlinear, homogeneous, and inhomogeneous differential equations (Geng et al. 2009; Odibat and Momani 2009; Odibat 2010; Khana et al. 2011). The principal benefit of this approach is its simplicity and ability to solve nonlinear equations. The approach is also valid in bounded and unbounded domains. It is based on the Lagrange multiplier method initiated by Inokuti et al. (1978). This approach is a modification of the general Lagrange multiplier approach to an iteration process, called functional correction (Momani and Odibat 2007). A substantial number of nonlinear problems are solved effectively by various researchers, generally with one or two iterations leading to good solutions. It is worth mentioning that almost all conventional perturbation methods are based on the assumption of small parameters. These small parameters are so sensitive that a slight change may influence the result. The right choice of small parameters yields optimal performance. However, an inappropriate choice of small parameters leads to poor, even significant effects. As such, no small parameter assumptions are made in this method, which may be the main advantage.

9.2 Procedure for VIM

In order to understand the procedure of VIM, let us consider the time-fractional differential equation as (Odibat and Momani 2009):

$$\frac{\partial^{\alpha}}{\partial t^{\alpha}} u(x,t) = N[x]\, u(x,t) + q(x,t), \quad t > 0, x \in \Re, \tag{9.1}$$

where $N[x]$ is the differential operator in x, subject to initial and boundary conditions (BCs):

$$u(x,0) = f(x), \quad 0 < \alpha \le 1,$$
$$u(x,t) \to 0 \ \text{as} \ |x| \to \infty, \quad t > 0, \tag{9.2}$$

and

$$u(x,0) = f(x), \quad \frac{\partial}{\partial t} u(x,0) = g(x), \quad 1 < \alpha \le 2,$$
$$u(x,t) \to 0 \ \text{as} \ |x| \to \infty, \quad t > 0, \tag{9.3}$$

Computational Fractional Dynamical Systems: Fractional Differential Equations and Applications, First Edition.
Snehashish Chakraverty, Rajarama Mohan Jena, and Subrat Kumar Jena.

where $f(x)$, $g(x)$, and $q(x, t)$ are all continuous functions, and α, $m - 1 < \alpha \leq m$, $m \in N$ is a parameter representing the order of fractional derivatives defined in the Caputo sense. As per the principle of VIM, the correction functional for Eq. (9.1) may be written as follows (Odibat and Momani 2009):

$$u_{k+1}(x,t) = u_k(x,t) + J_t^\beta \left[\lambda \left(\frac{\partial^\alpha}{\partial t^\alpha} u_k(x,t) - N[x]\tilde{u}_k(x,t) - q(x,t) \right) \right],$$

$$= u_k(x,t) + \frac{1}{\Gamma(\beta)} \int_0^t (t-\tau)^{\beta-1} \lambda(\tau) \left(\frac{\partial^\alpha}{\partial t^\alpha} u_k(x,\tau) - N[x]\tilde{u}_k(x,\tau) - q(x,\tau) \right) d\tau,$$

(9.4)

where J_t^β is the Reimann–Liouville integral operator of order $\beta = \alpha - \lfloor \alpha \rfloor$, $\lfloor \alpha \rfloor = \max \{m \in Z \mid m \leq \alpha\}$ which may be written as $\beta = \alpha - m + 1$ and λ is a general Lagrange multiplier that can be optimally defined through variational theory (Inokuti et al. 1978). Some approximations must be made to determine the Lagrange multiplier approximately. The functional correction Eq. (9.4) may be presented as (Odibat and Momani 2009):

$$u_{k+1}(x,t) = u_k(x,t) + \int_0^t \lambda(\tau) \left(\frac{\partial^m}{\partial t^m} u_k(x,\tau) - N[x]\tilde{u}_k(x,\tau) - q(x,\tau) \right) d\tau.$$

(9.5)

Here, we implement the restricted variations to the nonlinear term $N[x]\, u$, in which the multiplier can be calculated easily. Considering the aforementioned functional stationery, observing that $\delta \tilde{u}_k = 0$, we obtain

$$\delta u_{k+1}(x,t) = \delta u_k(x,t) + \delta \int_0^t \lambda(\tau) \left(\frac{\partial^m}{\partial t^m} u_k(x,\tau) - q(x,\tau) \right) d\tau.$$

(9.6)

From Eq. (9.6), we have the following Lagrange multipliers:

$$\lambda = -1, \quad \text{for } m = 1,$$ (9.7)

$$\lambda = \tau - t, \quad \text{for } m = 2.$$ (9.8)

Hence, for $m = 1$ $(0 < \alpha \leq 1)$, we substitute $\lambda = -1$ in Eq. (9.4) to achieve the following iteration expression:

$$u_{k+1}(x,t) = u_k(x,t) - J_t^\alpha \left[\frac{\partial^\alpha}{\partial t^\alpha} u_k(x,t) - N[x]u_k(x,t) - q(x,\tau) \right].$$

(9.9)

For $m = 2$ $(1 < \alpha \leq 2)$, substituting $\lambda = \tau - t$ in the functional Eq. (9.4), we obtain the following iteration formula:

$$u_{k+1}(x,t) = u_k(x,t) + \frac{1}{\Gamma(\alpha-1)} \int_0^t (t-\tau)^{\alpha-2}(\tau-t) \left(\frac{\partial^\alpha}{\partial t^\alpha} u_k(x,t) - N[x]u_k(x,t) - q(x,\tau) \right),$$

$$= u_k(x,t) - \frac{\alpha-1}{\Gamma(\alpha)} \int_0^t (t-\tau)^{\alpha-1} \left(\frac{\partial^\alpha}{\partial t^\alpha} u_k(x,t) - N[x]u_k(x,t) - q(x,\tau) \right),$$

(9.10)

Rewriting Eq. (9.10), we get

$$u_{k+1}(x,t) = u_k(x,t) - (\alpha-1)J_t^\alpha \left[\frac{\partial^\alpha}{\partial t^\alpha} u_k(x,t) - N[x]u_k(x,t) - q(x,\tau) \right].$$

(9.11)

Generally, the following Lagrange multipliers are used for $m \geq 1$ (Singh and Kumar 2017):

$$\lambda(t) = \frac{(-1)^m}{\Gamma(m)} (\tau-t)^{m-1}, \quad m \geq 1.$$

(9.12)

The initial approximation u_0 can be freely selected if it fulfills the initial and BCs of the problem. However, the effectiveness of this approach depends on the correct selection of the initial approximation u_0. Finally, we approximate the solution $u(x,t) = \lim_{k \to \infty} u_k(x,t)$ by the kth term solution.

9.3 Examples

We use the aforementioned method to solve a linear one-dimensional heat-like model and a nonlinear fractional advection equation, respectively, in Examples 9.1 and 9.2. It is worth noticing that VIM can also be used to handle linear and nonlinear fractional differential equations.

Example 9.1 Let us consider the one-dimensional heat-like model (Inokuti et al. 1978; Sadighi et al. 2008) as:

$$D_t^\alpha u(x,t) = \frac{1}{2}x^2 u_{xx}(x,t), \quad 0 < x < 1, t > 0, 0 < \alpha \le 1, \tag{9.13}$$

with the BCs:

$$u(0,t) = 0, \quad u(1,t) = e^t, \tag{9.14}$$

and initial condition (IC):

$$u(x,0) = x^2. \tag{9.15}$$

Solution

Comparing $m-1 < \alpha \le m$ with $0 < \alpha \le 1$ of Eq. (9.13), we get $m = 1$.

According to VIM and Eq. (9.9), the iteration formula for Eq. (9.13) may be written as:

$$u_{k+1}(x,t) = u_k(x,t) - J_t^\alpha \left[\frac{\partial^\alpha}{\partial t^\alpha} u_k(x,t) - \frac{1}{2}x^2 \frac{\partial^2}{\partial x^2} u_k(x,t) \right]. \tag{9.16}$$

By the aforemetioned variational iteration formula and choosing $u_0 = x^2$, we may find the following approximations:

$$k = 0 : u_1(x,t) = u_0 - J_t^\alpha \left[\frac{\partial^\alpha}{\partial t^\alpha} u_0 - \frac{1}{2}x^2 \frac{\partial^2}{\partial x^2} u_0 \right]$$

$$= x^2 - J_t^\alpha \left[\frac{\partial^\alpha}{\partial t^\alpha} x^2 - \frac{1}{2}x^2 \frac{\partial^2}{\partial x^2} x^2 \right]$$

$$= x^2 - J_t^\alpha [0 - x^2] = x^2 + x^2 \frac{t^\alpha}{\Gamma(1+\alpha)}$$

$$k = 1 : u_2(x,t) = u_1 - J_t^\alpha \left[\frac{\partial^\alpha}{\partial t^\alpha} u_1 - \frac{1}{2}x^2 \frac{\partial^2}{\partial x^2} u_1 \right]$$

$$= x^2 + x^2 \frac{t^\alpha}{\Gamma(1+\alpha)} - J_t^\alpha \left[\begin{array}{c} D_t^\alpha \left(x^2 + x^2 \frac{t^\alpha}{\Gamma(1+\alpha)} \right) \\ -\frac{1}{2}x^2 \frac{\partial^2}{\partial x^2} \left(x^2 + x^2 \frac{t^\alpha}{\Gamma(1+\alpha)} \right) \end{array} \right]$$

$$= x^2 + x^2 \frac{t^\alpha}{\Gamma(1+\alpha)} + x^2 \frac{t^{2\alpha}}{\Gamma(1+2\alpha)}$$

Similarly,

$$k = 2 : u_3(x,t) = x^2 + x^2 \frac{t^\alpha}{\Gamma(1+\alpha)} + x^2 \frac{t^{2\alpha}}{\Gamma(1+2\alpha)} + x^2 \frac{t^{3\alpha}}{\Gamma(1+3\alpha)}$$

So, the solution of the fractional heat-like Eq. (9.13) may be obtained as:

$$u(x,t) = \lim_{n \to \infty} u_n(x,t) = x^2 + x^2 \frac{t^\alpha}{\Gamma(1+\alpha)} + x^2 \frac{t^{2\alpha}}{\Gamma(1+2\alpha)} + x^2 \frac{t^{3\alpha}}{\Gamma(1+3\alpha)} + \cdots,$$

$$= x^2 \left(1 + \frac{t^\alpha}{\Gamma(1+\alpha)} + \frac{t^{2\alpha}}{\Gamma(1+2\alpha)} + \frac{t^{3\alpha}}{\Gamma(1+3\alpha)} + \cdots \right) = x^2 E(t^\alpha), \tag{9.17}$$

where $E(t^\alpha)$ is called the Mittag-Leffler function, which is given in Chapter 1.

In particular, at $\alpha = 1$, Eq. (9.17) reduces to $u(x,t) = x^2 e^t$, which is same as the solution of Sadighi et al. (2008). In Figure 9.1, the present solution is compared with the exact solution by increasing the order of approximation. Figures 9.2–9.5 give the fourth-order approximate solution plots of Example 9.1 at various fractional-order values.

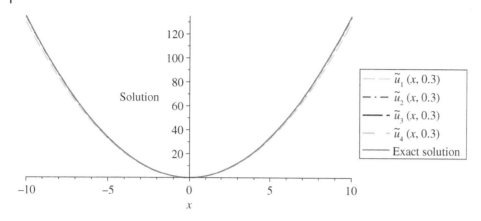

Figure 9.1 Comparison plot of the present solution with the exact solution of Example 9.1.

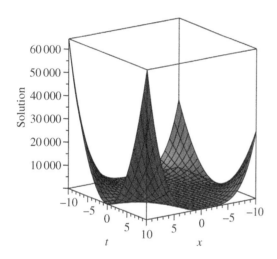

Figure 9.2 Fourth-order approximate solution plot of Example 9.1 at $\alpha = 1$.

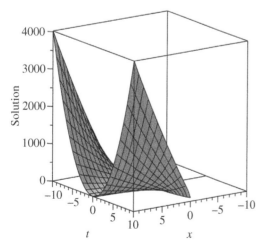

Figure 9.3 Fourth-order approximate solution plot of Example 9.1 at $\alpha = 0.35$.

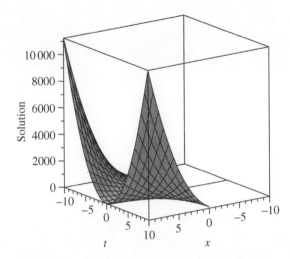

Figure 9.4 Fourth-order approximate solution plot of Example 9.1 at α = 0.55.

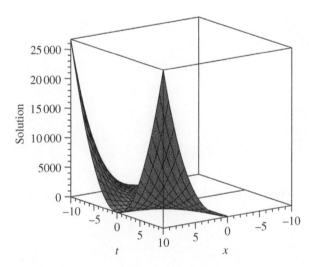

Figure 9.5 Fourth-order approximate solution plot of Example 9.1 at α = 0.75.

Example 9.2 Consider the following nonlinear advection equation (Wazwaz 2007):

$$\frac{\partial^{\alpha} u}{\partial t^{\alpha}} + u \frac{\partial u}{\partial x} = 0, \quad 0 < \alpha \leq 1, \tag{9.18}$$

with IC:

$$u(x, 0) = -x. \tag{9.19}$$

Solution

In this problem, $m = 1$ and $\lambda = -1$. Using VIM and Eq. (9.9), the iteration formula for this model may be expressed as:

$$u_{k+1}(x, t) = u_k(x, t) - J_t^{\alpha}\left[\frac{\partial^{\alpha}}{\partial t^{\alpha}} u_k(x, t) + u_k(x, t) \frac{\partial}{\partial x} u_k(x, t)\right]. \tag{9.20}$$

By the aforemetioned variational iteration formula and considering $u_0 = -x$, we have

$$k = 0 : u_1(x,t) = u_0 - J_t^\alpha \left[\frac{\partial^\alpha}{\partial t^\alpha} u_0 + u_0 \frac{\partial}{\partial x} u_0 \right]$$

$$= -x - J_t^\alpha \left[\frac{\partial^\alpha}{\partial t^\alpha}(-x) + (-x)\frac{\partial}{\partial x}(-x) \right]$$

$$= -x - J_t^\alpha x = -x - x\frac{t^\alpha}{\Gamma(1+\alpha)}$$

$$k = 1 : u_2(x,t) = u_1 - J_t^\alpha \left[\frac{\partial^\alpha}{\partial t^\alpha} u_1 + u_1 \frac{\partial}{\partial x} u_1 \right]$$

$$= -x - x\frac{t^\alpha}{\Gamma(1+\alpha)} - J_t^\alpha \left[\frac{\partial^\alpha}{\partial t^\alpha}\left(-x - x\frac{t^\alpha}{\Gamma(1+\alpha)}\right) + \left(-x - x\frac{t^\alpha}{\Gamma(1+\alpha)}\right)\frac{\partial}{\partial x}\left(-x - x\frac{t^\alpha}{\Gamma(1+\alpha)}\right) \right]$$

$$= -x - x\frac{t^\alpha}{\Gamma(1+\alpha)} - J_t^\alpha \left[-x + \left(-x - x\frac{t^\alpha}{\Gamma(1+\alpha)}\right)\left(-1 - \frac{t^\alpha}{\Gamma(1+\alpha)}\right) \right]$$

$$= -x - x\frac{t^\alpha}{\Gamma(1+\alpha)} - J_t^\alpha \left[2x\frac{t^\alpha}{\Gamma(1+\alpha)} + x\frac{t^{2\alpha}}{(\Gamma(1+\alpha))^2} \right]$$

$$= -x - x\frac{t^\alpha}{\Gamma(1+\alpha)} - 2x\frac{t^{2\alpha}}{\Gamma(1+2\alpha)} - x\frac{\Gamma(1+2\alpha)t^{3\alpha}}{(\Gamma(1+\alpha))^2\Gamma(1+3\alpha)}$$

So, the solution to Eq. (9.18) may be obtained as:

$$u(x,t) = \lim_{n\to\infty} u_n(x,t)$$

$$= -x - x\frac{t^\alpha}{\Gamma(1+\alpha)} - 2x\frac{t^{2\alpha}}{\Gamma(1+2\alpha)} - x\frac{\Gamma(1+2\alpha)t^{3\alpha}}{(\Gamma(1+\alpha))^2\Gamma(1+3\alpha)} - \cdots \qquad (9.21)$$

Particularly at $\alpha = 1$, Eq. (9.21) reduces to a closed-form solution $u(x,t) = \dfrac{x}{t-1}$, which is same as the solution of Wazwaz (2007). The comparison plot of the present solution with the exact solution is given in Figure 9.6. The third-order approximate solution is plotted in Figures 9.7–9.10 at different values of α.

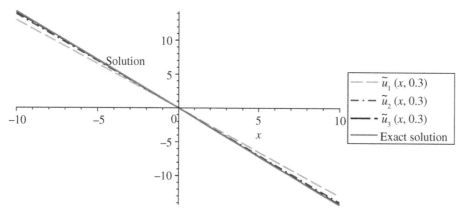

Figure 9.6 Comparison plot of the present solution with the exact solution of Example 9.2.

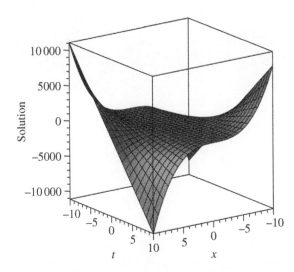

Figure 9.7   Third-order approximate solution plot of Example 9.2 at α = 1.

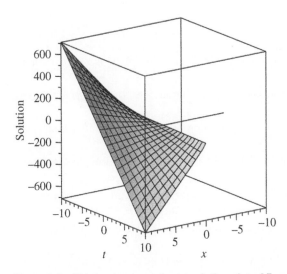

Figure 9.8   Third-order approximate solution plot of Example 9.2 at α = 0.35.

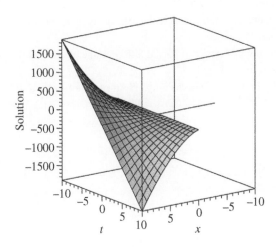

Figure 9.9   Third-order approximate solution plot of Example 9.2 at α = 0.55.

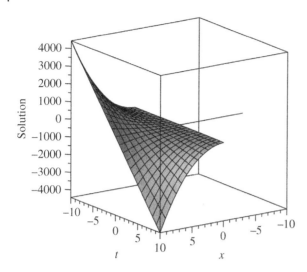

Figure 9.10 Third-order approximate solution plot of Example 9.2 at $\alpha = 0.75$.

References

Geng, F., Lin, Y., and Cui, M. (2009). A piecewise variational iteration method for Riccati differential equations. *Computers and Mathematics with Applications* 58 (11–12): 2518–2522.

He, J.H. (1998). Approximate analytical solution for seepage flow with fractional derivatives in porous media. *Computer Methods in Applied Mechanics and Engineering* 167 (1–2): 57–68.

He, J.H. and Wu, X.H. (2007). Variational iteration method: new development and applications. *Computers and Mathematics with Applications* 54 (7/8): 881–894.

Inokuti, M., Sekine, H., and Mura, T. (1978). General use of the Lagrange multiplier in nonlinear mathematical physics. In: *Variational Methods in the Mechanics of Solids* (ed. S. Nemat-Nasser), 156–162. New York: Pergamon Press.

Khana, Y., Faraz, N., Yildirim, A., and Wua, Q. (2011). Fractional variational iteration method for fractional initial-boundary value problems arising in the application of nonlinear science. *Computers and Mathematics with Applications* 62: 2273–2278.

Momani, S. and Odibat, Z. (2007). Numerical comparison of the methods for solving linear differential equations of fractional order. *Chaos Solitons Fractals* 31: 1248–1255.

Odibat, Z.M. (2010). A study on the convergence of variational iteration method. *Mathematical and Computer Modelling* 51 (9–10): 1181–1192.

Odibat, Z. and Momani, S. (2009). The variational iteration method: an efficient scheme for handling fractional partial differential equations in fluid mechanics. *Computers and Mathematics with Applications* 58: 2199–2208.

Sadighi, A., Ganji, D.D., Gorji, M., and Tolou, N. (2008). Numerical simulation of heat-like models with variable coefficients by the variational iteration method. *Journal of Physics: Conference Series* 96: 012083.

Singh, B.K. and Kumar, P. (2017). Fractional variational iteration method for solving fractional partial differential equations with proportional delay. *International Journal of Differential Equations* 5206380 (2017): 1–11.

Wazwaz, A.M. (2007). A comparison between the variational iteration method and adomian decomposition method. *Journal of Computational and Applied Mathematics* 207: 129–136.

10

Weighted Residual Methods

10.1 Introduction

The weighted residual is another effective approach for solving fractional differential equations (FDEs) with boundary conditions (BCs), often known as boundary value problems (BVPs). The weighted residual method (WRM) is an approximation approach that uses a linear combination of trial or shape functions with unknown coefficients to estimate the solution of FDEs. The approximate solution is then replaced in the governing FDE, yielding error or residual. Finally, in order to determine the unknown coefficients, the residual is made to zero at average points or made as minimal as possible based on the weight function. Gerald and Wheatley (2004) addressed WRMs, including collocation and Galerkin approaches. More information on various WRMs may be found in classic books (Baluch et al. 1983; Finlayson 2013; Hatami 2017). Locker (1971) has provided a least-square method for solving BVPs. Gerald and Wheatley (2004), Lindgren (2009), Finlayson (2013), Hatami (2017), and Logan (2011) address weighted residual-based finite-element algorithms. Boundary characteristic orthogonal polynomials (Bhat and Chakraverty 2004) utilized as trial or shape functions are sometimes advantageous. As a result, Chapter 11 focuses on solving BVP using boundary characteristic orthogonal polynomials integrated into Galerkin and least-square techniques.

In this regard, this chapter includes several WRMs, namely collocation, Galerkin method, and least-square techniques used for solving FDEs subject to BCs, and these are demonstrated in Sections 10.2, 10.3, and 10.4, respectively. Comparative results for specific FDEs with respect to WRMs are also included to understand the methods better.

Let us consider a FDE:

$$Lu(x) = f(x), \tag{10.1}$$

subject to BCs in the domain Ω. Here, L is the fractional operator acting on $u(x)$, and $f(x)$ is the applied force. In WRM, the solution of Eq. (10.1) is approximately considered as $\hat{u}(x) = \sum_{i=0}^{n} c_i \phi_i(x)$ satisfying the BCs. Here, c_i are the unknown coefficients to be determined for the trial functions $\phi_i(x)$ for $i = 0, 1, 2, ..., n$, where ϕ_i's are linearly independent functions. The assumed solution is substituted in Eq. (10.1), resulting in an error or residual. This residue is subsequently reduced or made to vanish in the domain, yielding a system of algebraic equations with unknown coefficients c_i. As such, WRMs mainly consist of the following steps:

Step (i) Let us assume an approximate solution:

$$u(x) \approx \hat{u}(x) = \sum_{i=0}^{n} c_i \phi_i(x), \tag{10.2}$$

involving linear independent trial functions $\phi_i(x)$ such that $\hat{u}(x)$ satisfies the BCs. Alternatively,

$$\hat{u}(x) = u_0(x) + \sum_{i=0}^{n} c_i \phi_i(x), \tag{10.3}$$

may also be considered where $u_0(x)$ is the function satisfying the BCs.

In general, the trial functions are chosen to interpolate the desired solution under BCs as given in Eq. (10.2) or Eq. (10.3). Various examples of trial function assumptions are discussed in Finlayson (2013). One possible assumption for the shape functions may be $\phi_0 = 1$, $\phi_1 = x$, and $\phi_i = x^{i-1}(x - X)$, for $i = 2, 3, ..., n - 1$ over the domain $\Omega = [0, X]$.

Computational Fractional Dynamical Systems: Fractional Differential Equations and Applications, First Edition.
Snehashish Chakraverty, Rajarama Mohan Jena, and Subrat Kumar Jena.
© 2023 John Wiley & Sons, Inc. Published 2023 by John Wiley & Sons, Inc.

Step (ii) Substitute $\hat{u}(x)$ in $L\hat{u}(x) - f(x) \neq 0$, which yields an error.

Step (iii) The measure of error is considered as residual, which is written as:

$$R(x) = L\hat{u}(x) - f(x). \tag{10.4}$$

Step (iv) An arbitrary weight function $w_i(x)$ is then multiplied with Eq. (10.4) and integrated over Ω, which gives

$$\int_\Omega w_i(x)(L\hat{u}(x) - f(x))dx = \int_\Omega w_i(x)R(x)dx \neq 0, \tag{10.5}$$

for $i = 0, 1, ..., n$. Making every residue zero in the entire domain is not always possible. The integral is either turned to disappear at finite points or created as small as possible depending on the weight function.

Step (v) Forcing the integral to zero over the entire domain Ω using

$$\int_\Omega w_i(x)R(x)dx = 0, \tag{10.6}$$

we obtain $n + 1$ independent algebraic equations with $n + 1$ unknown coefficients c_i, for $i = 0, 1, ..., n$. Solving these algebraic equations, we get the values of c_i. Substituting these obtained values of c_i in Eq. (10.2), we achieve the desired approximate solution of the original equation. It is worth mentioning that in Eq. (10.2) as $n \to \infty$, $u(x) = \hat{u}(x)$. Further, there exist other forms of WRMs, such as collocation, least-square, and Galerkin methods depending on the weight function. In this regard, different types of WRMs are addressed in Sections 10.2–10.4.

10.2 Collocation Method

In the collocation method (Gerald and Wheatley 2004; Logan, 2011; Finlayson 2013; Hatami 2017), the weight function is considered in terms of the Dirac delta function as:

$$w_k(x) = \delta(x - x_k) = \begin{cases} 1, & x = x_k \\ 0, & \text{otherwise} \end{cases} \tag{10.7}$$

where $k = 0, 1, ..., n$. This strategy is also known as the point collocation method in the literature. Now using Eq. (10.7), the integrand Eq. (10.6) may be written as:

$$\int_\Omega w_k(x)R(x)dx = R(x_k). $$

Generally, the residual is set to zero at $n + 1$ distinct points x_k within the domain Ω:

$$R(x_k) = L\left(\sum_{i=0}^{n} c_i \phi_i(x_k)\right) - f(x_k) = 0 \Rightarrow \sum_{i=0}^{n} c_i L(\phi_i(x_k)) - f(x_k), \tag{10.8}$$

for computing the unknown coefficients c_i for $n - 1$ collocating points.

10.3 Least-Square Method

In the least-square method (Locker, 1971; Finlayson 2013; Hatami 2017), the residue given in Eq. (10.4) is squared and integrated over the entire domain Ω:

$$I = \int_\Omega R^2(x, c_i)\,dx. \tag{10.9}$$

The integrand I is then minimized using $\dfrac{\partial}{\partial c_i}\displaystyle\int_{\Omega} R^2(x, c_i)\,dx = 0$, where c_i are the unknown coefficients of the approximate

solution $\hat{u}(x) = \displaystyle\sum_{i=0}^{n} c_i\phi_i(x)$, which further reduces to

$$2\int_{\Omega} R(x; c_i)\frac{\partial R}{\partial c_i}\,dx = 0, \quad \text{for } i = 0, 1, 2, ..., n.$$

$$\Rightarrow \int_{\Omega} R(x; c_i)\frac{\partial R}{\partial c_i}\,dx = 0. \tag{10.10}$$

From Eq. (10.10), the weight function for the least-square method is considered as:

$$W(x) = \frac{\partial R}{\partial c_i}, \quad \text{for } i = 0, 1, 2, ..., n. \tag{10.11}$$

Next, we discuss the Galerkin method.

10.4 Galerkin Method

In the Galerkin method (Gerald and Wheatley 2004; Logan 2011; Finlayson 2013; Hatami 2017), the weight function is considered in terms of trial functions:

$$w_k = \phi_k(x), \quad k = 0, 1, ..., n, \tag{10.12}$$

such that ϕ_k's are $(n+1)$ basis functions of x satisfying the BCs.

As such, Eq. (10.6) reduces to

$$\int_{\Omega} w_k(x)R(x)dx = \int_{\Omega} \phi_k(x)R(x)dx = 0. \tag{10.13}$$

Then, using Eq. (10.13), we obtain the system of equations in terms of the unknown coefficients.

10.5 Numerical Examples

Example 10.1 Consider the following FDE (Chakraverty et al. 2019):

$$D^{\alpha}u(x) + u(x) = x, \quad 1 < \alpha \le 2, \tag{10.14}$$

subject to the BCs $u(0) = 0$ and $u\left(\dfrac{\pi}{4}\right) = 2$ with the domain $x \in \left[0, \dfrac{\pi}{4}\right]$.

Solution

Here, $Lu(x) = D^{\alpha}u(x) + u(x)$ and $f(x) = x$. Let us consider the approximate solution of Eq. (10.14) as Eq. (10.2) at $n = 3$ be $\hat{u}(x) = \hat{u}_3(x) = c_0\phi_0(x) + c_1\phi_1(x) + c_2\phi_2(x) + c_3\phi_3(x)$, where trial functions $\phi_0(x) = 1, \phi_1(x) = x, \phi_2(x) = x\left(x - \dfrac{\pi}{4}\right)$, and $\phi_3(x) = x^2\left(x - \dfrac{\pi}{4}\right)$. Since the approximate solution satisfies the given BCs that help in the computation of unknown coefficients as:

$$\hat{u}(0) = c_0\phi_0(0) + c_1\phi_1(0) + c_2\phi_2(0) + c_3\phi_3(0) = 0 \Rightarrow c_0 = 1, \tag{10.15}$$

$$\hat{u}\left(\frac{\pi}{4}\right) = c_0\phi_0\left(\frac{\pi}{4}\right) + c_1\phi_1\left(\frac{\pi}{4}\right) + c_2\phi_2\left(\frac{\pi}{4}\right) + c_3\phi_3\left(\frac{\pi}{4}\right) = 2 \Rightarrow c_1 = \frac{4}{\pi}. \tag{10.16}$$

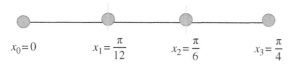

Figure 10.1 Collocation points for the domain $\Omega = \left[0, \frac{\pi}{4}\right]$.

Accordingly, the approximate solution may now be written as:

$$\hat{u}(x) = 1 + \frac{4}{\pi}x + c_2 x \left(x - \frac{\pi}{4}\right) + c_3 x^2 \left(x - \frac{\pi}{4}\right). \qquad (10.17)$$

The Caputo fractional derivatives of Eq. (10.17) for $1 < \alpha \leq 2$ may be obtained as:

$$D^\alpha \hat{u}(x) = \left(c_2 - c_3 \frac{\pi}{4}\right) \frac{2x^{2-\alpha}}{\Gamma(3-\alpha)} + \frac{6c_3 x^{3-\alpha}}{\Gamma(4-\alpha)}. \qquad (10.18)$$

So, the residual function reduces to

$$R(x) = D^\alpha \hat{u}(x) + \hat{u}(x) - x$$

$$= \left(c_2 - c_3 \frac{\pi}{4}\right) \frac{2x^{2-\alpha}}{\Gamma(3-\alpha)} + \frac{6c_3 x^{3-\alpha}}{\Gamma(4-\alpha)} + 1 + \frac{4}{\pi}x + c_2 x \left(x - \frac{\pi}{4}\right) + c_3 x^2 \left(x - \frac{\pi}{4}\right). \qquad (10.19)$$

Using Collocation Method

Now, the remaining two unknown coefficients c_2 and c_3 are determined for two collocating points x_1 and x_2 as shown in Figure 10.1.

Considering the residual function $R(x_k)$ for $\alpha = 2$ to zero at collocating points $x_1 = \frac{\pi}{12}$ and $x_2 = \frac{\pi}{6}$, we obtain the algebraic equations as follows:

$$R(x_1) = -0.1436c_3 + 1.4512c_2 + 5.0714, \qquad (10.20)$$

and

$$R(x_2) = 1.2840c_3 + 1.4513c_2 + 6.1429. \qquad (10.21)$$

On solving the aforementioned two equations, we get $c_2 = -3.5689$ and $c_3 = -0.75031$. Substituting the values of c_2 and c_3 in Eq. (10.17), we obtain the approximate solution of Eq. (10.14) at $\alpha = 2$ as:

$$\hat{u}_3(x) = 1 + 4.0769x - 2.9793x^2 - 0.75044x^3. \qquad (10.22)$$

Similarly, by following the aforementioned procedure, we can obtain the approximate solution of Eq. (10.14) at different values of fractional order as:

For $\alpha = \frac{4}{3}$, $\hat{u}_3(x) = 1 + 6.3429x - 11.5114x^2 + 6.4380x^3, \qquad (10.23)$

For $\alpha = \frac{5}{3}$, $\hat{u}_3(x) = 1 + 5.3871x - 6.0518x^2 + 1.0377x^3, \qquad (10.24)$

and so on.

Using Least-Square Method

In order to compute the remaining unknown coefficients c_2 and c_3 given in Eq. (10.17) using the least-square method, we have to minimize the residual function Eq. (10.19) at $\alpha = 2$ in the domain $\Omega = \left[0, \frac{\pi}{4}\right]$ as:

$$\int_0^{\pi/4} R(x; c_2, c_3) \frac{\partial R}{\partial c_2} dx = 0.7892793032c_3 + 2.009074592c_2 + 6.998030335 = 0, \qquad (10.25)$$

and

$$\int_0^{\pi/4} R(x; c_2, c_3) \frac{\partial R}{\partial c_3} dx = 1.648990455c_3 + 0.7892793032c_2 + 3.700556933 = 0. \qquad (10.26)$$

On solving Eqs. (10.25) and (10.26), we obtain $c_2 = -3.204076831$ and $c_3 = -0.7105228541$. The approximate solution up to four decimal places is obtained as:

$$\hat{u}_3(x) = 1 + 3.7902x - 2.6458x^2 - 0.7105x^3. \tag{10.27}$$

Likewise, using the aforementioned procedure, we may find the approximate solution of Eq. (10.14) for various fractional-order values as follows:

$$\text{For } \alpha = \frac{4}{3}, \quad \hat{u}_3(x) = 1 + 6.8302\,x - 11.7456\,x^2 + 5.9468\,x^3, \tag{10.28}$$

$$\text{For } \alpha = \frac{5}{3}, \quad \hat{u}_3(x) = 1 + 5.3130\,x - 6.2731\,x^2 + 1.4393\,x^3, \tag{10.29}$$

and so on for other values of α.

Using Galerkin Method

In order to calculate the values of c_2 and c_3 in Eq. (10.17) using the Galerkin method, we have to use Eq. (10.13) and residual function Eq. (10.19) at $\alpha = 2$ in the domain $\Omega = \left[0, \frac{\pi}{4}\right]$ as:

$$\int_0^{\pi/4} \phi_2(x)R(x; c_2, c_3)\, dx = -0.04783410291c_3 - 0.1217595348c_2 - 0.453298539 = 0, \tag{10.30}$$

and

$$\int_0^{\pi/4} \phi_3(x)R(x; c_2, c_3)\, dx = -0.03288412695c_3 - 0.04783410295c_2 - 0.1882900665 = 0. \tag{10.31}$$

On solving Eqs. (10.30) and (10.31), we obtain $c_2 = -3.438310155$ and $c_3 = -0.724409822$. So, we get the approximate solution to the original equation up to four decimal places as:

$$\hat{u}_3(x) = 1 + 3.9743\,x - 2.8691\,x^2 - 0.7244\,x^3, \tag{10.32}$$

Similarly, we may find the approximate solution of Eq. (10.14) for various fractional-order values as follows:

$$\text{For } \alpha = \frac{4}{3}, \quad \hat{u}_3(x) = 1 + 6.5680x - 12.0425x^2 + 6.7494x^3, \tag{10.33}$$

$$\text{For } \alpha = \frac{5}{3}, \quad \hat{u}_3(x) = 1 + 5.3384x - 6.1382x^2 + 1.2265x^3, \tag{10.34}$$

and so on for other values of α.

To validate the correctness of these methods, the integer-order ($\alpha = 2$) approximate solutions obtained from these three methods are compared with the exact solution (Chakraverty et al. 2019) $u(x) = \cos(2x) + \left(2 - \frac{\pi}{16}\right)\sin(2x) + \frac{x}{4}$, which is illustrated in Figure 10.2. Solution plots of Example 10.1 using three methods are depicted in Figures 10.3 and 10.4 for different values of fractional order. Further, the error $|u(x) - \hat{u}_3(x)|$ for each method is calculated, and then the respective errors for WRMs are shown in Figure 10.5.

Example 10.2 Let us consider the following FDE (Chakraverty et al. 2019):

$$D^\alpha u(x) + u'(x) + u(x) = x, \quad 1 < \alpha \le 2, \tag{10.35}$$

subject to the BCs $u(0) = 0$ and $u(1) = 0$ with the domain $x \in [0, 1]$.

Solution

Here, $Lu(x) = D^\alpha u(x) + u'(x) + u(x)$ and $f(x) = x$. Let us consider a cubic approximate solution of Eq. (10.35) as $\hat{u}(x) = \hat{u}_3(x) = c_0\phi_0(x) + c_1\phi_1(x) + c_2\phi_2(x) + c_3\phi_3(x)$, where trial functions $\phi_0(x) = 1$, $\phi_1(x) = x$, $\phi_2(x) = x(1-x)$, and $\phi_3(x) = x^2(1-x)$. Using the given BCs, we get

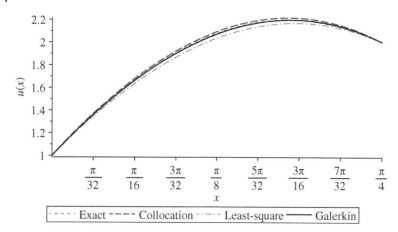

Figure 10.2 Comparison of the approximate solutions of Example 10.1 with the exact solution when $\alpha = 2$.

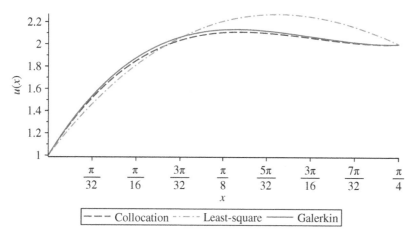

Figure 10.3 Comparison of the approximate solutions of Example 10.1 when $\alpha = \dfrac{4}{3}$.

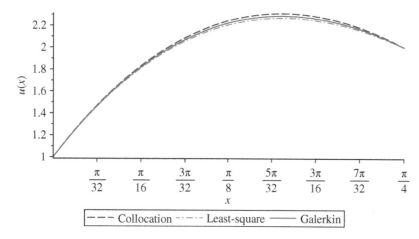

Figure 10.4 Comparison of the approximate solutions of Example 10.1 when $\alpha = \dfrac{5}{3}$.

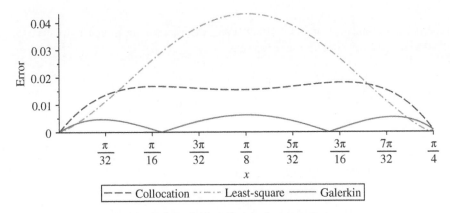

Figure 10.5 Error plot of approximate solutions for Example 10.1.

$$\hat{u}(0) = c_0\phi_0(0) + c_1\phi_1(0) + c_2\phi_2(0) + c_3\phi_3(0) = 0 \Rightarrow c_0 = 0,$$
(10.36)

$$\hat{u}(1) = c_0\phi_0(1) + c_1\phi_1(1) + c_2\phi_2(1) + c_3\phi_3(1) = 0 \Rightarrow c_1 = 0.$$
(10.37)

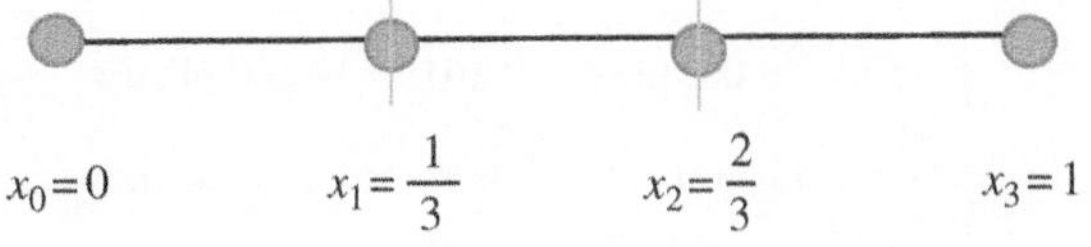

Figure 10.6 Collocation points for the domain $\Omega = [0, 1]$.

So, the approximate solution reduces to

$$\hat{u}(x) = c_2\, x\,(1-x) + c_3\, x^2\,(1-x).$$
(10.38)

As such, the residual for $1 < \alpha \le 2$ is obtained as:

$$R(x) = D^\alpha \hat{u}(x) + \hat{u}'(x) + \hat{u}(x) - x$$

$$= (c_3 - c_2)\frac{2x^{2-\alpha}}{\Gamma(3-\alpha)} - \frac{6c_3 x^{3-\alpha}}{\Gamma(4-\alpha)} + 2(c_3 - c_2)x - 3c_3 x^2 + c_2 x\,(1-x) + c_3 x^2\,(1-x) - x.$$
(10.39)

Using Collocation Method

Now, the two unknown coefficients c_2 and c_3 are determined for two collocating points $x_1 = \frac{1}{3}$ and $x_2 = \frac{2}{3}$, as shown in Figure 10.6.

Assuming the residual function $R(x)$ for $\alpha = 2$ to zero at collocating points $x = x_1$ and $x = x_2$, we obtain the algebraic equations as follows:

$$R(x_1) = -1.444444444c_2 + 0.4074074074c_3 - 0.3333333333 = 0,$$
(10.40)

and

$$R(x_2) = -2.111111111c_2 - 1.851851852c_3 - 0.6666666667 = 0.$$
(10.41)

On solving Eqs. (10.40) and (10.41), we get $c_2 = -0.2514551805$ and $c_3 = -0.07334109427$. Putting the values of c_2 and c_3 in Eq. (10.38), we obtain the approximate solution of Eq. (10.35) at $\alpha = 2$ as:

$$\hat{u}_3(x) = -0.2514\,x + 0.1781\,x^2 + 0.0733\,x^3$$
(10.42)

Likewise, we may obtain the approximate solution of Eq. (10.35) at various values of α by using the aforementioned procedure:

$$\text{For } \alpha = \frac{4}{3}, \quad \hat{u}_3(x) = -0.3872x + 0.5504x^2 - 0.1632x^3,$$
(10.43)

$$\text{For } \alpha = \frac{5}{3}, \quad \hat{u}_3(x) = -0.3264x + 0.3362x^2 - 0.0098x^3,$$
(10.44)

and so on.

Using Least-Square Method

To compute the remaining unknown coefficients c_2 and c_3 given in Eq. (10.38) using the least-square approach, we need to minimize the residual function Eq. (10.39) at $\alpha = 2$ in the domain $\Omega = [0, 1]$ as:

$$\int_0^1 R(x; c_2, c_3)\, \frac{\partial R}{\partial c_2}\, dx = 2.85c_3 + 3.7c_2 + 1.083333333 = 0, \tag{10.45}$$

and

$$\int_0^1 R(x; c_2, c_3)\, \frac{\partial R}{\partial c_3}\, dx = 4.876190476c_3 + 2.850000000c_2 + 1.033333333 = 0. \tag{10.46}$$

From Eqs. (10.45) and (10.46), we obtain $c_2 = -0.2356532209$ and $c_3 = -0.07418119845$. So, the approximate solution to four decimal places is written as:

$$\hat{u}_3(x) = -0.2357x + 0.1615x^2 + 0.0742x^3 \tag{10.47}$$

Following the aforementioned procedure likewise, one can find the approximate solution of Eq. (10.35) for various fractional-order values as follows:

$$\text{For } \alpha = \frac{4}{3}, \quad \hat{u}_3(x) = -0.3966x + 0.5334x^2 - 0.1368x^3, \tag{10.48}$$

$$\text{For } \alpha = \frac{5}{3}, \quad \hat{u}_3(x) = -0.3296x + 0.3495x^2 - 0.0199x^3. \tag{10.49}$$

Using Galerkin Method

In order to compute the values of c_2 and c_3 in Eq. (10.38) using the Galerkin method at $\alpha = 2$ in the domain $\Omega = [0, 1]$, we use

$$\int_0^1 \varphi_2(x) R(x; c_2, c_3)\, dx = -0.1333333333c_3 - 0.3c_2 - 0.08333333333 = 0, \tag{10.50}$$

and

$$\int_0^1 \varphi_3(x) R(x; c_2, c_3)\, dx = -0.1238095238c_3 - 0.1666666667c_2 - 0.05 = 0. \tag{10.51}$$

On solving Eqs. (10.50) and (10.51), we get $c_2 = -0.2446808511$ and $c_3 = -0.07446808497$. So, the approximate solution to Eq. (10.35) is obtained as:

$$\hat{u}_3(x) = -0.2447x + 0.1702x^2 + 0.0745x^3, \tag{10.52}$$

Similarly,

$$\text{For } \alpha = \frac{4}{3}, \quad \hat{u}_3(x) = -0.4040x + 0.5998x^2 - 0.1957x^3, \tag{10.53}$$

$$\text{For } \alpha = \frac{5}{3}, \quad \hat{u}_3(x) = -0.3295x + 0.3498x^2 - 0.0203x^3. \tag{10.54}$$

In order to validate these methods, the approximate solutions produced by these three approaches are compared with the exact solution (Chakraverty et al. 2019) $u(x) = e^{-x/2}\left(-\sin\left(\frac{\sqrt{3}}{2}x\right)\cot\left(\frac{\sqrt{3}}{2}\right) + \cos\left(\frac{\sqrt{3}}{2}x\right)\right) + x - 1$ at $\alpha = 2$, which is shown in Figure 10.7. Figures 10.8 and 10.9 provide approximate solution plots of Example 10.2 for various values of fractional order. Also, the absolute error of each method is determined, and the corresponding errors for WRMs are displayed in Figure 10.10.

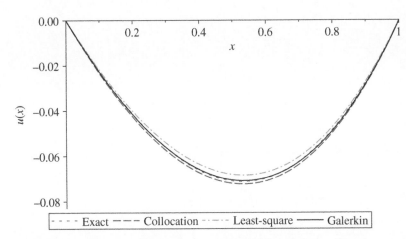

Figure 10.7 Comparison of the approximate solutions of Example 10.2 with the exact solution when $\alpha = 2$.

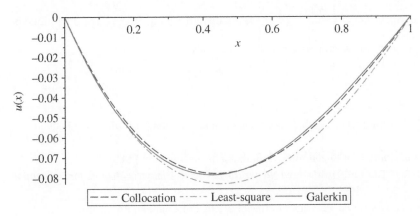

Figure 10.8 Comparison of the approximate solutions of Example 10.2 when $\alpha = \dfrac{4}{3}$.

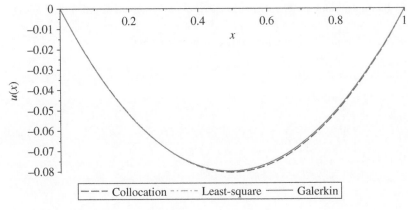

Figure 10.9 Comparison of the approximate solutions of Example 10.2 when $\alpha = \dfrac{5}{3}$.

From the aforementioned test examples, It is concluded that

- One can improve the accuracy of the solutions by considering higher-order approximation.
- As shown in Figures 10.5 and 10.10, the Galerkin method gives the highest accuracy for both the examples since the error produced by this method is minimum compared to the collocation and least-square methods.

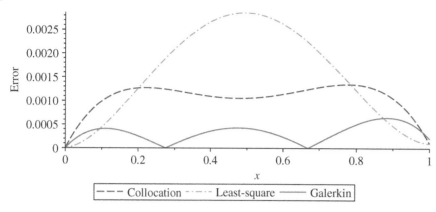

Figure 10.10 Error plot of approximate solutions for Example 10.2.

- It is worth mentioning that if the trial functions are chosen in such a way that they form the orthogonal basis, the Galerkin method is found to be more efficient since $\phi_i(x)\phi_j(x) = \begin{cases} \phi_i^2(x), & i = j \\ 0 & i \neq j \end{cases}$ is obtained from Eq. (10.13). This technique is discussed in Chapter 11 in terms of boundary characteristic orthogonal polynomials.

References

Baluch, M.H., Mohsen, M.F.N., and Ali, A.I. (1983). Method of weighted residuals as applied to nonlinear differential equations. *Applied Mathematical Modelling* 7 (5): 362–365.

Bhat, R.B. and Chakraverty, S. (2004). *Numerical Analysis in Engineering*. Oxford: Alpha Science Int'l Ltd.

Chakraverty, S., Mahato, N., Karunakar, P., and Rao, T.D. (2019). *Advanced Numerical and Semi-Analytical Methods for Differential Equations*. Wiley.

Finlayson, B.A. (2013). *The Method of Weighted Residuals and Variational Principles*, 73e. Philadelphia, PA: SIAM.

Gerald, C.F. and Wheatley, P.O. (2004). *Applied Numerical Analysis*, 7e. New Delhi: Pearson Education Inc.

Hatami, M. (2017). *Weighted Residual Methods: Principles, Modifications and Applications*. London: Academic Press.

Lindgren, L.E. (2009). From Weighted Residual Methods to Finite Element Methods. Technical Report.

Locker, J. (1971). The method of least squares for boundary value problems. *Transactions of the American Mathematical Society* 154: 57–68.

Logan, D.L. (2011). *A First Course in the Finite Element Method*, 5ee. Stamford, CT: Cengage Learning.

11

Boundary Characteristic Orthogonal Polynomials

11.1 Introduction

Bhat (1985, 1986) proposed boundary characteristic orthogonal polynomials (BCOPs) in 1985, which have been applied in various scientific and engineering fields. Several authors have employed BCOPs in many problems, such as Bhat and Chakraverty (2004) and Singh and Chakraverty (1994a) for two-dimensional BCOPs first time in a systematic manner. BCOPs have been beneficial in well-known approaches like Rayleigh–Ritz, Galerkin, and collocation. The Gram–Schmidt orthogonalization approach (Johnson 2014) can be used to generate BCOPs. The resulting BCOPs have to satisfy some of the boundary conditions of the considered models (Singh and Chakraverty 1994b; Bhat and Chakraverty 2004). Initially, the general approximation solution to the problem is assumed to be a linear combination of BCOPs. The residual can be obtained by replacing the approximate solution in the boundary value problem (Singh and Chakraverty 1994a; Chakraverty et al. 2008). A linear system of equations may be developed by employing the residual. Finally, the resultant linear system may be handled using any known analytical/numerical method. The orthogonal nature of BCOPs makes them straightforward to analyze.

Torvik and Bagley (1984) proposed a fractional model after moving a rigid plate dipped in a Newtonian fluid as follows:

$$a\frac{d^2u(x)}{dx^2} + bD^\alpha u(x) + cu(x) = f(x), \quad \alpha = \frac{3}{2}, \tag{11.1}$$

where a, b, and c are constants depend on mass, area of the plate, stiffness of spring, fluid density, and viscosity $f(x)$ is the external force and $u(x)$ stands for the displacement of the plate. Equation (11.1) has been generalized, and numerical techniques have been developed for the solution of this equation with initial value conditions (Diethelm and Ford 2002; Cenesiz et al. 2010; Wang and Wang 2010; Mekkaoui and Hammouch 2012). One may compute the movement of the plate at any other points in Newtonian fluid and model multi-point boundary value problems of the B-T equation. As regards, the generalized nonlinear B-T equation with two-point boundary conditions has been studied by Stanek (Stanek 2013). Moreover, the coefficients a, b, and c may change with the changes of fluid density and viscosity. So a, b, and c are treated as functions of x. So, the three-point boundary value problems of the B-T equation with variable coefficients is written as follows:

$$\frac{d^2u(x)}{dx} + p(x)D^\alpha u(x) + q(x)u(x) = f(x), \quad 0 < \alpha < 2, x \in [a, b], \tag{11.2}$$

with boundary conditions:

$$u(a) = a_1, \quad u(b) + \lambda u(c) = b_1, \quad c \in (a, b), \tag{11.3}$$

or

$$u(a) + \mu u(c) = a_1, \quad u(b) = b_1, \quad c \in (a, b), \tag{11.4}$$

where $p(x)$, $q(x)$, and $f(x)$ are known functions and a_1, b_1, μ, λ, c are known constants.

The Gram–Schmidt orthogonalization procedure for producing orthogonal polynomials is described in the next section.

Computational Fractional Dynamical Systems: Fractional Differential Equations and Applications, First Edition.
Snehashish Chakraverty, Rajarama Mohan Jena, and Subrat Kumar Jena.
© 2023 John Wiley & Sons, Inc. Published 2023 by John Wiley & Sons, Inc.

11.2 Gram–Schmidt Orthogonalization Procedure

Let us consider a set of linearly independent functions $(f_i(x) = x^i, i = 0, 1, 2, ...)$ in $[a, b]$. From these set of functions, we can construct appropriate orthogonal functions by using the Gram–Schmidt orthogonalization process (Bhat and Chakraverty 2004; Johnson 2014) as follows:

$$\phi_0 = f_0,$$
$$\phi_1 = f_1 - \alpha_{10}\phi_0,$$
$$\phi_2 = f_2 - \alpha_{20}\phi_0 - \alpha_{21}\phi_1,$$
$$\vdots$$

where, $\alpha_{10} = \dfrac{\langle f_1, \phi_0 \rangle}{\langle \phi_0, \phi_0 \rangle}$, $\alpha_{20} = \dfrac{\langle f_2, \phi_0 \rangle}{\langle \phi_0, \phi_0 \rangle}$, $\alpha_{21} = \dfrac{\langle f_2, \phi_1 \rangle}{\langle \phi_1, \phi_1 \rangle}$, etc., and $\langle \rangle$ defines the inner product of the respective polynomials.

In general, we can write the aforementioned procedure as:

$$\phi_0 = f_0,$$
$$\phi_i = f_i - \sum_{j=0}^{i-1} \alpha_{ij}\phi_j,$$

where,

$$\alpha_{ij} = \frac{\left\langle f_i, \phi_j \right\rangle}{\left\langle \phi_j, \phi_j \right\rangle} = \frac{\displaystyle\int_a^b W(x)f_i(x)\phi_j(x)dx}{\displaystyle\int_a^b W(x)\phi_j(x)\phi_j(x)dx}.$$

The aforementioned procedure is valid only when the inner product exists for the interval $[a, b]$ with respect to the weight function $W(x)$. Throughout the paper, we have considered $W(x) = 1$ for simplicity.

11.3 Generation of BCOPs

The first member of BCOPs set, viz. $\phi_0(x)$, is chosen as the simplest polynomial of the least order, which satisfies the boundary conditions of the considered problem. The other members of the orthogonal set in the interval $a \leq x \leq b$ are generated by using the Gram–Schmidt orthogonalization process (Singh and Chakraverty 1994a; Bhat and Chakraverty 2004) as follows:

$$\phi_1(x) = (x - l_1)\phi_0(x),$$
$$\vdots$$
$$\phi_k(x) = (x - l_k)\phi_{k-1}(x) - m_k\phi_{k-2}(x),$$

where $l_k = \dfrac{\displaystyle\int_a^b x\phi_{k-1}^2(x)dx}{\displaystyle\int_a^b \phi_{k-1}^2(x)dx}$ and $m_k = \dfrac{\displaystyle\int_a^b x\phi_{k-1}(x)\phi_{k-2}(x)dx}{\displaystyle\int_a^b \phi_{k-2}^2(x)dx}$.

Here, we consider $W(x) = 1$. The polynomials $\phi_k(x)$ satisfy the orthogonality condition:

$$\int_a^b \phi_i(x)\phi_j(x)dx = 0, \quad \text{if } i \neq j.$$

Next, we discuss the Galerkin method-based BCOPs for solving the B-T equation with variable coefficient.

11.4 Galerkin Method with BCOPs

Let us consider three-point boundary value problems Eq. (11.2) subject to boundary conditions of Eqs. (11.3) and (11.4). We assume an approximate solution for the aforementioned differential equation satisfying the boundary conditions and involving unknown real constants c_0, c_1, c_2, ..., c_n as:

$$\hat{u}(x) = r(x) + h(x)\left(\sum_{i=0}^{n} c_i \phi_i(x)\right), \tag{11.5}$$

where $r(x)$ and $h(x)$ control the boundary conditions, and ϕ_i's are the BCOPs that satisfy the given boundary conditions.

Substituting Eq. (11.5) in Eq. (11.2), we may obtain the residual R as:

$$R(x; c_0, c_1, c_2, ..., c_n) = \frac{d^2}{dx^2}\left(r(x) + h(x)\left(\sum_{i=0}^{n} c_i \phi_i(x)\right)\right) + p(x)D^\alpha\left(r(x) + h(x)\left(\sum_{i=0}^{n} c_i \phi_i(x)\right)\right)$$
$$+ q(x)\left(r(x) + \sum_{i=0}^{n} c_i h(x)\phi_i(x)\right) - f(x). \tag{11.6}$$

The residual R is orthogonalized to the $(n+1)$ BCOPs functions ϕ_0, ϕ_1, ..., ϕ_n, which gives

$$\int_a^b R(x; c_0, c_1, c_2, ..., c_n)\,\phi_j(x)\,dx = 0, \quad j = 0, 1, 2, ..., n,$$

$$\int_a^b \left[\begin{array}{l} \dfrac{d^2}{dx^2}\left(r(x) + h(x)\left(\sum_{i=0}^{n} c_i \phi_i(x)\right)\right) + p(x)\,D^\alpha\left(r(x) + h(x)\left(\sum_{i=0}^{n} c_i \phi_i(x)\right)\right) \\ + q(x)\left(r(x) + \sum_{i=0}^{n} c_i h(x)\phi_i(x)\right) - f(x) \end{array}\right]\phi_j(x)\,dx = 0,$$
$$\text{for } j = 0, 1, 2, ..., n \tag{11.7}$$

where $\int_a^b \phi_i(x)\phi_j(x)dx = 0$, if $i \neq j$.

Eq. (11.7) gives $(n+1)$ simultaneous equations in $(n+1)$ unknowns, which can be solved by any standard method. Further substituting the evaluated constants c_0, c_1, ..., c_n in Eq. (11.5), we may get the approximate solution to the original Eq. (11.2). One may note that the terms containing ϕ_0, ϕ_1, ..., ϕ_n will vanish due to the orthogonal property. This makes the method efficient.

Section 11.5 presents another approach, viz. least-square method (LSM), with BCOPs.

11.5 Least-Square Method with BCOPs

In the LSM (Locker, 1971; Finlayson 2013; Hatami 2017), the residue given in Eq. (11.6) is squared and integrated over the entire domain $[a, b]$:

$$I = \int_{[a,b]} R^2(x, c_i)\, dx. \tag{11.8}$$

The integrand I is then minimized using $\dfrac{\partial}{\partial c_i}\displaystyle\int_{[a,b]} R^2(x, c_i)\, dx = 0$, where c_i are the unknown coefficients of the approximate

solution $\hat{u}(x) = r(x) + h(x)\left(\sum_{i=0}^{n} c_i \phi_i(x)\right)$, which further reduces to

$$2 \int\limits_{[a,b]} R(x; c_i) \frac{\partial R}{\partial c_i} dx = 0, \quad \text{for } i = 0, 1, 2, ..., n.$$

$$\Rightarrow \int\limits_{[a,b]} R(x; c_i) \frac{\partial R}{\partial c_i} dx = 0. \tag{11.9}$$

From Eq. (11.9), the weight function for the LSM is considered as:

$$W(x) = \frac{\partial R}{\partial c_i}, \quad \text{for } i = 0, 1, 2, ..., n. \tag{11.10}$$

11.6 Application Problems

This section implements the present methods to solve two Bagley–Torvik (B-T) equation examples. Solutions of these examples are also compared with the exact solution.

Example 11.1 Let us consider the three-point boundary value problem for the generalized B-T equation (Huang et al. 2016):

$$D^2 u(x) + \sqrt{\pi} x^2 D^{3/2} u(x) + \left(1 - 4x^{1/2}\right) u(x) = x^2 + 2, \tag{11.11}$$

subject to boundary conditions:

$$u(0) = 0, \quad u\left(\frac{1}{5}\right) + u\left(\frac{1}{10}\right) = \frac{1}{20}, \tag{11.12}$$

with $x \in \left[0, \frac{1}{5}\right]$.

The exact solution of the Eq. (11.11) subject to boundary conditions Eq. (11.12) is $u(x) = x^2$.

Using Galerkin Method Based on BCOPs

Let us consider two terms guess solution of Eqs. (11.11) and (11.12) as:

$$\hat{u}(x) = r(x) + h(x)[c_0\phi_0(x) + c_1\phi_1(x)] = r(x) + h(x)\left(\sum_{i=0}^{1} c_i\phi_i(x)\right), \tag{11.13}$$

where $r(x) = x^2, h(x) = x\left(x - \frac{1}{5}\right)\left(x - \frac{1}{10}\right)$, which controls the boundary conditions.

The residual function R may be written as:

$$R(x; c_0, c_1) = D^2\left(r(x) + h(x)\left(\sum_{i=0}^{1} c_i\phi_i\right)\right) + \sqrt{\pi} x^2 D^{3/2}\left(r(x) + h(x)\left(\sum_{i=0}^{1} c_i\phi_i\right)\right)$$
$$+ (1 - 4x^{0.5})\left(r(x) + h(x)\left(\sum_{i=0}^{1} c_i\phi_i\right)\right) - x^2 - 2, \tag{11.14}$$

where $\phi_0(x) = 1$ and $\phi_1(x) = x - 0.1$ are the BOCPs in the domain $\left[0, \frac{1}{5}\right]$.

Substituting the values of the functions $r(x), h(x), \phi_0$, and ϕ_1 in Eq. (11.14), we have

$$R = (0.000804916 - 0.000241475c_0 + 0.0000402458c_1)x^{5/2} + (-0.08c_0 + 0.008c_1)x^{3/2}$$
$$+ (4.001609832c_0 - 1.600643933c_1)x^{7/2} + 8.80257573x^{9/2}c_1 + (x^3 - 0.3x^2 + 6.02x - 0.6)c_0 \tag{11.15}$$
$$(x^4 - 0.4x^3 + 12.05x^2 - 2.4020x + 1)c_1.$$

Now by using the Galerkin method with BCOPs, we have

$$\int\limits_0^{1/5} (R\phi_0)dx = 0, \quad \text{and} \quad \int\limits_0^{1/5} (R\phi_1)dx = 0,$$ (11.16)

which gives the two equations as:

$$0.00006361285324c_0 + 0.004029140957c_1 + 8.227871507 \times 10^{-7} = 0,$$ (11.17)

and

$$0.004013278139c_0 + 0.000002115661834c_1 + 4.571039726 \times 10^{-8} = 0.$$ (11.18)

Solving the aforementioned system of equations, we have

$$\begin{aligned} c_0 &= -0.00001128223243 \\ c_1 &= -0.0002040309496 \end{aligned}$$ (11.19)

So, the solution to Eqs. (11.11) and (11.12) may now be written as:

$$\begin{aligned} \hat{u}(x) &= r(x) + h(x)[c_0\phi_0(x) + c_1\phi_1(x)] \\ &= -0.0002040309496x^4 + 0.00007033014741x^3 + x^2 + 1.824172506 \times 10^{-7}x. \end{aligned}$$ (11.20)

The numerical results of the present solution and the exact solution are shown in Table 11.1. The plot of the exact and present solutions has also been presented in Figure 11.1. One may see that the exact solution of Eq. (11.11) agrees precisely by taking two terms only.

Using the Least-Square Method Based on BCOPs

According to the LSM, taking the square of the residual equation, we have

$$\begin{aligned} S(x;c_0,c_1) &= 0.02416010589c_0^2 + (0.00004870485423c_1 + 5.516997793 \times 10^{-7})c_0 \\ &\quad + 0.0003394335925c_1^2 + 5.348123125 \times 10^{-8}c_1 + 6.910824182 \times 10^{-12}. \end{aligned}$$ (11.21)

From Eq. (11.8), we have

$$\frac{\partial S}{\partial c_0} = 0, \quad \text{and} \quad \frac{\partial S}{\partial c_1} = 0.$$ (11.22)

Eq. (11.22) gives two equations as:

$$0.04832021178c_0 + 0.00004870485423c_1 + 5.516997793 \times 10^{-7} = 0,$$ (11.23)

$$0.00004870485423c_0 + 0.0006788671850c_1 + 5.348123125 \times 10^{-8} = 0.$$ (11.24)

Table 11.1 Comparison of the present solution with the exact solution.

x	Present solution $n = 1$	Exact solution (Huang et al. 2016)
0.00	0.00	0.00
0.02	4.000014×10^{-4}	4×10^{-4}
0.04	16.000003×10^{-4}	16×10^{-4}
0.06	35.999989×10^{-4}	36×10^{-4}
0.08	63.999986×10^{-4}	64×10^{-4}
0.10	1×10^{-2}	1×10^{-2}
0.12	$144.000029 \times 10^{-4}$	144×10^{-4}

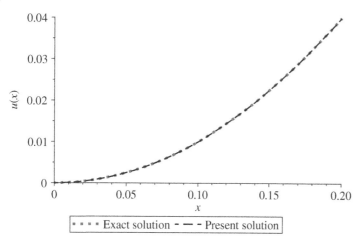

Figure 11.1 Behavior of the exact and present solution of Example 11.1.

Table 11.2 Comparison of the present solution with the exact solution.

x	Present solution $n = 1$	Exact solution (Huang et al. 2016)
0.00	0.00	0.00
0.02	3.999985×10^{-4}	4×10^{-4}
0.04	15.999974×10^{-4}	16×10^{-4}
0.06	35.999972×10^{-4}	36×10^{-4}
0.08	63.999981×10^{-4}	64×10^{-4}
0.10	1×10^{-2}	1×10^{-2}
0.12	$144.000024 \times 10^{-4}$	144×10^{-4}

Solving the aforementioned linear system of equations, we have

$$c_0 = -0.0000113389906$$

$$c_1 = -0.00007796660162 \tag{11.25}$$

The solution of original Eq. (11.11) may be written as:

$$\hat{u}(x) = r(x) + h(x)[c_0\phi_0(x) + c_1\phi_1(x)]$$

$$= -0.00007796660162x^4 + 0.00001984765005x^3 + x^2 - 7.08466088 \times 10^{-8}x, \tag{11.26}$$

Again the numerical results of the present solution and the exact solutions are given in Table 11.2. The exact and present solution plot has been depicted in Figure 11.2. One may see that the exact solution of Eq. (11.11) is same as the present method solution by taking two terms.

Example 11.2   Next, we consider the following three-point boundary value problem for the generalized B-T equation of variable coefficient (Huang et al. 2016):

$$D^2u(x) - 5\sqrt{\pi}xD^{1/2}u(x) + 16x^{1/2}u(x) = 6x, \tag{11.27}$$

subject to the boundary conditions:

$$u(0) + 2u\left(\frac{1}{10}\right) = \frac{1}{500}, \quad u\left(\frac{1}{5}\right) = \frac{1}{125}, \tag{11.28}$$

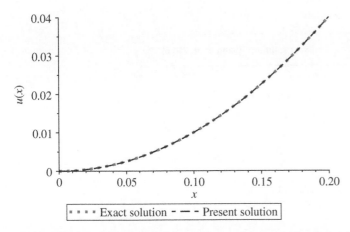

Figure 11.2 Behavior of the exact and present solution of Example 11.1.

with $x \in \left[0, \dfrac{1}{5}\right]$.

The exact solution of the Eqs. (11.27) and (11.28) is $u(x) = x^3$.

Using Galerkin Method Based on BCOPs

We again consider two terms guess solution as:

$$\hat{u}(x) = r(x) + h(x)[c_0\phi_0(x) + c_1\phi_1(x)] = r(x) + h(x)\left(\sum_{i=0}^{1} c_i\phi_i(x)\right), \tag{11.29}$$

where $r(x) = x^3, h(x) = x\left(x - \dfrac{1}{5}\right)\left(x - \dfrac{1}{10}\right)$, which control the boundary conditions.

The residual function is written as"

$$R(x, c_0, c_1) = D^2\left(r(x) + h(x)\left(\sum_{i=0}^{1} c_i\phi_i\right)\right) - 5\sqrt{\pi}xD^{1/2}\left(r(x) + h(x)\left(\sum_{i=0}^{1} c_i\phi_i\right)\right)$$
$$+ \left(16x^{\frac{1}{2}}\right)\left(r(x) + h(x)\left(\sum_{i=0}^{1} c_i\phi_i\right)\right) - 6x, \tag{11.30}$$

where $\phi_0(x) = 1$ and $\phi_1(x) = x - 0.1$ are the BCOPs in the domain $\left[0, \dfrac{1}{5}\right]$.

Using the values of the functions $r(x)$, $h(x)$, ϕ_0, and ϕ_1 in Eq. (11.30), we obtain

$$R = (-0.00321968 + 0.001287871c_1 - 0.00321968c_0)x^{7/2} + (0.1199597541c_0 - 0.01199597541c_1)x^{3/2}$$
$$+ (0.1331991804c_1 - 0.799195082c_0)x^{5/2} - 2.28939391\,x^{9/2}c_1 + \left(-2.4x + 0.1 + 12x^2\right)c_1 + (6x - 0.6)c_0. \tag{11.31}$$

With the help of the Galerkin method with BCOPs, we have

$$\int_0^{1/5} (R\phi_0)dx = 0, \quad \text{and} \quad \int_0^{1/5} (R\phi_1)dx = 0. \tag{11.32}$$

Solving Eq. (11.32) gives the two equations as:

$$0.00004091095414c_0 + 0.003990955764c_1 - 5.119589935 \times 10^{-7} = 0, \tag{11.33}$$

and

$$0.003991368857c_0 - 2.254426072 \times 10^{-7}c_1 - 3.257920868 \times 10^{-8} = 0. \tag{11.34}$$

Table 11.3 Comparison of the present solution with the exact solution.

x	Present solution $n = 1$	Exact solution (Huang et al. 2016)
0.00	0.00	0.00
0.02	7.999399×10^{-6}	8×10^{-6}
0.04	64.000183×10^{-6}	64×10^{-6}
0.06	$216.001022 \times 10^{-6}$	216×10^{-6}
0.08	$512.001076 \times 10^{-6}$	512×10^{-6}
0.10	1×10^{-3}	1×10^{-3}
0.12	$1727.99794 \times 10^{-6}$	1728×10^{-6}

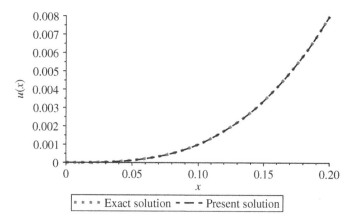

Figure 11.3 Behavior of the exact and present solution of Example 11.2.

Solving the aforementioned two linear systems of equations, we get

$$c_0 = 0.000008169655751$$
$$c_1 = 0.0001281960501 \tag{11.35}$$

Accordingly, the solution of original Eq. (11.27) may now be obtained as:

$$\hat{u}(x) = 0.0001281960501x^4 + x^3 + 0.00000395890578x^2 - 9.29989852 \times 10^{-8}x, \tag{11.36}$$

The numerical results of the present solution and the exact solution are given in Table 11.3. The plot of the exact and present solutions has been presented in Figure 11.3. One may see that the exact solution of Eq. (11.27) is in good agreement with the present solution by taking two terms.

Using the Least-Square Method Based on BCOPs

Taking the square of the residual equation Eq. (11.31), we have

$$S(x; c_0, c_1) = 0.02389672501c_0^2 + \left(-0.0001433717427c_1 - 3.897731001 \times 10^{-7} \right)c_0$$
$$+ 3.317228577 \times 10^{-12} + 0.0003352149388c_1^2 - 4.043552417 \times 10^{-8}c_1, \tag{11.37}$$

From Eq. (11.8), we have

$$\frac{\partial S}{\partial c_0} = 0, \quad \text{and} \quad \frac{\partial S}{\partial c_1} = 0. \tag{11.38}$$

Eq. (11.38) provides two equations as:

$$0.04779345002c_0 - 0.00001433717427c_1 - 3.897731001 \times 10^{-7} = 0, \tag{11.39}$$

$$-0.00001433717427c_0 + 0.0006704298776c_1 - 4.043552417 \times 10^{-8} = 0. \tag{11.40}$$

Table 11.4 Comparison of the present solution with the exact solution.

x	Present solution $n = 1$	Exact solution (Huang et al. 2016)
0.00	0.0	0.0
0.02	8.000960×10^{-6}	8×10^{-6}
0.04	64.001745×10^{-6}	64×10^{-6}
0.06	$216.001933 \times 10^{-6}$	216×10^{-6}
0.08	$512.001337 \times 10^{-6}$	512×10^{-6}
0.10	1×10^{-3}	1×10^{-3}
0.12	$1727.998200 \times 10^{-6}$	1728×10^{-6}

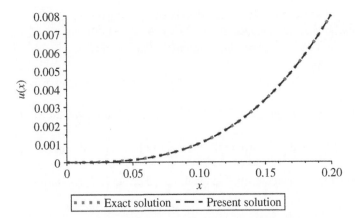

Figure 11.4 Behavior of the exact and present solution of Example 11.2.

Solving Eqs. (11.39) and (11.40), we get

$$c_0 = 0.000008173511675$$
$$c_1 = 0.00006048762233 \tag{11.41}$$

So, we obtain the approximate solution of Eq. (11.27) as:

$$\hat{u}(x) = 0.00006048762233x^4 + x^3 + 5.72327614 \times 10^{-7}x^2 + 4.24949888 \times 10^{-8}x. \tag{11.42}$$

One may see that the exact solution of Eq. (11.27) agrees precisely by taking two terms. The numerical results of the present solution and the exact solution are given in Table 11.4. The plot of the exact and present solutions has been illustrated in Figure 11.4.

From the results obtained from Examples 11.1 and 11.2, we can draw the following conclusions:

- The present solutions are in good agreement with the exact solutions.
- The accuracy of present methods may be improved by taking more terms of BCOPs.
- The methods are used in linear and nonlinear fractional differential equations, and the solutions are validated.

References

Bhat, R.B. (1985). Natural frequencies of rectangular plates using characteristic orthogonal polynomials in Rayleigh-Ritz method. *Journal of Sound and Vibration* 102 (4): 493–499.

Bhat, R.B. (1986). Transverse vibrations of a rotating uniform cantilever beam with tip mass as predicted by using beam characteristic orthogonal polynomials in the Rayleigh-Ritz method. *Journal of Sound and Vibration* 105 (2): 199–210.

Bhat, R.B. and Chakraverty, S. (2004). *Numerical Analysis in Engineering*. Oxford: Alpha Science Int'l Ltd.

Cenesiz, Y., Keskin, Y., and Kurnaz, A. (2010). The solution of the Bagley–Torvik equation with the generalized Taylor collocation method. *Journal of the Franklin Institute* 347: 452–466.

Chakraverty, S., Saini, H., and Panigrahi, S.K. (2008). Prediction of product parameters of fly ash cement bricks using two dimensional orthogonal polynomials in the regression analysis. *Computers and Concrete* 5 (5): 449–459.

Diethelm, K. and Ford, J. (2002). Numerical solution of the Bagley–Torvik equation. *BIT Numerical Mathematics* 42 (3): 490–507.

Finlayson, B.A. (2013). *The Method of Weighted Residuals and Variational Principles*, 73e. Philadelphia, PA: SIAM.

Hatami, M. (2017). *Weighted Residual Methods: Principles, Modifications and Applications*. London: Academic Press.

Huang, Q.A., Zhong, X.C., and Guo, B.L. (2016). Approximate solution of Bagley–Torvik equations with variable coefficients and three-point boundary-value conditions. *International Journal of Applied and Computational Mathematics* 2: 327–347.

Johnson, P.S. (2014). Gram-Schmidt orthogonalization process.

Locker, J. (1971). The method of least squares for boundary value problems. *Transactions of the American Mathematical Society* 154: 57–68.

Mekkaoui, T. and Hammouch, Z. (2012). Approximate analytical solutions to the Bagley–Torvik equation by the fractional iteration method. *Annals of the University of Craiova, Mathematics and Computer Science Series* 39 (2): 251–256.

Singh, B. and Chakraverty, S. (1994a). Boundary characteristic orthogonal polynomials in numerical approximation. *Communications in Numerical Methods in Engineering* 10 (12): 1027–1043.

Singh, B. and Chakraverty, S. (1994b). Flexural vibration of skew plates using boundary characteristic orthogonal polynomials in two variables. *Journal of Sound and Vibration* 173 (2): 157–178.

Stanek, S. (2013). Two-point boundary value problems for the generalized Bagley-Torvik fractional differential equation. *Central European Journal of Mathematics* 11 (3): 574–593.

Torvik, P.J. and Bagley, R.L. (1984). On the appearance of the fractional derivative in the behavior of real materials. *Journal of Applied Mechanics* 51: 294–298.

Wang, Z.H. and Wang, X. (2010). General solution of the Bagley–Torvik equation with fractional-order derivative. *Communications in Nonlinear Science and Numerical Simulation* 15: 1279–1285.

12

Residual Power Series Method

12.1 Introduction

In Chapter 11, we have already discussed boundary characteristics orthogonal polynomials-based solution of linear and nonlinear ordinary/partial/fractional differential equations where the orthogonal polynomials are generated by using the Gram–Schmidt orthogonalization procedure. In this chapter, we will discuss about residual power series method (RPSM). The RPSM was first developed in 2013 by Omar Abu Arqub, a Jordanian mathematician, to determine the values of the coefficients of power series solution for the first- and second-order fuzzy differential equations (Arqub 2013). RPSM is an intuitive and reliable to construct power series solutions for linear and nonlinear equations without linearization, perturbation, or discretization. Over the last few years, the RPSM has been used to solve different nonlinear ordinary and partial differential equations (PDEs) of various forms, classifications, and orders. One may find the successful implementation of this method in the solutions of the generalized Lane–Emden equation (Arqub et al. 2013), higher-order ordinary differential equations (Arqub et al. 2013), fractional coupled physical equations arising in fluids flow (Arafa and Elmahdy 2018), solitary pattern solutions for nonlinear time-fractional dispersive PDEs (Arqub et al. 2015), and predicting and representing the multiplicity of solutions to boundary value problems of fractional order (Arqub et al. 2014), etc. The RPSM distinguishes itself from many other analytical and numerical methods in some significant respects (El-Ajou et al. 2015). Firstly, the coefficients of the corresponding terms do not need to be compared, and a recursion relation is not required. Secondly, by minimizing the associated residual error, the RPSM provides a straightforward way to ensure the convergence of the series solution. Thirdly, computational round-off errors do not impact the RPSM and do not take substantial computer memory and time. Fourthly, while shifting from the low-order to the higher-order and from simple linearity to complex nonlinearity, the RPSM does not involve any conversion. Consequently, the method can be employed directly to the given problem by choosing the suitable initial guess approximation. Before explaining the detailed procedure of this method, let us define some useful theorems and lemmas in the following sections.

12.2 Theorems and Lemma Related to RPSM

Definition 12.1 (Arafa and Elmahdy 2018)
A series of the form:

$$\sum_{k=0}^{\infty} a_k(t-t_0)^{k\alpha} = a_0 + a_1(t-t_0)^{\alpha} + a_2(t-t_0)^{2\alpha} + \cdots, \quad \text{for } 0 \leq n-1 < \alpha \leq n, t \geq t_0, \tag{12.1}$$

is called fractional power series expansion (FPSE) at $t = t_0$, where a_k is the coefficient of series.

Theorem 12.1 (Hira and Ghazala 2017; Jena and Chakraverty 2019a)
If $f(t) = \sum_{k=0}^{\infty} a_k(t-t_0)^{k\alpha}$ and $D^{k\alpha}f(t) \in C(t_0, t_0 + R)$ for $k = 0, 1, 2, \ldots$ then the value of a_k in Eq. (12.1) is given by

$$a_k = \frac{D^{k\alpha}f(t_0)}{\Gamma(k\alpha + 1)}.$$

Computational Fractional Dynamical Systems: Fractional Differential Equations and Applications, First Edition.
Snehashish Chakraverty, Rajarama Mohan Jena, and Subrat Kumar Jena.
© 2023 John Wiley & Sons, Inc. Published 2023 by John Wiley & Sons, Inc.

Definition 12.2 (Jena and Chakraverty 2019b)

An FPSE of the form $\sum_{k=0}^{\infty} b_k(x)(t - t_0)^{k\alpha}$ is called multiple FPSE about $t = t_0$, where b_k's are coefficients of the series.

12.3 Basic Idea of RPSM

In order to provide the approximate solution for nonlinear fractional-order differential equation using RPSM, a general nonlinear fractional differential equation is considered as:

$$D_t^{\alpha}\psi(x,t) = N(\psi) + R(\psi), \tag{12.2}$$

subject to the initial condition (IC):

$$\psi(x,0) = b(x), \tag{12.3}$$

where $N(\psi)$ is a nonlinear term and $R(\psi)$ is the linear term.

Step 1. Let us assume that the solution of Eq. (12.2) in terms of fractional power series about the initial point $t = t_0$ is written as (Arafa and Elmahdy 2018; Jena and Chakraverty 2019a):

$$\psi(x,t) = \sum_{k-0}^{\infty} b_k(x)\frac{t^{k\alpha}}{\Gamma(1 + k\alpha)}, \quad 0 < \alpha \le 1, -\infty < x < \infty, 0 \le t < R. \tag{12.4}$$

To evaluate the value of $\psi(x, t)$, let $\psi_m(x, t)$ signifies the mth truncated series of $\psi(x, t)$ as:

$$\psi_m(x,t) = \sum_{k=0}^{m} b_k(x)\frac{t^{ak}}{\Gamma(ak + 1)}, \quad 0 < \alpha \le 1, 0 \le t. \tag{12.5}$$

For $m = 0$, the 0th residual power series (RPS) solution of $\psi(x, t)$ may be written as:

$$\psi_0(x,t) = b_0(x). \tag{12.6}$$

Using Eq. (12.6), Eq. (12.5) can be modified as:

$$\psi_m(x,t) = b_0(x) + \sum_{k=1}^{m} b_k(x)\frac{t^{ak}}{\Gamma(ak + 1)}, \quad 0 < \alpha \le 1, 0 \le t, m = 1, 2, 3, \ldots \tag{12.7}$$

So mth RPS solution can be evaluated after obtaining all $b_k(x)$, $k = 1, 2, \ldots, m$.

Step 2. Let us consider the residual function (RF) of Eq. (12.2) as (Arafa and Elmahdy 2018):

$$\mathrm{Res}_\psi(x,t) = D_t^{\alpha}\psi(x,t) - N(\psi) - R(\psi), \tag{12.8}$$

and mth RF may be written as (Arafa and Elmahdy 2018):

$$\mathrm{Res}_{\psi,m}(x,t) = D_t^{\alpha}\psi_m(x,t) - N(\psi_m) - R(\psi_m), \quad m = 1, 2, 3, \ldots \tag{12.9}$$

Some useful results about $\mathrm{Res}_{\psi, m}(x, t)$ have been included in (Jena and Chakraverty 2019a, 2019b), which are given as follows:

$$\mathrm{Res}_\psi(x,t) = 0.$$

$$\lim_{m \to \infty} \mathrm{Res}_{\psi,m}(x,t) = \mathrm{Res}_\psi(x,t). \tag{12.10}$$

$$D_t^{i\alpha}\mathrm{Res}_\psi(x,0) = D_t^{i\alpha}\mathrm{Res}_{\psi,m}(x,0) = 0, \quad i = 0, 1, 2, \ldots, m.$$

Step 3. Putting Eq. (12.7) into Eq. (12.9) and calculating $D_t^{(k-1)\alpha}\mathrm{Res}_{\psi,m}(x,t), k = 1, 2, \ldots$ at $t = 0$, together with the aforementioned three results, we obtain the following algebraic equations:

$$D_t^{(k-1)\alpha}\mathrm{Res}_{\psi,m}(x,0) = 0, \quad 0 < \alpha \le 1, k = 1, 2, \ldots \tag{12.11}$$

Step 4. By solving Eq. (12.11), we can get the coefficients $b_k(x)$, $k = 1, 2, \ldots m$. Thus m^{th} RPS approximate solution is derived.

12.4 Convergence Analysis

Lemma 12.1 If $f(x)$ is a continuous function and $\alpha, \beta > 0$ then

$$I_c^\alpha I_c^\beta f(x) = I_c^{\alpha+\beta} f(x) = I_c^\beta I_c^\alpha f(x).$$

Theorem 12.2 (Hira and Ghazala 2017; Jena and Chakraverty 2019a, 2019b)

(a) If the FPSE of the form $\sum\limits_{n=0}^{\infty} a_n x^{n\alpha}, x \geq 0$ converges at $x = x_1$, then it converges absolutely $\forall x$ satisfying $|x| < |x_1|$.

(b) If the FPSE diverges at $x = x_1$, then it will diverge $\forall x$ such that $|x| > |x_1|$.

Theorem 12.3 (Hira and Ghazala 2017; Jena and Chakraverty 2019a, 2019b)

Suppose $D_t^{r+k\alpha}, D_t^{r+(k+1)\alpha} \in C[R, t_0] \times [R, t_0 + R]$, then

$$\left(I_t^{r+k\alpha} D_t^{r+k\alpha} u\right)(x,t) - \left(I_t^{r+(k+1)\alpha} D_t^{r+(k+1)\alpha} u\right)(x,t) = \frac{(t-t_0)^{r+k\alpha}}{\Gamma(r+k\alpha+1)} D_t^{r+k\alpha} u(x,t_0),$$

where $D_t^{r+k\alpha} = \underbrace{(D_t.D_t.D_t...)}_{r-times}\underbrace{\left(D_t^\alpha.D_t^\alpha.D_t^\alpha...\right)}_{k-times}$ and $0 \leq n-1 < \alpha \leq n$.

Theorem 12.4 (Hira and Ghazala 2017; Jena and Chakraverty 2019a, 2019b)

Let $w(x,t), D_t^{k\alpha} w(x,t) \in C[R, t_0] \times [R, t_0 + R]$ where $k = 0, 1, 2, ..., N+1$ and $j = 0, 1, 2, ..., n-1$. Also $D_t^{k\alpha} w(x,t)$ may be differentiated $n-1$ times with respect to t. Then,

$$w(x,t) \cong \sum_{j=0}^{n-1}\sum_{i=0}^{N} W_{j+i\alpha}(x)(t-t_0)^{j+i\alpha},$$

where $W_{j+i\alpha}(x) = \dfrac{D_t^{j+i\alpha} w(x,t_0)}{\Gamma(j+i\alpha+1)}$. Also, $\exists$ a value ε, $0 \leq \varepsilon \leq t$, the error term has the term as follows:

$$\|E_N(x,t)\| = \sup_{t\in[0,T]}\left|\sum_{j=0}^{n-1}\left[\frac{D^{j+(N+1)\alpha} w(x,\varepsilon)}{\Gamma((N+1)\alpha+j+1)} t^{(N+1)\alpha+j}\right]\right|.$$

Interested authors may see the proofs and details of the aforementioned theorems in the references (Hira and Ghazala 2017; Jena and Chakraverty 2019a, 2019b).

12.5 Examples

In order to demonstrate the fundamental concept of RPSM, we consider the following two time-fractional one-dimensional heat-like and nonlinear advection equations, respectively, in Examples 12.1 and 12.2.

Example 12.1 Let us consider the one-dimensional heat-like model (Özis and Agirseven 2008; Sadighi et al. 2008) as:

$$D_t^\alpha \psi(x,t) = \frac{1}{2}x^2 \psi_{xx}(x,t), \quad 0 < x < 1, t > 0, 0 < \alpha \leq 1, \tag{12.12}$$

with the boundary conditions (BCs):

$$\psi(0,t) = 0, \quad \psi(1,t) = e^t. \tag{12.13}$$

and IC:

$$\psi(x,0) = x^2. \tag{12.14}$$

Solution

According to the RPSM, $\psi_0(x, t) = x^2$ and the infinite series solution of Eq. (12.12) can be written as:

$$\psi(x, t) = x^2 + \sum_{k=1}^{\infty} b_k(x) \frac{t^{\alpha k}}{\Gamma(\alpha k + 1)}. \tag{12.15}$$

The *m*th truncated series solution of $\psi(x, t)$ becomes

$$\psi_m(x, t) = x^2 + \sum_{k=1}^{m} b_k(x) \frac{t^{\alpha k}}{\Gamma(\alpha k + 1)}, \quad m = 1, 2, 3, \dots \tag{12.16}$$

For $m = 1$, first RPS solution for Eq. (12.7) may be written as:

$$\psi_1(x, t) = x^2 + b_1(x) \frac{t^{\alpha}}{\Gamma(\alpha + 1)}. \tag{12.17}$$

To determine the value of $b_1(x)$, we substitute Eq. (12.17) in the first RF of Eq. (12.12) $\mathrm{Res}_{\psi,1}(x, t) = D_t^{\alpha} \psi_1(x, t) - \frac{1}{2} x^2 (\psi_1(x, t))_{xx}$, this gives

$$\mathrm{Res}_{\psi,1}(x, t) = b_1(x) - \frac{1}{2} x^2 \left(2 + b_1''(x) \frac{t^{\alpha}}{\Gamma(\alpha + 1)} \right). \tag{12.18}$$

Using (iii) of Eq. (12.10) for $i = 0$, that is, $\mathrm{Res}_{\psi}(x, 0) = \mathrm{Res}_{\psi,1}(x, 0) = 0$, we get

$$\mathrm{Res}_{\psi,1}(x, 0) = b_1(x) - \frac{1}{2} x^2 (2 + b_1''(x) \times 0) = 0.$$

$$b_1(x) = x^2. \tag{12.19}$$

For $m = 2$, second RPS solution for Eq. (12.7) may be written as:

$$\psi_2(x, t) = x^2 + x^2 \frac{t^{\alpha}}{\Gamma(\alpha + 1)} + b_2(x) \frac{t^{2\alpha}}{\Gamma(2\alpha + 1)}. \tag{12.20}$$

To find the value of $b_2(x)$, Eq. (12.20) is substituted in the second RF of Eq. (12.12) $\mathrm{Res}_{\psi,2}(x, t) = D_t^{\alpha} \psi_2(x, t) - \frac{1}{2} x^2 (\psi_2(x, t))_{xx}$. Then, we have

$$\begin{aligned} \mathrm{Res}_{\psi,2}(x, t) &= x^2 + \frac{b_2(x) t^{\alpha}}{\Gamma(2\alpha + 1)} \frac{\Gamma(2\alpha + 1)}{\Gamma(\alpha + 1)} - \frac{1}{2} x^2 \left(2 + \frac{2t^{\alpha}}{\Gamma(\alpha + 1)} + b_2''(x) \frac{t^{2\alpha}}{\Gamma(2\alpha + 1)} \right) \\ &= \frac{b_2(x) t^{\alpha}}{\Gamma(\alpha + 1)} - \frac{x^2 t^{\alpha}}{\Gamma(\alpha + 1)} - b_2''(x) \frac{t^{2\alpha}}{\Gamma(2\alpha + 1)} \frac{x^2}{2} \end{aligned} \tag{12.21}$$

Using (iii) of Eq. (12.10) for $i = 1$ that is $D_t^{\alpha} \mathrm{Res}_{\psi}(x, 0) = D_t^{\alpha} \mathrm{Res}_{\psi,2}(x, 0) = 0$, we get

$$D_t^{\alpha} \mathrm{Res}_{\psi,2}(x, 0) = b_2(x) - x^2 - \left. \frac{b_2''(x) x^2}{2\Gamma(2\alpha + 1)} \frac{\Gamma(2\alpha + 1) t^{\alpha}}{\Gamma(\alpha + 1)} \right|_{t=0} = 0,$$

$$b_2(x) = x^2. \tag{12.22}$$

Similarly, for $m = 3$, third RPS solution for the Eq. (12.7) is expressed as:

$$\psi_3(x, t) = x^2 + x^2 \frac{t^{\alpha}}{\Gamma(\alpha + 1)} + x^2 \frac{t^{2\alpha}}{\Gamma(2\alpha + 1)} + b_3(x) \frac{t^{3\alpha}}{\Gamma(3\alpha + 1)}, \tag{12.23}$$

Putting Eq. (12.23) in the third RF of Eq. (12.12) $\operatorname{Re} s_{\psi,3}(x,t) = D_t^\alpha \psi_3(x,t) - \frac{1}{2}x^2(\psi_3(x,t))_{xx}$, we have

$$\operatorname{Res}_{\psi,3}(x,t) = x^2 + x^2\frac{t^\alpha}{\Gamma(\alpha+1)} + b_3(x)\frac{t^{2\alpha}}{\Gamma(2\alpha+1)}$$
$$-\left(\frac{1}{2}x^2\right)\left(2 + \frac{2t^\alpha}{\Gamma(\alpha+1)} + \frac{2t^{2\alpha}}{\Gamma(2\alpha+1)} + b_3''(x)\frac{t^{3\alpha}}{\Gamma(3\alpha+1)}\right) \tag{12.24}$$

Using Eq. (12.10) for $i = 2$, that is $D_t^{2\alpha} \operatorname{Re} s_\psi(x,0) = D_t^{2\alpha} \operatorname{Re} s_{\psi,3}(x,0) = 0$, it follows that

$$b_3(x) = x^2. \tag{12.25}$$

Continuing this way, one may find the values of $b_4(x)$, $b_5(x)$, ...
So, the solution of the fractional heat-like Eq. (12.12) may be obtained as:

$$\psi(x,t) = x^2 + x^2\frac{t^\alpha}{\Gamma(1+\alpha)} + x^2\frac{t^{2\alpha}}{\Gamma(1+2\alpha)} + x^2\frac{t^{3\alpha}}{\Gamma(1+3\alpha)} + \cdots,$$
$$= x^2\left(1 + \frac{t^\alpha}{\Gamma(1+\alpha)} + \frac{t^{2\alpha}}{\Gamma(1+2\alpha)} + \frac{t^{3\alpha}}{\Gamma(1+3\alpha)} + \cdots\right) = x^2 E(t^\alpha), \tag{12.26}$$

where $E(t^\alpha)$ is called the Mittag-Leffler function, which is given in Chapter 1.

In particular, at $\alpha = 1$, Eq. (12.26) reduces to $\psi(x,t) = x^2 e^t$ which is same as the solution of Sadighi et al. (2008). The present solution is compared with the exact solution with an increasing number of terms of solution at $\alpha = 1$, which is given in Figure 12.1. Figures 12.2–12.5 exhibit the fourth-order approximate solution plots of Example 12.1 for various α values.

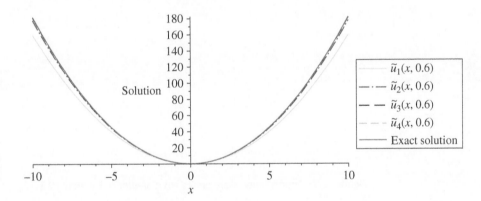

Figure 12.1 Comparison plot of the present solution with the exact solution at $\alpha = 1$ and $t = 0.6$.

Figure 12.2 Fourth-order approximate solution plot of Example 12.1 at $\alpha = 1$.

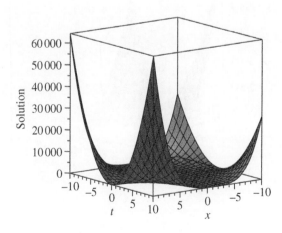

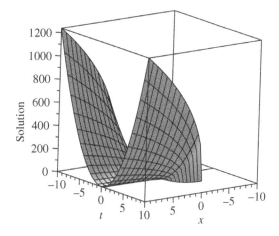

Figure 12.3 Fourth-order approximate solution plot of Example 12.1 at $\alpha = 0.15$.

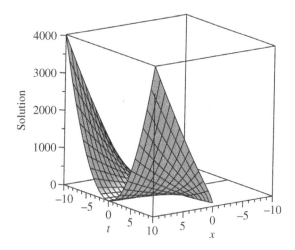

Figure 12.4 Fourth-order approximate solution plot of Example 12.1 at $\alpha = 0.35$.

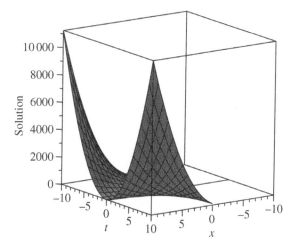

Figure 12.5 Fourth-order approximate solution plot of Example 12.1 at $\alpha = 0.55$.

Example 12.2 Consider the following nonlinear advection equation (Wazwaz 2007):

$$\frac{\partial^\alpha \psi}{\partial t^\alpha} + \psi \frac{\partial \psi}{\partial x} = 0, \quad 0 < \alpha \le 1, \tag{12.27}$$

with IC:

$$\psi(x, 0) = -x. \tag{12.28}$$

Solution

Here, $\psi_0(x, t) = -x$ and the infinite series solution of Eq. (12.27) is written as:

$$\psi(x, t) = -x + \sum_{k=1}^{\infty} b_k(x) \frac{t^{\alpha k}}{\Gamma(\alpha k + 1)}. \tag{12.29}$$

The mth truncated series solution of $\psi(x, t)$ becomes

$$\psi_m(x, t) = -x + \sum_{k=1}^{m} b_k(x) \frac{t^{\alpha k}}{\Gamma(\alpha k + 1)}, \quad m = 1, 2, 3, \dots \tag{12.30}$$

For $m = 1$, first RPS solution for Eq. (12.7) may be written as:

$$\psi_1(x, t) = -x + b_1(x) \frac{t^\alpha}{\Gamma(\alpha + 1)}. \tag{12.31}$$

Plugging Eq. (12.31) in the first RF of Eq. (12.27) $\text{Res}_{\psi,1}(x, t) = \frac{\partial^\alpha \psi_1}{\partial t^\alpha} + \psi_1 \frac{\partial \psi_1}{\partial x}$, it gives

$$\text{Res}_{\psi,1}(x, t) = b_1(x) + \left(-x + b_1(x) \frac{t^\alpha}{\Gamma(\alpha + 1)} \right) \left(-1 + b_1'(x) \frac{t^\alpha}{\Gamma(\alpha + 1)} \right). \tag{12.32}$$

Applying (iii) of Eq. (12.10) for $i = 0$, that is $\text{Res}_\psi(x, 0) = \text{Res}_{\psi,\,1}(x, 0) = 0$, we obtain

$$\text{Res}_{\psi,1}(x, 0) = b_1(x) + x = 0.$$
$$b_1(x) = -x. \tag{12.33}$$

For $m = 2$, second RPS solution for Eq. (12.7) may be written as:

$$\psi_2(x, t) = -x + -x \frac{t^\alpha}{\Gamma(\alpha + 1)} + b_2(x) \frac{t^{2\alpha}}{\Gamma(2\alpha + 1)}. \tag{12.34}$$

To find the value of $b_2(x)$, substituting Eq. (12.34) in the second RF of Eq. (12.27) $\text{Res}_{\psi,2}(x, t) = \frac{\partial^\alpha \psi_2}{\partial t^\alpha} + \psi_2 \frac{\partial \psi_2}{\partial x}$, we have

$$\text{Res}_{\psi,2}(x, t) = \left(-x + b_2(x) \frac{t^\alpha}{\Gamma(\alpha + 1)} \right) + \left(-x + -x \frac{t^\alpha}{\Gamma(\alpha + 1)} + b_2(x) \frac{t^{2\alpha}}{\Gamma(2\alpha + 1)} \right)$$
$$\left(-1 - \frac{t^\alpha}{\Gamma(\alpha + 1)} + b_2'(x) \frac{t^{2\alpha}}{\Gamma(2\alpha + 1)} \right) \tag{12.35}$$

Using (iii) of Eq. (12.10) for $i = 1$, that is, $D_t^\alpha \text{Res}_\psi(x, 0) = D_t^\alpha \text{Res}_{\psi,2}(x, 0) = 0$, we get

$$D_t^\alpha \text{Res}_{\psi,2}(x, 0) = \left. \begin{matrix} b_2(x) + \left(-x + b_2(x) \frac{t^\alpha}{\Gamma(\alpha + 1)} \right) \left(-1 - \frac{t^\alpha}{\Gamma(\alpha + 1)} + b_2'(x) \frac{t^{2\alpha}}{\Gamma(2\alpha + 1)} \right) + \\ \left(-x + -x \frac{t^\alpha}{\Gamma(\alpha + 1)} + b_2(x) \frac{t^{2\alpha}}{\Gamma(2\alpha + 1)} \right) \left(-1 + b_2'(x) \frac{t^\alpha}{\Gamma(\alpha + 1)} \right) \end{matrix} \right|_{t=0} = 0,$$
$$b_2(x) = -2x. \tag{12.36}$$

Similarly, for $m = 3$, third RPS solution for the Eq. (12.7) is expressed as:

$$\psi_3(x, t) = -x - x \frac{t^\alpha}{\Gamma(\alpha + 1)} - 2x \frac{t^{2\alpha}}{\Gamma(2\alpha + 1)} + b_3(x) \frac{t^{3\alpha}}{\Gamma(3\alpha + 1)}, \tag{12.37}$$

Putting Eq. (12.37) in the third RF of Eq. (12.27), we have

$$\text{Res}_{\psi,3}(x,t) = -x - 2x\frac{t^\alpha}{\Gamma(\alpha+1)} + b_3(x)\frac{t^{2\alpha}}{\Gamma(2\alpha+1)} + \left(-x - x\frac{t^\alpha}{\Gamma(\alpha+1)} - 2x\frac{t^{2\alpha}}{\Gamma(2\alpha+1)} + b_3(x)\frac{t^{3\alpha}}{\Gamma(3\alpha+1)}\right)$$

$$\left(-1 - \frac{t^\alpha}{\Gamma(\alpha+1)} - 2\frac{t^{2\alpha}}{\Gamma(2\alpha+1)} + b_3'(x)\frac{t^{3\alpha}}{\Gamma(3\alpha+1)}\right).$$

$$= -x - 2x\frac{t^\alpha}{\Gamma(\alpha+1)} + b_3(x)\frac{t^{2\alpha}}{\Gamma(2\alpha+1)} + x + 2x\frac{t^\alpha}{\Gamma(\alpha+1)} + \left(\frac{4x}{\Gamma(2\alpha+1)} + \frac{x}{(\Gamma(\alpha+1))^2}\right)t^{2\alpha}$$

$$+ \left(-b_3'(x)\frac{x}{\Gamma(3\alpha+1)} + \frac{4x}{\Gamma(2\alpha+1)\Gamma(\alpha+1)} - \frac{b_3(x)}{\Gamma(3\alpha+1)}\right)t^{3\alpha} + \left(\frac{-xb_3'(x) - b_3(x)}{\Gamma(\alpha+1)\Gamma(3\alpha+1)} + \frac{4x}{(\Gamma(2\alpha+1))^2}\right)t^{4\alpha}$$

$$+ \left(\frac{-2b_3(x) - 2xb_3'(x)}{\Gamma(2\alpha+1)\Gamma(3\alpha+1)}\right)t^{5\alpha} + b_3(x)b_3'(x)\frac{t^{6\alpha}}{(\Gamma(3\alpha+1))^2}.$$

Using Eq. (12.9) for $i = 2$ that is $D_t^{2\alpha}\,\text{Re}\,s_\psi(x,0) = D_t^{2\alpha}\,\text{Re}\,s_{\psi,3}(x,0) = 0$, it follows that

$$D_t^{2\alpha}\,\text{Re}\,s_{\psi,3}(x,0) = \left. \begin{array}{l} b_3(x) + \Gamma(2\alpha+1)\left(\dfrac{4x}{\Gamma(2\alpha+1)} + \dfrac{x}{(\Gamma(\alpha+1))^2}\right) + \\[2mm] \left(-b_3'(x)\dfrac{x}{\Gamma(3\alpha+1)} + \dfrac{4x}{\Gamma(2\alpha+1)\Gamma(\alpha+1)} - \dfrac{b_3(x)}{\Gamma(3\alpha+1)}\right)\dfrac{\Gamma(3\alpha+1)t^\alpha}{\Gamma(\alpha+1)} + \\[2mm] \left(\dfrac{-xb_3'(x) - b_3(x)}{\Gamma(\alpha+1)\Gamma(3\alpha+1)} + \dfrac{4x}{(\Gamma(2\alpha+1))^2}\right)\dfrac{\Gamma(4\alpha+1)t^{2\alpha}}{\Gamma(2\alpha+1)} \\[2mm] + \left(\dfrac{-2b_3(x) - 2xb_3'(x)}{\Gamma(2\alpha+1)\Gamma(3\alpha+1)}\right)\dfrac{\Gamma(5\alpha+1)t^{3\alpha}}{\Gamma(3\alpha+1)} + \dfrac{b_3(x)b_3'(x)}{(\Gamma(3\alpha+1))^2}\dfrac{\Gamma(6\alpha+1)t^{4\alpha}}{\Gamma(4\alpha+1)} \end{array} \right|_{t=0} = 0,$$

$$b_3(x) = -4x - x\frac{\Gamma(2\alpha+1)}{(\Gamma(\alpha+1))^2}. \tag{12.38}$$

Continuing the process likewise, one may find the values of $b_4(x)$, $b_5(x)$, ...
So, the solution to Eq. (12.27) may be obtained as:

$$\psi(x,t) = -x - x\frac{t^\alpha}{\Gamma(1+\alpha)} - 2x\frac{t^{2\alpha}}{\Gamma(1+2\alpha)} - 4x\frac{t^{3\alpha}}{\Gamma(1+3\alpha)} - \frac{x\Gamma(1+2\alpha)}{(\Gamma(1+\alpha))^2}\frac{t^{3\alpha}}{\Gamma(1+3\alpha)} - \cdots \tag{12.39}$$

Particularly at $\alpha = 1$, Eq. (12.39) reduces to a solution $\psi(x,t) = \dfrac{x}{t-1}$, which is same as the solution of Wazwaz (2007). The comparison of the present solution with the exact solution at $\alpha = 1$ is shown in Figure 12.6. Plots of the third-order approximate solutions of Example 12.2 for a wide range of alpha values are presented in Figures 12.7–12.10.

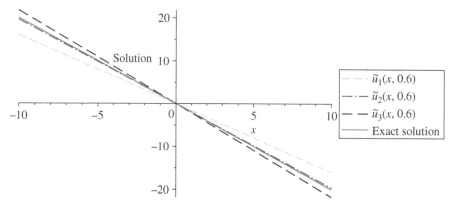

Figure 12.6 Comparison plot of the present solution with the exact solution at $\alpha = 1$ and $t = 0.6$.

Figure 12.7 Third-order approximate solution plot of Example 12.2 at $\alpha = 1$.

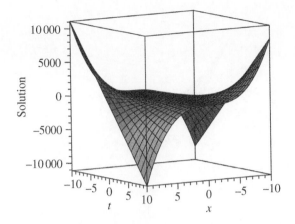

Figure 12.8 Third-order approximate solution plot of Example 12.2 at $\alpha = 0.15$.

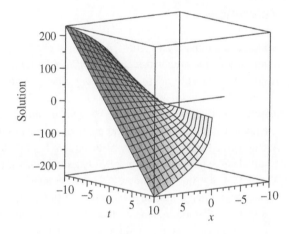

Figure 12.9 Third-order approximate solution plot of Example 12.2 at $\alpha = 0.35$.

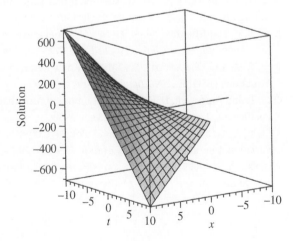

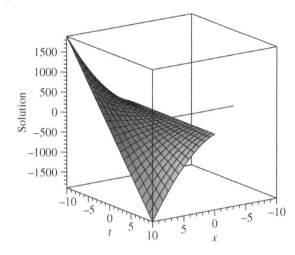

Figure 12.10 Third-order approximate solution plot of Example 12.2 at $\alpha = 0.55$.

References

Arafa, A. and Elmahdy, G. (2018). Application of residual power series method to fractional coupled physical equations arising in fluids flow. *International Journal of Differential Equations* 2018, 7692849: 10.

Arqub, O.A. (2013). Series solution of fuzzy differential equations under strongly generalized differentiability. *Journal of Advanced Research in Applied Mathematics* 5 (1): 31–52.

Arqub, O.A., El-Ajou, A., Bataineh, A.S. and Hashim, I. (2013). A representation of the exact solution of generalized Lane-Emden equations using a new analytical method. *Abstract and Applied Analysis* 2013378593: 10.

Arqub, O.A., Abo-Hammour, Z., Al-Badarneh, R. and Momani, S. (2013). A reliable analytical method for solving higher-order initial value problems. *Discrete Dynamics in Nature and Society* 2013, 673829: 12.

Arqub, O.A., El-Ajou, A., Zhour, Z.A., and Momani, S. (2014). Multiple solutions of nonlinear boundary value problems of fractional order: a new analytic iterative technique. *Entropy* 16 (1): 471–493.

Arqub, O.A., El-Ajou, A., and Momani, S. (2015). Constructing and predicting solitary pattern solutions for nonlinear time fractional dispersive partial differential equations. *Journal of Computational Physics* 293: 385–399.

El-Ajou, A., Arqub, O.A., Momani, S. et al. (2015). A novel expansion iterative method for solving linear partial differential equations of fractional order. *Applied Mathematics and Computation* 257: 119–133.

Hira, T. and Ghazala, A. (2017). Residual power series method for solving time-space-fractional Benney-Lin equation arising in falling film problems. *Journal of Applied Mathematics and Computing* 55: 683–708.

Jena, R.M. and Chakraverty, S. (2019a). Residual power series method for solving time-fractional model of vibration equation of large membranes. *Journal of Applied and Computational Mechanics* 5 (4): 603–615.

Jena, R.M. and Chakraverty, S. (2019b). A new iterative method based solution for fractional Black-Scholes Option Pricing Equations (BSOPE). *SN Applied Sciences* 1 (1): 95.

Özis, T. and Agırseven, D. (2008). He's homotopy perturbation method for solving heat-like and wave-like equations with variable coefficients. *Physics Letters A* 372: 5944–5950.

Sadighi, A., Ganji, D.D., Gorji, M., and Tolou, N. (2008). Numerical simulation of heat-like models with variable coefficients by the variational iteration method. *Journal of Physics: Conference Series* 96: 012083.

Wazwaz, A.M. (2007). A comparison between the variational iteration method and adomian decomposition method. *Journal of Computational and Applied Mathematics* 207: 129–136.

13

Homotopy Analysis Method

13.1 Introduction

In Chapter 12, we have discussed the residual power series method (RPSM) based on the generalized Taylor's series formula and the residual error function for solving linear and nonlinear ordinary/partial/fractional differential equations. In this chapter, we will address the homotopy analysis method (HAM). Various computationally efficient methods have been developed recently to solve fractional differential equations. However, neither perturbation nor non-perturbation approaches can provide an easy way to conveniently modify and monitor the convergence region and rate of the series. Liao (1995, 2005b) suggested the homotopy analysis approach in 1992. Based on the homotopy of topology, the validity of the HAM is independent of whether there exist small parameters in the considered equation (Liao 2003, 2004, 2009) or not. Hence, HAM can overcome the previous restrictions and limitations of the perturbation techniques to allow us to examine highly nonlinear problems (Liao 2005a). This approach includes a certain auxiliary parameter $\hbar \neq 0$ and an auxiliary linear operator L, which gives us an easy way to monitor and adjust the rate of convergence of the series solution (Jena et al. 2019; Srivastava et al. 2020). Several researchers have successfully applied HAM to solve various types of nonlinear physical problems arising in science and engineering (Zhang et al. 2011; Sakar and Erdogan 2013).

13.2 Theory of Homotopy Analysis Method

To illustrate the idea of HAM, we consider the following fractional differential equation (Liao 1995, 2003, 2004, 2005a, 2005b, 2009; Jena et al. 2019; Srivastava et al. 2020) in general:

$$FD(u(x,t)) = 0, \quad (x,t) \in \Omega, \tag{13.1}$$

where FD is a fractional operator and $u(x,t)$ is the unknown function in the domain Ω. Generalizing the traditional homotopy method (Liao 1995, 2003, 2004, 2005a, 2005b, 2009), the zero-order deformation equation is constructed and can be written as:

$$(1-p)L[\phi(x,t;p) - u_0(x,t)] = p\hbar H(x,t)FD(\phi(x,t;p)), \tag{13.2}$$

subject to the initial conditions (ICs):

$$\phi^{(k)}(x,0;p) = g_k(x), k = 0,1,2,...,m-1, \tag{13.3}$$

where $p \in [0,1]$ is the embedding parameter, $\hbar \neq 0$ is a nonzero auxiliary parameter, $H(x,t) \neq 0$ is a nonzero auxiliary function, $u_0(x,t)$ is the IC of $u(x,t)$, $\phi(x,t;p)$ is the unknown function, and $L = D_t^\alpha(n-1 < \alpha \leq n)$ is an auxiliary linear operator with the property (Sakar and Erdogan 2013):

$$L[\phi(x,t)] = 0 \quad \text{when} \quad \phi(x,t) = 0. \tag{13.4}$$

Computational Fractional Dynamical Systems: Fractional Differential Equations and Applications, First Edition.
Snehashish Chakraverty, Rajarama Mohan Jena, and Subrat Kumar Jena.
© 2023 John Wiley & Sons, Inc. Published 2023 by John Wiley & Sons, Inc.

It may be noted that one has the freedom to choose auxiliary parameters such as $\hbar$ and L in the HAM. By substituting $p = 0$ and $p = 1$ in Eq. (13.2), we obtain

$$\phi(x,t;0) = u_0(x,t), \quad \text{and} \quad \phi(x,t;1) = u(x,t),$$

(13.5)

respectively. Thus, as p increases from 0 to 1, the solution $\phi(x,t;p)$ varies from $u_0(x,t)$ to the solution $u(x,t)$. Expanding $\phi(x, t; p)$ using Taylor's series with respect to p, we have

$$\phi(x,t;p) = u_0(x,t) + \sum_{m=1}^{\infty} u_m(x,t)p^m,$$

(13.6)

where

$$u_m(x,t) = \frac{1}{m!}\frac{\partial^m \phi(x,t;p)}{\partial p^m}\bigg|_{p=0}.$$

(13.7)

Equation (13.6) converges at $p = 1$ if the auxiliary linear operator, the initial guess, the auxiliary operator $\hbar$, and the auxiliary function are properly chosen, and then we get

$$u(x,t) = u_0(x,t) + \sum_{m=1}^{\infty} u_m(x,t).$$

(13.8)

Substituting $\hbar = -1$, Eq. (13.2) reduces to

$$(1-p)L[\phi(x,t;p) - u_0(x,t)] + pH(x,t)FD(\phi(x,t;p)) = 0.$$

(13.9)

The aforementioned type of equation may generally be obtained in HPM. So HPM is a particular case of HAM. Now, let us define the vector as:

$$\vec{u}_n = \{u_0(x,t), u_1(x,t), ..., u_n(x,t)\}.$$

(13.10)

Differentiating Eq. (13.2) m times with respect to p, putting $p = 0$ and then dividing them by m!, we obtain the mth-order deformation equation with the assumption $H(x, t) = 1$ as follows (Liao 2003):

$$L[u_m(x,t) - \chi_m u_{m-1}(x,t)] = \hbar R_m\left(u_{m-1}\vec{}(x,t)\right),$$

(13.11)

subject to the following ICs:

$$u_m^{(k)}(x,0) = 0, \quad k = 0,1,2,...,m-1,$$

(13.12)

where

$$R_m\left(u_{m-1}\vec{}(x,t)\right) = \frac{1}{(m-1)!}\frac{\partial^{m-1}FD(\phi(x,t;p))}{\partial p^{m-1}}\bigg|_{p=0},$$

(13.13)

and

$$\chi_m = \begin{cases} 0, & m \le 1 \\ 1, & m > 1 \end{cases}$$

(13.14)

Applying an integral operator J_t^α on both sides of Eq. (13.11), we find

$$J_t^\alpha D_t^\alpha[u_m(x,t) - \chi_m u_{m-1}(x,t)] = \hbar J_t^\alpha R_m\left(u_{m-1}\vec{}(x,t)\right)$$

$$u_m(x,t) = \chi_m u_{m-1}(x,t) + \hbar J_t^\alpha R_m\left(u_{m-1}\vec{}(x,t)\right).$$

(13.15)

The mth-order approximate solution is then written as

$$u(x,t) = \sum_{k=0}^{m-1} u_k(x,t).$$

(13.16)

13.3 Convergence Theorem of HAM

Theorem 13.1 (Sakar and Erdogan 2013)

As long as the series $u(x,t) = u_0(x,t) + \sum_{m=1}^{\infty} u_m(x,t)$ converges, where $u_m(x,t)$ is governed by Eq. (13.11) under the definitions Eqs. (13.13) and (13.14), it must be a solution of Eq. (13.1).

Proof: One may see reference (Liao 2003) for the proof of this theorem.

13.4 Test Examples

Here, we apply the present method to solve a time-fractional one-dimensional heat-like equation in Example 13.1 and fractional nonlinear advection equation in Example 13.2.

Example 13.1 Let us consider the one-dimensional heat-like model (Özis and Agırseven 2008; Sadighi et al. 2008) as:

$$D_t^\alpha u(x,t) = \frac{1}{2}x^2 u_{xx}(x,t), \quad 0 < x < 1, \quad t > 0, \quad 0 < \alpha \leq 1, \tag{13.17}$$

with the boundary conditions (BCs):

$$u(0,t) = 0, \quad u(1,t) = e^t, \tag{13.18}$$

and IC:

$$u(x,0) = x^2. \tag{13.19}$$

Solution

To solve Eqs. (13.17)–(13.19) using HAM, we chose the initial approximation as:

$$u_0(x,t) = u(x,0) = x^2. \tag{13.20}$$

Equation (13.17) suggests the nonlinear operator as:

$$FD[\phi(x,t;p)] = D_t^\alpha \phi(x,t;p) - \frac{1}{2}x^2 \phi_{xx}(x,t;p), \tag{13.21}$$

and the linear operator:

$$L[\phi(x,t;p)] = D_t^\alpha \phi(x,t;p), \tag{13.22}$$

with the property $L[c] = 0$ where c is the integration constant. Using the aforementioned definition with the assumption $H(x,t) = 1$, the zeroth-order deformation equation may be constructed as:

$$(1-p)L[\phi(x,t;p) - u_0(x,t)] = p\hbar FD(\phi(x,t;p)). \tag{13.23}$$

Obviously, at $p = 0$ and $p = 1$ in Eq. (13.23) gives

$$\phi(x,t;0) = u_0(x,t), \quad \text{and} \quad \phi(x,t;1) = u(x,t), \tag{13.24}$$

respectively. So, as p increases from 0 to 1, the solution $\phi(x,t;p)$ varies from the initial guess $u_0(x,t)$ to the solution $u(x,t)$. Now, the mth-order deformation equation can be written as:

$$L[u_m(x,t) - \chi_m u_{m-1}(x,t)] = \hbar R_m\left(\overrightarrow{u_{m-1}}(x,t)\right), \tag{13.25}$$

where

$$R_m\left(\overrightarrow{u_{m-1}}(x,t)\right) = D_t^\alpha u_{m-1}(x,t) - \frac{1}{2}x^2(u_{m-1}(x,t))_{xx}. \tag{13.26}$$

The solution of mth-order deformation Eq. (13.25) for $m \geq 1$ becomes

$$u_m(x,t) = \chi_m u_{m-1}(x,t) + \hbar J_t^\alpha R_m\left(\overrightarrow{u_{m-1}}(x,t)\right). \tag{13.27}$$

From Eqs. (13.20) and (13.27), we have

$$u_0(x,t) = x^2, \tag{13.28}$$

$$
\begin{aligned}
u_1(x,t) &= \chi_1 u_0(x,t) + \hbar J_t^\alpha R_1\left(\overrightarrow{u_0}(x,t)\right) = \hbar J_t^\alpha\left[D_t^\alpha u_0(x,t) - \frac{1}{2}x^2(u_0(x,t))_{xx}\right] \\
&= \hbar J_t^\alpha\left(0 - x^2\right) = \frac{-\hbar x^2 t^\alpha}{\Gamma(\alpha+1)},
\end{aligned} \tag{13.29}
$$

$$
\begin{aligned}
u_2(x,t) &= \chi_2 u_1(x,t) + \hbar J_t^\alpha R_2\left(\overrightarrow{u_1}(x,t)\right) = \frac{-\hbar x^2 t^\alpha}{\Gamma(\alpha+1)} + \hbar J_t^\alpha\left[D_t^\alpha u_1(x,t) - \frac{1}{2}x^2(u_1(x,t))_{xx}\right] \\
&= \frac{-\hbar x^2 t^\alpha}{\Gamma(\alpha+1)} + \hbar J_t^\alpha\left(-\hbar x^2 + \frac{\hbar x^2 t^\alpha}{\Gamma(\alpha+1)}\right) = \frac{-\hbar x^2 t^\alpha}{\Gamma(\alpha+1)} + \hbar\left(\frac{-\hbar x^2 t^\alpha}{\Gamma(\alpha+1)} + \frac{\hbar x^2 t^{2\alpha}}{\Gamma(2\alpha+1)}\right) \\
&= \frac{-x^2 t^\alpha}{\Gamma(\alpha+1)}\hbar(\hbar+1) + \frac{\hbar^2 x^2 t^{2\alpha}}{\Gamma(2\alpha+1)},
\end{aligned} \tag{13.30}
$$

$$u_3(x,t) = \chi_3 u_2(x,t) + \hbar J_t^\alpha R_3\left(\overrightarrow{u_2}(x,t)\right) = \frac{-x^2 t^\alpha}{\Gamma(\alpha+1)}\hbar(\hbar+1)^2 - \frac{\hbar^3 x^2 t^{3\alpha}}{\Gamma(3\alpha+1)}, \tag{13.31}$$

and so on. Therefore, the four-term approximate solution of Eq. (13.17) is given by:

$$u_{HAM}(x,t) = \sum_{i=0}^{3} u_i(x,t). \tag{13.32}$$

By substituting $\hbar = -1$, the solution of the fractional heat-like equation may be obtained as:

$$
\begin{aligned}
u(x,t) &= x^2 + x^2\frac{t^\alpha}{\Gamma(1+\alpha)} + x^2\frac{t^{2\alpha}}{\Gamma(1+2\alpha)} + x^2\frac{t^{3\alpha}}{\Gamma(1+3\alpha)} + \cdots, \\
&= x^2\left(1 + \frac{t^\alpha}{\Gamma(1+\alpha)} + \frac{t^{2\alpha}}{\Gamma(1+2\alpha)} + \frac{t^{3\alpha}}{\Gamma(1+3\alpha)} + \cdots\right) = x^2 E(t^\alpha),
\end{aligned} \tag{13.33}
$$

where $E(t^\alpha)$ is called the Mittag-Leffler function. In particular, at $\alpha = 1$, Eq. (13.33) reduces to $u(x,t) = x^2 e^t$, which is same as the solution of Sadighi et al. (2008). The third-order approximate solution plot is depicted in Figure 13.1 for $\hbar = -0.3, -0.6, -0.9, -1.0$ at $\alpha = 1$ and $x = 1$. In this figure, one may see the comparison of the present results with the exact solution. From the results presented in Figure 13.1, we observe that at $\hbar = -1.0$, the HAM solution is the same as the exact solution.

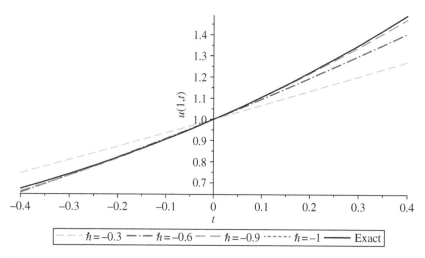

Figure 13.1 Solution of Eq. (13.17) for different values of $\hbar$.

Example 13.2 Consider the following nonlinear advection equation (Wazwaz 2007):

$$\frac{\partial^\alpha u}{\partial t^\alpha} + u\frac{\partial u}{\partial x} = 0, \quad 0 < \alpha \le 1, \tag{13.34}$$

with IC:

$$u(x,0) = -x. \tag{13.35}$$

Solution

Here, we chose the initial approximation as;

$$u_0(x,t) = u(x,0) = -x. \tag{13.36}$$

From Eq. (13.34), the nonlinear operator is written as:

$$FD[\phi(x,t;p)] = D_t^\alpha\phi(x,t;p) + \phi(x,t;p)\phi_x(x,t;p), \tag{13.37}$$

and the linear operator as:

$$L[\phi(x,t;p)] = D_t^\alpha\phi(x,t;p), \tag{13.38}$$

Using the aforementioned definition with an assumption $H(x,t) = 1$, the zeroth-order deformation equation can be constructed as:

$$(1-p)L[\phi(x,t;p) - u_0(x,t)] = p\hbar FD(\phi(x,t;p)). \tag{13.39}$$

The mth-order deformation equation is

$$L[u_m(x,t) - \chi_m u_{m-1}(x,t)] = \hbar R_m\left(\overrightarrow{u_{m-1}(x,t)}\right), \tag{13.40}$$

where

$$R_m\left(\overrightarrow{u_{m-1}(x,t)}\right) = D_t^\alpha u_{m-1}(x,t) + \sum_{k=0}^{m-1} u_k(x,t)(u_{m-1-k}(x,t))_x. \tag{13.41}$$

The solution of mth-order deformation Eq. (13.40) for $m \ge 1$ becomes

$$u_m(x,t) = \chi_m u_{m-1}(x,t) + \hbar J_t^\alpha R_m\left(\overrightarrow{u_{m-1}(x,t)}\right). \tag{13.42}$$

From Eqs. (13.36) and (13.42), we obtain

$$u_0(x,t) = -x, \tag{13.43}$$

$$u_1(x,t) = \chi_1 u_0(x,t) + \hbar J_t^\alpha R_1\left(\overrightarrow{u_0(x,t)}\right) = \hbar J_t^\alpha\left[D_t^\alpha u_0(x,t) + u_0(x,t)(u_0(x,t))_x\right]$$
$$= \hbar J_t^\alpha(0 + (-x)(-1)) = \frac{\hbar xt^\alpha}{\Gamma(\alpha+1)}, \tag{13.44}$$

$$u_2(x,t) = \chi_2 u_1(x,t) + \hbar J_t^\alpha R_2\left(\overrightarrow{u_1(x,t)}\right) = \frac{\hbar xt^\alpha}{\Gamma(\alpha+1)} + \hbar J_t^\alpha\begin{bmatrix} D_t^\alpha u_1(x,t) + u_0(x,t)(u_1(x,t))_x + \\ u_1(x,t)(u_0(x,t))_x \end{bmatrix}$$
$$= \frac{\hbar xt^\alpha}{\Gamma(\alpha+1)} + \hbar J_t^\alpha\left(\hbar x - \frac{\hbar xt^\alpha}{\Gamma(\alpha+1)} - \frac{\hbar xt^\alpha}{\Gamma(\alpha+1)}\right) = \frac{\hbar xt^\alpha}{\Gamma(\alpha+1)} + \hbar J_t^\alpha\left(\hbar x - \frac{2\hbar xt^\alpha}{\Gamma(\alpha+1)}\right)$$
$$= \frac{\hbar xt^\alpha}{\Gamma(\alpha+1)} + \hbar\left(\frac{\hbar xt^\alpha}{\Gamma(\alpha+1)} - \frac{2\hbar xt^{2\alpha}}{\Gamma(2\alpha+1)}\right) = \frac{xt^\alpha}{\Gamma(\alpha+1)}\left(\hbar^2 + \hbar\right) - \frac{2\hbar^2 xt^{2\alpha}}{\Gamma(2\alpha+1)}, \tag{13.45}$$

$$u_3(x,t) = \chi_3 u_2(x,t) + \hbar J_t^\alpha R_3\left(\overrightarrow{u_2(x,t)}\right) = \frac{xt^\alpha}{\Gamma(\alpha+1)}\left(\hbar^2+\hbar\right) - \frac{2\hbar^2 xt^{2\alpha}}{\Gamma(2\alpha+1)} + \hbar J_t^\alpha$$

$$\begin{bmatrix} D_t^\alpha u_2(x,t) + u_0(x,t)(u_2(x,t))_x + \\ u_1(x,t)(u_1(x,t))_x + u_2(x,t)(u_0(x,t))_x \end{bmatrix} = \frac{xt^\alpha}{\Gamma(\alpha+1)}\left(\hbar^2+\hbar\right) - \frac{2\hbar^2 xt^{2\alpha}}{\Gamma(2\alpha+1)} + \hbar J_t^\alpha$$

(13.46)

$$\left\{\frac{t^\alpha}{\Gamma(\alpha+1)}\left(-4x\hbar^2 - 2x\hbar\right) + x\left(\hbar^2+\hbar\right) + \frac{x\hbar^2 t^{2\alpha}}{(\Gamma(\alpha+1))^2} + \frac{4\hbar^2 xt^{2\alpha}}{\Gamma(2\alpha+1)}\right\} = \frac{xt^\alpha}{\Gamma(\alpha+1)}\left(2\hbar^2 + \hbar + \hbar^3\right) +$$

$$\frac{xt^{2\alpha}}{\Gamma(2\alpha+1)}\left(-4\hbar^3 - 4\hbar^2\right) + \frac{x\hbar^3\Gamma(2\alpha+1)t^{3\alpha}}{(\Gamma(\alpha+1))^2\Gamma(3\alpha+1)} + \frac{4\hbar^3 xt^{3\alpha}}{\Gamma(3\alpha+1)}$$

and so on. The fourth-order series solution of Eq. (13.34) by HAM can be written in the form:

$$u_{HAM}(x,t) = \sum_{i=0}^{3} u_i(x,t).$$

or, especially when $\hbar = -1$,

$$u_{HAM}(x,t) = -x - x\frac{t^\alpha}{\Gamma(1+\alpha)} - 2x\frac{t^{2\alpha}}{\Gamma(1+2\alpha)} - 4x\frac{t^{3\alpha}}{\Gamma(1+3\alpha)} - \frac{x\Gamma(1+2\alpha)}{(\Gamma(1+\alpha))^2}\frac{t^{3\alpha}}{\Gamma(1+3\alpha)} - \cdots$$

(13.47)

Particularly at $\alpha = 1$, Eq. (13.47) reduces to a solution $u(x,t) = \frac{x}{t-1}$, which is same as the solution of Wazwaz (2007). Figure 13.2 shows that the HAM solution converges to the exact solution at $\hbar = -1.0$ for $\alpha = 1$ and $x = 1$. It may be noted that it is not always true to achieve a better result at $\hbar = -1.0$. In this case, one may obtain a closed-form solution at $\hbar = -1.2$. It can be concluded from the aforementioed discussion that the proper values of the control parameter $\hbar$ should be selected to provide better results with less errors. It is worth noting that the HAM produces not only the approximate convergent series solution but also an exact solution depending on the considered problem with the proper $\hbar$ value.

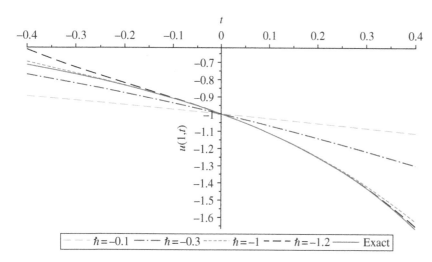

Figure 13.2 Solution of Eq. (13.34) for different values of $\hbar$. *Source:* Based on Wazwaz (2007).

References

Jena, R.M., Chakraverty, S., and Jena, S.K. (2019). Dynamic response analysis fractionally damped beams subjected to external loads using Homotopy Analysis Method (HAM). *Journal of Applied and Computational Mechanics* 5 (2): 355–366.

Liao, S.J. (1995). An approximate solution technique which does not depend upon small parameters: a special example. *International Journal of Non-Linear Mechanics* 30: 371–380.

Liao, S.J. (2003). *Beyond Perturbation: Introduction to the Homotopy Analysis Method*. Boca Raton: Chapman and Hall/CRC Press.

Liao, S.J. (2004). On the homotopy analysis method for nonlinear problems. *Applied Mathematics and Computation* 147: 499–513.

Liao, S.J. (2005a). An analytic approach to solve multiple solutions of a strongly nonlinear problem. *Applied Mathematics and Computation* 169: 854–865.

Liao, S.J. (2005b). Comparison between the homotopy analysis method and Homotopy perturbation method. *Applied Mathematics and Computation* 169: 1186–1194.

Liao, S.J. (2009). Notes on the homotopy analysis method: some definitions and theorems. *Communications in Nonlinear Science and Numerical Simulation* 14: 983–997.

Özis, T. and Agırseven, D. (2008). He's homotopy perturbation method for solving heat-like and wave-like equations with variable coefficients. *Physics Letters A* 372: 5944–5950.

Sadighi, A., Ganji, D.D., Gorji, M., and Tolou, N. (2008). Numerical simulation of heat-like models with variable coefficients by the variational iteration method. *Journal of Physics Conference Series* 96: 012083.

Sakar, M.G. and Erdogan, F. (2013). The homotopy analysis method for solving the time-fractional Fornberg–Whitham equation and comparison with Adomian's decomposition method. *Applied Mathematical Modelling* 37: 8876–8885.

Srivastava, H.M., Jena, R.M., Chakraverty, S., and Jena, S.K. (2020). Dynamic response analysis of fractionally-damped generalized Bagley–Torvik equation subject to external loads. *Russian Journal of Mathematical Physics* 27 (2): 254–268.

Wazwaz, A.M. (2007). A comparison between the variational iteration method and adomian decomposition method. *Journal of Computational and Applied Mathematics* 207: 129–136.

Zhang, X., Tang, B., and He, Y. (2011). Homotopy analysis method for higher-order fractional integro-differential equations. *Computers and Mathematics with Applications* 62: 3194–3203.

14

Homotopy Analysis Transform Method

14.1 Introduction

A Chinese mathematician, Liao, proposed the homotopy analysis method (HAM) (Liao 2003, 2004) by using the basic definition of differential geometry and topology. Homotopy analysis transform method (HATM) is a combination of HAM and the different transform methods. This method monitors and manipulates the series solution, which converges easily to the exact solution. As a result, several authors have recently studied different phenomena using HATM (Mohamed et al. 2014; Saad and Al-Shomrani 2016; Ziane and Cherif 2017; Maitama and Zhao 2020; Saratha et al. 2020). The HAM takes a longer time for computing and needs large computer memory. There has been a need to combine this approach with other transformation techniques to reduce the computing time and overcome other limitations. This method provides powerful features, including a nonlocal effect, a simple solution mechanism, a broad convergence region free from assumptions, discretization, and perturbation. The HATM solution involves auxiliary parameters, $\hbar$, which helps us to adjust and control the convergence of the series solution.

14.2 Transform Methods for the Caputo Sense Derivative

Definition 14.1 The Laplace transform of the Caputo fractional derivative is defined as (Baleanu and Jassim 2019):

$$L\left[D_t^\alpha u(x,t)\right] = s^\alpha L[u(x,t)] - \sum_{k=0}^{n-1} s^{(\alpha-k-1)} u^{(k)}(x,0), \quad n-1 < \alpha \le 1, n \in \mathbb{N}. \tag{14.1}$$

Definition 14.2 The Sumudu transform of the Caputo fractional derivative is defined as (Jena and Chakraverty 2019a):

$$S\left[D_t^\alpha u(x,t)\right] = s^{-\alpha} S[u(x,t)] - \sum_{k=0}^{n-1} s^{-\alpha+k} u^{(k)}(x,0), \quad n-1 < \alpha \le 1, n \in \mathbb{N}. \tag{14.2}$$

Definition 14.3 The Elzaki transform of the Caputo fractional derivative is defined as (Jena and Chakraverty 2019b):

$$E\left[D_t^\alpha u(x,t)\right] = \frac{E[u(x,t)]}{s^\alpha} - \sum_{k=0}^{n-1} s^{k-\alpha+2} u^{(k)}(x,0), \quad n-1 < \alpha \le n, \quad n \in \mathbb{N}. \tag{14.3}$$

Definition 14.4 The Aboodh transform of the Caputo fractional derivative is defined as (Aboodh 2013; Aboodh et al. 2017):

$$A\left[D_t^\alpha u(x,t)\right] = s^\alpha A[u(x,t)] - \sum_{k=0}^{n-1} s^{-k+\alpha-2} u^{(k)}(x,0), \quad n-1 < \alpha \le n, \quad n \in \mathbb{N}. \tag{14.4}$$

Computational Fractional Dynamical Systems: Fractional Differential Equations and Applications, First Edition.
Snehashish Chakraverty, Rajarama Mohan Jena, and Subrat Kumar Jena.
© 2023 John Wiley & Sons, Inc. Published 2023 by John Wiley & Sons, Inc.

Table 14.1 Transforms of some essential functions.

Functions	Laplace transform	Sumudu transform	Elzaki transform	Aboodh transform
Definitions	$L[f(t)] = f(s)$	$S[g(t)] = g(s)$	$E[h(t)] = h(s)$	$A[p(t)] = p(s)$
	$= \int_0^\infty e^{-st} f(t)dt$	$= \int_0^\infty e^{-t} f(st)dt$	$= s \int_0^\infty e^{\frac{-t}{s}} f(t)dt$	$= \frac{1}{s} \int_0^\infty e^{-st} f(t)dt$
		$= \frac{1}{s} f\left(\frac{1}{s}\right)$	$= s f\left(\frac{1}{s}\right)$	$= \frac{1}{s} f(s)$
1	$\dfrac{1}{s}$	1	s^2	$\dfrac{1}{s^2}$
t^α	$\dfrac{\Gamma(1+\alpha)}{s^{\alpha+1}}$	$s^\alpha \Gamma(1+\alpha)$	$s^{\alpha+2}\Gamma(1+\alpha)$	$\dfrac{\Gamma(1+\alpha)}{s^{\alpha+2}}$
e^{at}	$\dfrac{1}{s-a}$	$\dfrac{1}{1-as}$	$\dfrac{s^2}{1-as}$	$\dfrac{1}{s(1-s)}$
$\sin(at)$	$\dfrac{a}{s^2+a^2}$	$\dfrac{as}{1+a^2s^2}$	$\dfrac{as^3}{1+a^2s^2}$	$\dfrac{a}{s(s^2+a^2)}$
$\cos(at)$	$\dfrac{s}{s^2+a^2}$	$\dfrac{1}{1+s^2a^2}$	$\dfrac{s}{1+s^2a^2}$	$\dfrac{1}{s^2+a^2}$

Table 14.1 shows the transformations of several standard functions with respect to the four transformation methods described earlier, along with their definitions.

The following section addresses a systematic analysis of the homotopy analysis Laplace transform method (HALTM), homotopy analysis Sumudu transform method (HASTM), homotopy analysis Elzaki transform method (HAETM), and homotopy analysis Aboodh transform method (HAATM) methods.

14.3 Homotopy Analysis Laplace Transform Method (HALTM)

To illustrate the basic concept of HALTM, we consider the nonlinear nonhomogenous fractional partial differential equation (PDE) as follows (Saad and Al-Shomrani 2016):

$$D_t^\alpha u(x,t) + Ru(x,t) + Nu(x,t) = h(x,t), n-1 < \alpha \le n, \tag{14.5}$$

where D_t^α is the Caputo fractional derivative, R and N are the linear and nonlinear differential operators, and $h(x, t)$ is the source term. Applying Laplace transform on both sides of Eq. (14.5) and using Eq. (14.1), we have

$$L[u(x,t)] = \sum_{k=0}^{n-1} s^{-k-1} u^{(k)}(x,0) + s^{-\alpha} L[h(x,t) - Ru(x,t) - Nu(x,t)]. \tag{14.6}$$

Let us define a nonlinear operator as:

$$N[\phi(x,t;p)] = L[\phi(x,t;p)] - \sum_{k=0}^{n-1} s^{-k-1} \phi^{(k)}(x,t;p)(0^+) + s^{-\alpha} L\left[\begin{matrix} R\phi(x,t;p) + N\phi(x,t;p) \\ -h(x,t) \end{matrix}\right], \tag{14.7}$$

where $p \in [0, 1]$ is the embedding parameter, and $\phi(x, t; p)$ is the unknown function. Constructing a homotopy, we get

$$(1-p)L[\phi(x,t;p) - u_0(x,t)] = \hbar p H(x,t) N[\phi(x,t;p)], \tag{14.8}$$

where $H(x, t)$ represents the nonzero auxiliary function, $\hbar \neq 0$ is an auxiliary parameter, and $u_0(x, t)$ is the initial value of $u(x, t)$.

If $p = 0$ and $p = 1$, then we obtain

$$\varphi(x, t; 0) = u_0(x, t), \quad \text{and} \quad \varphi(x, t; 1) = u(x, t), \tag{14.9}$$

respectively. As p increases from 0 to 1 then $\varphi(x, t; p)$ varies from $u_0(x, t)$ to the solution of Eq. (14.5). Expanding $\varphi(x, t; p)$ in Taylor's series with respect to p, we get

$$\varphi(x, t; p) = u_0(x, t) + \sum_{m=1}^{\infty} u_m(x, t) p^m, \tag{14.10}$$

where

$$u_m(x, t) = \frac{1}{m!} \frac{\partial^m u(x, t; p)}{\partial p^m} \bigg|_{p=0}. \tag{14.11}$$

If R, $u_0(x, t)$, $H(x, t)$, and $\hbar$ are properly chosen, then Eq. (14.10) converges at $p = 1$. So we have

$$u(x, t) = u_0(x, t) + \sum_{m=1}^{\infty} u_m(x, t). \tag{14.12}$$

Let us define the vector:

$$\overrightarrow{u_m}(t) = \{u_0(x, t), u_1(x, t), ..., ...u_m(x, t)\}. \tag{14.13}$$

Differentiating Eq. (14.8) m-times with respect to p, setting $p = 0$ and then dividing by $m!$, we have the mth-order deformation equation (Liao 2003, 2004) as follows:

$$L[u_m(x, t) - \chi_m u_{m-1}(x, t)] = \hbar H(x, t) \Re_m \left(\overrightarrow{u_{m-1}}(x, t) \right), \tag{14.14}$$

By applying inverse Laplace transform on both sides of Eq. (14.14), we obtain

$$u_m(x, t) = \chi_m u_{m-1}(x, t) + L^{-1} \left[\hbar H(x, t) \Re_m \left(\overrightarrow{u_{m-1}}(x, t) \right) \right], \tag{14.15}$$

where

$$\Re_m \left(\overrightarrow{u_{m-1}}(x, t) \right) = \frac{1}{(m-1)!} \frac{\partial^{m-1} N(\varphi(x, t; p))}{\partial p^{m-1}} \bigg|_{p=0}, \tag{14.16}$$

and

$$\chi_m = \begin{cases} 0, & m \leq 1, \\ 1, & m > 1. \end{cases} \tag{14.17}$$

In this way, one can easily obtain $u_m(x, t)$ for $m \geq 1$, which is written as:

$$u(x, t) = \sum_{i=0}^{m} u_i(x, t). \tag{14.18}$$

when $m \to \infty$ then, we get an exact approximation of the original Eq. (14.5).

14.4 Homotopy Analysis Sumudu Transform Method (HASTM)

Taking Sumudu transform on both sides of Eq. (14.5) and using Eq. (14.2), we have

$$S[u(x, t)] = \sum_{k=0}^{n-1} s^k u^{(k)}(x, 0) + s^\alpha S[h(x, t) - Ru(x, t) - Nu(x, t)], \tag{14.19}$$

The nonlinear operator is

$$N[\varphi(x,t;p)] = S[\varphi(x,t;p)] - \sum_{k=0}^{n-1} s^k \varphi^{(k)}(x,t;p)(0^+) + s^\alpha S \begin{bmatrix} R\varphi(x,t;p) + N\varphi(x,t;p) \\ -h(x,t) \end{bmatrix}. \tag{14.20}$$

Constructing the homotopy, we get

$$(1-p)S[\varphi(x,t;p) - u_0(x,t)] = \hbar p H(x,t) N[\varphi(x,t;p)]. \tag{14.21}$$

Substituting $p = 0$ and $p = 1$ in Eq. (14.21), we obtain Eq. (14.9). Using Taylor's series expansion on $\varphi(x, t; p)$, one may get Eqs. (14.10) and (14.11). If R, $u_0(x, t)$, $H(x, t)$, and $\hbar$ are correctly chosen, then Eq. (14.10) converges at $p = 1$.

Let us define the vector Eq. (14.13). Differentiating Eq. (14.21) m-times with respect to p, setting $p = 0$ and then dividing by m!, we have the mth-order deformation equation as follows:

$$S[u_m(x,t) - \chi_m u_{m-1}(x,t)] = \hbar H(x,t) \Re_m \left(\overrightarrow{u_{m-1}}(x,t) \right), \tag{14.22}$$

By applying inverse Sumudu transform on both sides of Eq. (14.22), we obtain

$$u_m(x,t) = \chi_m u_{m-1}(x,t) + S^{-1} \left[\hbar H(x,t) \Re_m \left(\overrightarrow{u_{m-1}}(x,t) \right) \right], \tag{14.23}$$

with Eqs. (14.16) and (14.17).

14.5 Homotopy Analysis Elzaki Transform Method (HAETM)

Taking Elzaki transform on both sides of Eq. (14.5) and with the help of Eq. (14.3), we have

$$E[u(x,t)] = \sum_{k=0}^{n-1} s^{k+2} u^{(k)}(x,0) + s^\alpha E[h(x,t) - Ru(x,t) - Nu(x,t)], \tag{14.24}$$

Define the nonlinear operator as:

$$N[\varphi(x,t;p)] = E[\varphi(x,t;p)] - \sum_{k=0}^{n-1} s^{k+2} \varphi^{(k)}(x,t;p)(0^+) + s^\alpha E \begin{bmatrix} R\varphi(x,t;p) + N\varphi(x,t;p) \\ -h(x,t) \end{bmatrix}, \tag{14.25}$$

The homotopy may be constructed as:

$$(1-p)E[\varphi(x,t;p) - u_0(x,t)] = \hbar p H(x,t) N[\varphi(x,t;p)]. \tag{14.26}$$

Let us define the vector Eq. (14.13). Differentiating Eq. (14.26) m-times with respect to p, setting $p = 0$ and then dividing by m!, the mth-order deformation equation is obtained as follows:

$$E[u_m(x,t) - \chi_m u_{m-1}(x,t)] = \hbar H(x,t) \Re_m \left(\overrightarrow{u_{m-1}}(x,t) \right). \tag{14.27}$$

By applying inverse Elzaki transform on both sides of Eq. (14.27), we find

$$u_m(x,t) = \chi_m u_{m-1}(x,t) + E^{-1} \left[\hbar H(x,t) \Re_m \left(\overrightarrow{u_{m-1}}(x,t) \right) \right], \tag{14.28}$$

with Eqs. (14.16) and (14.17).

14.6 Homotopy Analysis Aboodh Transform Method (HAATM)

Applying Aboodh transform on both sides of Eq. (14.5) and using Eq. (14.4), we have

$$A[u(x,t)] = \sum_{k=0}^{n-1} s^{-k-2} u^{(k)}(x,0) + s^{-\alpha} A[h(x,t) - Ru(x,t) - Nu(x,t)], \tag{14.29}$$

Define a nonlinear operator as:

$$N[\varphi(x,t;p)] = A[\varphi(x,t;p)] - \sum_{k=0}^{n-1} s^{-k-2}\varphi^{(k)}(x,t;p)(0^+) + s^{-\alpha}A\begin{bmatrix} R\varphi(x,t;p) + N\varphi(x,t;p) \\ -h(x,t) \end{bmatrix}, \tag{14.30}$$

where $p \in [0, 1]$, and $\varphi(x, t; p)$ is the unknown function.

Constructing a homotopy, we get

$$(1-p)A[\varphi(x,t;p) - u_0(x,t)] = \hbar p H(x,t)N[\varphi(x,t;p)], \tag{14.31}$$

where $H(x, t)$ represent the nonzero auxiliary function, $\hbar \neq 0$ is an auxiliary parameter, and $u_0(x, t)$ is the initial value. If $p = 0$ and $p = 1$ then, we obtain Eq. (14.9). As p increases from 0 to 1, then $\varphi(x, t; p)$ varies from $u_0(x, t)$ to the solution of Eq. (14.5). Using Taylor's series expansion on $\varphi(x, t; p)$, one obtains Eq. (14.10) with Eq. (14.11). If $R, u_0(x, t), H(x, t)$, and $\hbar$ are properly chosen, then Eq. (14.10) converges at $p = 1$. So we get Eq. (14.12).

Let us define the vector Eq. (14.13). Differentiating Eq. (14.31) m-times with respect to p, setting $p = 0$ and then dividing by m!, we have the mth-order deformation equation as:

$$A[u_m(x,t) - \chi_m u_{m-1}(x,t)] = \hbar H(x,t)\mathfrak{R}_m\left(\overrightarrow{u_{m-1}}(x,t)\right), \tag{14.32}$$

Applying inverse Aboodh transform on both sides of Eq. (14.32), we obtain

$$u_m(x,t) = \chi_m u_{m-1}(x,t) + A^{-1}\left[\hbar H(x,t)\mathfrak{R}_m\left(\overrightarrow{u_{m-1}}(x,t)\right)\right], \tag{14.33}$$

with Eqs. (14.16) and (14.17).

Next, we solve two test problems to demonstrate the present methods.

14.7 Numerical Examples

14.7.1 Implementation of HALTM

Example 14.1 Let us consider the one-dimensional heat-like model (Özis and Agırseven 2008; Sadighi et al. 2008) as:

$$D_t^\alpha u(x,t) = \frac{1}{2}x^2 u_{xx}(x,t), \quad 0 < x < 1, \quad t > 0, \quad 0 < \alpha \leq 1, \tag{14.34}$$

with the boundary conditions:

$$u(0,t) = 0, u(1,t) = e^t, \tag{14.35}$$

and initial condition:

$$u(x,0) = x^2. \tag{14.36}$$

Solution

In order to solve Eqs. (14.34)–(14.36) using HALTM, we chose the initial approximation as:

$$u_0(x,t) = u(x,0) = x^2. \tag{14.37}$$

Applying Laplace transform on both sides of Eq. (14.34) and simplifying, we obtain

$$L[u(x,t)] = \frac{x^2}{s} + \frac{s^{-\alpha}}{2}L\left[x^2\frac{\partial^2 u(x,t)}{\partial x^2}\right]. \tag{14.38}$$

The nonlinear operator may be written as:

$$N[\varphi(x,t;p)] = L[\varphi(x,t;p)] - \frac{x^2}{s}(1-\chi_m) - \frac{s^{-\alpha}}{2}L\left[x^2\frac{\partial^2 \varphi(x,t;p)}{\partial x^2}\right]. \tag{14.39}$$

Now, the *m*th-order deformation equation with the assumption $H(x, t) = 1$ can be written as:

$$L[u_m(x,t) - \chi_m u_{m-1}(x,t)] = \hbar R_m\left(\overrightarrow{u_{m-1}}(x,t)\right), \tag{14.40}$$

where

$$R_m\left(\overrightarrow{u_{m-1}}(x,t)\right) = L[u_{m-1}(x,t)] - \frac{x^2}{s}(1 - \chi_m) - \frac{s^{-\alpha}}{2}L\left[x^2\frac{\partial^2 u_{m-1}(x,t)}{\partial x^2}\right]. \tag{14.41}$$

Applying inverse Laplace transform in Eq. (14.40), we have

$$u_m(x,t) = \chi_m u_{m-1}(x,t) + L^{-1}\left\{\hbar\Re_m\left(\overrightarrow{u_{m-1}}(x,t)\right)\right\}. \tag{14.42}$$

From Eqs. (14.37) and (14.42), we have

$$u_0(x,t) = x^2, \tag{14.43}$$

$$u_1(x,t) = \chi_1 u_0(x,t) + L^{-1}\left[\hbar R_1\left(\overrightarrow{u_0}(x,t)\right)\right] = \frac{-\hbar x^2 t^\alpha}{\Gamma(\alpha+1)}, \tag{14.44}$$

$$\begin{aligned}
u_2(x,t) &= \chi_2 u_1(x,t) + L^{-1}\left[\hbar R_2\left(\overrightarrow{u_1}(x,t)\right)\right] = \frac{-\hbar x^2 t^\alpha}{\Gamma(\alpha+1)} + \hbar\left(\frac{-\hbar x^2 t^\alpha}{\Gamma(\alpha+1)} + \frac{\hbar x^2 t^{2\alpha}}{\Gamma(2\alpha+1)}\right) \\
&= \frac{-x^2 t^\alpha}{\Gamma(\alpha+1)}\hbar(\hbar+1) + \frac{\hbar^2 x^2 t^{2\alpha}}{\Gamma(2\alpha+1)},
\end{aligned} \tag{14.45}$$

$$\begin{aligned}
u_3(x,t) &= \chi_3 u_2(x,t) + L^{-1}\left[\hbar R_3\left(\overrightarrow{u_2}(x,t)\right)\right] = \frac{-x^2 t^\alpha}{\Gamma(\alpha+1)}\hbar(\hbar+1)^2 + \frac{x^2 t^{2\alpha}}{\Gamma(1+2\alpha)}2\hbar^2(1+\hbar) \\
&\quad - \frac{\hbar^3 x^2 t^{3\alpha}}{\Gamma(3\alpha+1)}
\end{aligned} \tag{14.46}$$

and so on. Therefore, the fourth-order approximate solution of Eq. (14.34) is given by:

$$u_{HALTM}(x,t;\hbar) = u_0(x,t;\hbar) + \sum_{i=1}^{3} u_i(x,t;\hbar). \tag{14.47}$$

By substituting $\hbar = -1$, the solution of the fractional heat-like equation may be obtained as:

$$\begin{aligned}
u(x,t) &= x^2 + x^2\frac{t^\alpha}{\Gamma(1+\alpha)} + x^2\frac{t^{2\alpha}}{\Gamma(1+2\alpha)} + x^2\frac{t^{3\alpha}}{\Gamma(1+3\alpha)} + \cdots, \\
&= x^2\left(1 + \frac{t^\alpha}{\Gamma(1+\alpha)} + \frac{t^{2\alpha}}{\Gamma(1+2\alpha)} + \frac{t^{3\alpha}}{\Gamma(1+3\alpha)} + \cdots\right) = x^2 E(t^\alpha),
\end{aligned} \tag{14.48}$$

where $E(t^\alpha)$ is called the Mittag-Leffler function. In particular, at $\alpha = 1$, Eq. (14.48) reduces to $u(x, t) = x^2 e^t$, which is same as the solution of Sadighi et al. (2008).

Example 14.2   Consider the following nonlinear advection equation (Wazwaz 2007):

$$\frac{\partial^\alpha u}{\partial t^\alpha} + u\frac{\partial u}{\partial x} = 0, \quad 0 < \alpha \le 1, \tag{14.49}$$

with initial condition:

$$u(x,0) = -x. \tag{14.50}$$

Solution

Here, we chose the initial approximation as:

$$u_0(x,t) = u(x,0) = -x. \tag{14.51}$$

Applying Laplace transform on both sides of Eq. (14.49) and simplifying, we obtain

$$L[u(x,t)] = \frac{-x}{s} - \frac{1}{s^\alpha} L\left[u\frac{\partial u(x,t)}{\partial x}\right]. \tag{14.52}$$

From Eq. (14.49), the nonlinear operator is written as:

$$N[\phi(x,t;p)] = L[\phi(x,t;p)] + \frac{x}{s}(1-\chi_m) + \frac{1}{s^\alpha}L\left[\phi(x,t;p)\frac{\partial\phi(x,t;p)}{\partial x}\right]. \tag{14.53}$$

The mth-order deformation equation when $H(x,t) = 1$ is

$$L[u_m(x,t) - \chi_m u_{m-1}(x,t)] = \hbar R_m\left(\overrightarrow{u_{m-1}}(x,t)\right), \tag{14.54}$$

where

$$R_m\left(\overrightarrow{u_{m-1}}(x,t)\right) = L[u_{m-1}(x,t)] + \frac{x}{s}(1-\chi_m) + \frac{1}{s^\alpha}L\left[\sum_{k=0}^{m-1} u_k(x,t)(u_{m-1-k}(x,t))_x\right]. \tag{14.55}$$

Taking inverse Laplace transform on both sides of Eq. (14.54), we get

$$u_m(x,t) = \chi_m u_{m-1}(x,t) + L^{-1}\left[\hbar R_m\left(\overrightarrow{u_{m-1}}(x,t)\right)\right]. \tag{14.56}$$

Now, from Eq. (14.56) for $m \geq 1$, we find

$$u_1(x,t) = \chi_1 u_0(x,t) + L^{-1}\left[\hbar R_1\left(\overrightarrow{u_0}(x,t)\right)\right] = \frac{\hbar x t^\alpha}{\Gamma(\alpha+1)}, \tag{14.57}$$

$$\begin{aligned} u_2(x,t) &= \chi_2 u_1(x,t) + L^{-1}\left[\hbar R_2\left(\overrightarrow{u_1}(x,t)\right)\right] = \frac{\hbar x t^\alpha}{\Gamma(\alpha+1)} + \hbar\left(\frac{\hbar x t^\alpha}{\Gamma(\alpha+1)} - \frac{2\hbar x t^{2\alpha}}{\Gamma(2\alpha+1)}\right) \\ &= \frac{x t^\alpha}{\Gamma(\alpha+1)}\hbar(\hbar+1) - \frac{2\hbar^2 x t^{2\alpha}}{\Gamma(2\alpha+1)}, \end{aligned} \tag{14.58}$$

$$\begin{aligned} u_3(x,t) &= \chi_3 u_2(x,t) + L^{-1}\left[\hbar R_3\left(\overrightarrow{u_2}(x,t)\right)\right] = \frac{x t^\alpha}{\Gamma(\alpha+1)}\hbar(\hbar+1)^2 - \frac{4x t^{2\alpha}}{\Gamma(2\alpha+1)}\hbar^2(\hbar+1) \\ &\quad + \frac{x\hbar^3\Gamma(2\alpha+1)t^{3\alpha}}{(\Gamma(\alpha+1))^2\Gamma(3\alpha+1)} + \frac{4\hbar^3 x t^{3\alpha}}{\Gamma(3\alpha+1)}, \end{aligned} \tag{14.59}$$

and so on. The fourth-order series solution of Eq. (14.49) by HALTM can be written in the form:

$$u_{HALTM}(x,t;\hbar) = u_0(x,t;\hbar) + \sum_{i=1}^{3} u_i(x,t;\hbar).$$

or, especially when $\hbar = -1$,

$$u_{HALTM}(x,t) = -x - x\frac{t^\alpha}{\Gamma(1+\alpha)} - 2x\frac{t^{2\alpha}}{\Gamma(1+2\alpha)} - 4x\frac{t^{3\alpha}}{\Gamma(1+3\alpha)} - \frac{x\Gamma(1+2\alpha)}{(\Gamma(1+\alpha))^2}\frac{t^{3\alpha}}{\Gamma(1+3\alpha)} - \cdots. \tag{14.60}$$

Particularly at $\alpha = 1$, Eq. (14.60) reduces to a solution $u(x,t) = \dfrac{x}{t-1}$, which is same as the solution of Wazwaz (2007).

14.7.2 Implementation of HASTM

In order to solve Eqs. (14.34)–(14.36), let us choose the initial approximation as Eq. (14.37). Applying Sumudu transform on both sides of Eq. (14.35) and substituting the initial condition, we obtain

$$S[u(x,t)] = x^2 + \frac{s^\alpha}{2}S\left[x^2\frac{\partial^2 u(x,t)}{\partial x^2}\right]. \tag{14.61}$$

The nonlinear operator is

$$N[\varphi(x,t;p)] = S[\varphi(x,t;p)] - x^2(1-\chi_m) - \frac{s^\alpha}{2} S\left[x^2 \frac{\partial^2 \varphi(x,t;p)}{\partial x^2}\right]. \tag{14.62}$$

Now, the mth-order deformation equation when $H(x,t) = 1$ can be written as:

$$S[u_m(x,t) - \chi_m u_{m-1}(x,t)] = \hbar R_m\left(\overrightarrow{u_{m-1}}(x,t)\right), \tag{14.63}$$

where

$$R_m\left(\overrightarrow{u_{m-1}}(x,t)\right) = S[u_{m-1}(x,t)] - x^2(1-\chi_m) - \frac{s^\alpha}{2} S\left[x^2 \frac{\partial^2 u_{m-1}(x,t)}{\partial x^2}\right]. \tag{14.64}$$

Applying inverse Sumudu transform on both sides of Eq. (14.63), it gives

$$u_m(x,t) = \chi_m u_{m-1}(x,t) + S^{-1}\left\{\hbar \mathfrak{R}_m\left(\overrightarrow{u_{m-1}}(x,t)\right)\right\}. \tag{14.65}$$

From Eqs. (14.37) and (14.65), we have

$$u_0(x,t) = x^2, \tag{14.66}$$

$$u_1(x,t) = \chi_1 u_0(x,t) + S^{-1}\left[\hbar R_1\left(\overrightarrow{u_0}(x,t)\right)\right] = \frac{-\hbar x^2 t^\alpha}{\Gamma(\alpha+1)}, \tag{14.67}$$

$$\begin{aligned} u_2(x,t) &= \chi_2 u_1(x,t) + S^{-1}\left[\hbar R_2\left(\overrightarrow{u_1}(x,t)\right)\right] = \frac{-\hbar x^2 t^\alpha}{\Gamma(\alpha+1)} + \hbar\left(\frac{-\hbar x^2 t^\alpha}{\Gamma(\alpha+1)} + \frac{\hbar x^2 t^{2\alpha}}{\Gamma(2\alpha+1)}\right) \\ &= \frac{-x^2 t^\alpha}{\Gamma(\alpha+1)} \hbar(\hbar+1) + \frac{\hbar^2 x^2 t^{2\alpha}}{\Gamma(2\alpha+1)}, \end{aligned} \tag{14.68}$$

$$\begin{aligned} u_3(x,t) &= \chi_3 u_2(x,t) + S^{-1}\left[\hbar R_3\left(\overrightarrow{u_2}(x,t)\right)\right] = \frac{-x^2 t^\alpha}{\Gamma(\alpha+1)} \hbar(\hbar+1)^2 + \frac{x^2 t^{2\alpha}}{\Gamma(1+2\alpha)} 2\hbar^2(1+\hbar) \\ &\quad - \frac{\hbar^3 x^2 t^{3\alpha}}{\Gamma(3\alpha+1)} \end{aligned} \tag{14.69}$$

and so on. Therefore, the fourth-order approximate solution of Eq. (14.34) is given by:

$$u_{HASTM}(x,t;\hbar) = u_0(x,t;\hbar) + \sum_{i=1}^{3} u_i(x,t;\hbar). \tag{14.70}$$

By substituting $\hbar = -1$, the solution of Eq. (14.34) may be written as:

$$u(x,t) = x^2\left(1 + \frac{t^\alpha}{\Gamma(1+\alpha)} + \frac{t^{2\alpha}}{\Gamma(1+2\alpha)} + \frac{t^{3\alpha}}{\Gamma(1+3\alpha)} + \cdots\right) = x^2 E(t^\alpha). \tag{14.71}$$

where $E(t^\alpha)$ is called the Mittag-Leffler function. In particular, at $\alpha = 1$, Eq. (14.71) reduces to $u(x,t) = x^2 e^t$, which is same as the solution of Sadighi et al. (Sadighi et al. 2008).

Further, taking Sumudu transform on both sides of Eq. (14.49) and simplifying, we have

$$S[u(x,t)] = -x - s^\alpha S\left[u\frac{\partial u(x,t)}{\partial x}\right]. \tag{14.72}$$

From Eq. (14.49), the nonlinear operator is written as:

$$N[\phi(x,t;p)] = S[\phi(x,t;p)] + x(1-\chi_m) + s^\alpha S\left[\phi(x,t;p)\frac{\partial \phi(x,t;p)}{\partial x}\right]. \tag{14.73}$$

The mth-order deformation equation with the assumption $H(x,t) = 1$ is Eq. (14.63),

where

$$R_m\left(\overrightarrow{u_{m-1}}(x,t)\right) = S[u_{m-1}(x,t)] + x(1-\chi_m) + s^\alpha S\left[\sum_{k=0}^{m-1} u_k(x,t)(u_{m-1-k}(x,t))_x\right]. \tag{14.74}$$

Taking inverse Sumudu transform on Eq. (14.63), it gives Eq. (14.65).

Now, from Eqs. (14.51) and (14.65) for $m \geq 1$, we find

$$u_0(x, t) = -x, \tag{14.75}$$

$$u_1(x, t) = \frac{\hbar x t^\alpha}{\Gamma(\alpha + 1)}, \tag{14.76}$$

$$u_2(x, t) = \frac{\hbar x t^\alpha}{\Gamma(\alpha + 1)} + \hbar\left(\frac{\hbar x t^\alpha}{\Gamma(\alpha + 1)} - \frac{2\hbar x t^{2\alpha}}{\Gamma(2\alpha + 1)}\right) = \frac{x t^\alpha}{\Gamma(\alpha + 1)}\hbar(\hbar + 1) - \frac{2\hbar^2 x t^{2\alpha}}{\Gamma(2\alpha + 1)}, \tag{14.77}$$

$$u_3(x, t) = \frac{x t^\alpha}{\Gamma(\alpha + 1)}\hbar(\hbar + 1)^2 - \frac{4 x t^{2\alpha}}{\Gamma(2\alpha + 1)}\hbar^2(\hbar + 1) + \frac{x\hbar^3 \Gamma(2\alpha + 1) t^{3\alpha}}{(\Gamma(\alpha + 1))^2 \Gamma(3\alpha + 1)} + \frac{4\hbar^3 x t^{3\alpha}}{\Gamma(3\alpha + 1)}, \tag{14.78}$$

and so on. The fourth-order series solution is written in the form:

$$u_{HASTM}(x, t; \hbar) = u_0(x, t; \hbar) + \sum_{i=1}^{3} u_i(x, t; \hbar). \tag{14.79}$$

or, especially when $\hbar = -1$,

$$u_{HASTM}(x, t) = -x - x\frac{t^\alpha}{\Gamma(1 + \alpha)} - 2x\frac{t^{2\alpha}}{\Gamma(1 + 2\alpha)} - 4x\frac{t^{3\alpha}}{\Gamma(1 + 3\alpha)} - \frac{x}{(\Gamma(1 + \alpha))^2}\frac{\Gamma(1 + 2\alpha)}{\Gamma(1 + 3\alpha)}\frac{t^{3\alpha}}{\Gamma(1 + 3\alpha)} - .... \tag{14.80}$$

Particularly at $\alpha = 1$, Eq. (14.80) reduces to a solution $u(x, t) = \dfrac{x}{t-1}$, which is same as the solution of Wazwaz (2007).

14.7.3 Implementation of HAETM

To solve Eqs. (14.34)–(14.36), we take the initial approximation as:

$$u_0(x, t) = u(x, 0) = x^2. \tag{14.81}$$

Applying Elzaki transform on both sides of Eq. (14.34) and using Eq. (14.81), we get

$$E[u(x, t)] = x^2 s^2 + \frac{s^\alpha}{2} E\left[x^2 \frac{\partial^2 u(x, t)}{\partial x^2}\right]. \tag{14.82}$$

The nonlinear operator may be written as:

$$N[\varphi(x, t; p)] = E[\varphi(x, t; p)] - x^2 s^2 (1 - \chi_m) - \frac{s^\alpha}{2} E\left[x^2 \frac{\partial^2 \varphi(x, t; p)}{\partial x^2}\right]. \tag{14.83}$$

Now, the mth-order deformation equation with the assumption $H(x, t) = 1$ may be written as:

$$E[u_m(x, t) - \chi_m u_{m-1}(x, t)] = \hbar R_m\left(\overrightarrow{u_{m-1}}(x, t)\right), \tag{14.84}$$

where

$$R_m\left(\overrightarrow{u_{m-1}}(x, t)\right) = E[u_{m-1}(x, t)] - x^2 s^2 (1 - \chi_m) - \frac{s^\alpha}{2} E\left[x^2 \frac{\partial^2 u_{m-1}(x, t)}{\partial x^2}\right]. \tag{14.85}$$

Applying inverse Elzaki transform on both sides of Eq. (14.84), we have

$$u_m(x, t) = \chi_m u_{m-1}(x, t) + E^{-1}\left\{\hbar \mathfrak{R}_m\left(\overrightarrow{u_{m-1}}(x, t)\right)\right\}. \tag{14.86}$$

From Eqs. (14.81) and (14.86), we have

$$u_0(x, t) = x^2, \tag{14.87}$$

$$u_1(x, t) = \chi_1 u_0(x, t) + E^{-1}\left[\hbar R_1\left(\overrightarrow{u_0}(x, t)\right)\right] = \frac{-\hbar x^2 t^\alpha}{\Gamma(\alpha + 1)}, \tag{14.88}$$

$$u_2(x,t) = \chi_2 u_1(x,t) + E^{-1}\left[\hbar R_2\left(u_1\vec{(x,t)}\right)\right] = \frac{-\hbar x^2 t^\alpha}{\Gamma(\alpha+1)} + \hbar\left(\frac{-\hbar x^2 t^\alpha}{\Gamma(\alpha+1)} + \frac{\hbar x^2 t^{2\alpha}}{\Gamma(2\alpha+1)}\right)$$
$$= \frac{-x^2 t^\alpha}{\Gamma(\alpha+1)}\hbar(\hbar+1) + \frac{\hbar^2 x^2 t^{2\alpha}}{\Gamma(2\alpha+1)}, \tag{14.89}$$

$$u_3(x,t) = \chi_3 u_2(x,t) + E^{-1}\left[\hbar R_3\left(u_2\vec{(x,t)}\right)\right] = \frac{-x^2 t^\alpha}{\Gamma(\alpha+1)}\hbar(\hbar+1)^2 + \frac{x^2 t^{2\alpha}}{\Gamma(1+2\alpha)}2\hbar^2(1+\hbar)$$
$$- \frac{\hbar^3 x^2 t^{3\alpha}}{\Gamma(3\alpha+1)}, \tag{14.90}$$

and so on. Therefore, the fourth-order approximate solution of Eq. (14.34) is given by:

$$u_{HAETM}(x,t;\hbar) = u_0(x,t;\hbar) + \sum_{i=1}^{3} u_i(x,t;\hbar). \tag{14.91}$$

By substituting $\hbar = -1$, the solution of the fractional heat-like equation may be obtained as:

$$u(x,t) = x^2\left(1 + \frac{t^\alpha}{\Gamma(1+\alpha)} + \frac{t^{2\alpha}}{\Gamma(1+2\alpha)} + \frac{t^{3\alpha}}{\Gamma(1+3\alpha)} + \cdots\right) = x^2 E(t^\alpha). \tag{14.92}$$

when $\alpha = 1$, Eq. (14.92) reduces to $u(x,t) = x^2 e^t$, which is same as the solution of Sadighi et al. (2008).

Again, applying Elzaki transform on both sides of Eq. (14.49) and using Eq. (14.51), we obtain

$$E[u(x,t)] = -xs^2 - s^\alpha E\left[u\frac{\partial u(x,t)}{\partial x}\right]. \tag{14.93}$$

From Eq. (14.49), the nonlinear operator is written as:

$$N[\phi(x,t;p)] = E[\phi(x,t;p)] + xs^2(1-\chi_m) + s^\alpha E\left[\phi(x,t;p)\frac{\partial\phi(x,t;p)}{\partial x}\right]. \tag{14.94}$$

The mth-order deformation equation when $H(x,t) = 1$ is

$$E[u_m(x,t) - \chi_m u_{m-1}(x,t)] = \hbar R_m\left(u_{m-1}\vec{(x,t)}\right), \tag{14.95}$$

where

$$R_m\left(\overrightarrow{u_{m-1}(x,t)}\right) = E[u_{m-1}(x,t)] + xs^2(1-\chi_m) + s^\alpha E\left[\sum_{k=0}^{m-1} u_k(x,t)(u_{m-1-k}(x,t))_x\right]. \tag{14.96}$$

Taking inverse Elzaki transform on both sides of Eq. (14.95), we get

$$u_m(x,t) = \chi_m u_{m-1}(x,t) + E^{-1}\left[\hbar R_m\left(u_{m-1}\vec{(x,t)}\right)\right]. \tag{14.97}$$

Now, from Eqs. (14.97) and (14.51) for $m \geq 1$, the following expressions are obtained as:

$$u_0(x,t) = -x, \tag{14.98}$$

$$u_1(x,t) = \frac{\hbar x t^\alpha}{\Gamma(\alpha+1)}, \tag{14.99}$$

$$u_2(x,t) = \frac{\hbar x t^\alpha}{\Gamma(\alpha+1)} + \hbar\left(\frac{\hbar x t^\alpha}{\Gamma(\alpha+1)} - \frac{2\hbar x t^{2\alpha}}{\Gamma(2\alpha+1)}\right) = \frac{x t^\alpha}{\Gamma(\alpha+1)}\hbar(\hbar+1) - \frac{2\hbar^2 x t^{2\alpha}}{\Gamma(2\alpha+1)}, \tag{14.100}$$

$$u_3(x,t) = \frac{x t^\alpha}{\Gamma(\alpha+1)}\hbar(\hbar+1)^2 - \frac{4x t^{2\alpha}}{\Gamma(2\alpha+1)}\hbar^2(\hbar+1) + \frac{x\hbar^3\Gamma(2\alpha+1)t^{3\alpha}}{(\Gamma(\alpha+1))^2\Gamma(3\alpha+1)} + \frac{4\hbar^3 x t^{3\alpha}}{\Gamma(3\alpha+1)}, \tag{14.101}$$

and so on. The fourth-order series solution of Eq. (14.49) by *HAETM* can be written in the form:

$$u_{HAETM}(x,t;\hbar) = u_0(x,t;\hbar) + \sum_{i=1}^{3} u_i(x,t;\hbar). \tag{14.102}$$

or, especially when $\hbar = -1$,

$$u_{HAETM}(x,t) = -x - x\frac{t^\alpha}{\Gamma(1+\alpha)} - 2x\frac{t^{2\alpha}}{\Gamma(1+2\alpha)} - 4x\frac{t^{3\alpha}}{\Gamma(1+3\alpha)} - \frac{x\,\Gamma(1+2\alpha)}{(\Gamma(1+\alpha))^2}\frac{t^{3\alpha}}{\Gamma(1+3\alpha)} - \dots \tag{14.103}$$

Particularly at $\alpha = 1$, Eq. (14.103) reduces to a solution $u(x,t) = \dfrac{x}{t-1}$, which is same as the solution of Wazwaz (2007).

14.7.4   Implementation of HAATM

To solve Eqs. (14.34)–(14.36) using *HAATM*, we have taken the initial approximation as:

$$u_0(x,t) = u(x,0) = x^2. \tag{14.104}$$

Applying Aboodh transform on both sides of Eq. (14.34) and using initial approximation Eq. (14.104), we get

$$A[u(x,t)] = \frac{x^2}{s^2} + \frac{s^{-\alpha}}{2}A\left[x^2\frac{\partial^2 u(x,t)}{\partial x^2}\right]. \tag{14.105}$$

The nonlinear operator is obtained as:

$$N[\varphi(x,t;p)] = A[\varphi(x,t;p)] - \frac{x^2}{s^2}(1-\chi_m) - \frac{s^{-\alpha}}{2}A\left[x^2\frac{\partial^2\varphi(x,t;p)}{\partial x^2}\right]. \tag{14.106}$$

Now, the mth-order deformation equation for $H(x,t) = 1$ may be written as:

$$A[u_m(x,t) - \chi_m u_{m-1}(x,t)] = \hbar R_m\left(\overrightarrow{u_{m-1}}(x,t)\right), \tag{14.107}$$

where

$$R_m\left(\overrightarrow{u_{m-1}}(x,t)\right) = A[u_{m-1}(x,t)] - \frac{x^2}{s^2}(1-\chi_m) - \frac{s^{-\alpha}}{2}A\left[x^2\frac{\partial^2 u_{m-1}(x,t)}{\partial x^2}\right]. \tag{14.108}$$

Applying inverse Aboodh transform, Eq. (14.107) reduces to

$$u_m(x,t) = \chi_m u_{m-1}(x,t) + A^{-1}\left\{\hbar\mathfrak{R}_m\left(\overrightarrow{u_{m-1}}(x,t)\right)\right\}. \tag{14.109}$$

From Eqs. (14.109) for $m \geq 1$, the following expressions are obtained:

$$u_1(x,t) = \chi_1 u_0(x,t) + A^{-1}\left[\hbar R_1\left(\overrightarrow{u_0}(x,t)\right)\right] = \frac{-\hbar x^2 t^\alpha}{\Gamma(\alpha+1)}, \tag{14.110}$$

$$u_2(x,t) = \chi_2 u_1(x,t) + A^{-1}\left[\hbar R_2\left(\overrightarrow{u_1}(x,t)\right)\right] = \frac{-\hbar x^2 t^\alpha}{\Gamma(\alpha+1)} + \hbar\left(\frac{-\hbar x^2 t^\alpha}{\Gamma(\alpha+1)} + \frac{\hbar x^2 t^{2\alpha}}{\Gamma(2\alpha+1)}\right)$$
$$= \frac{-x^2 t^\alpha}{\Gamma(\alpha+1)}\hbar(\hbar+1) + \frac{\hbar^2 x^2 t^{2\alpha}}{\Gamma(2\alpha+1)}, \tag{14.111}$$

$$u_3(x,t) = \chi_3 u_2(x,t) + A^{-1}\left[\hbar R_3\left(\overrightarrow{u_2}(x,t)\right)\right] = \frac{-x^2 t^\alpha}{\Gamma(\alpha+1)}\hbar(\hbar+1)^2 + \frac{x^2 t^{2\alpha}}{\Gamma(1+2\alpha)}2\hbar^2(1+\hbar)$$
$$- \frac{\hbar^3 x^2 t^{3\alpha}}{\Gamma(3\alpha+1)}, \tag{14.112}$$

and so on. Hence, the fourth-order approximate solution of Eq. (14.34) is given by:

$$u_{HAATM}(x,t;\hbar) = u_0(x,t;\hbar) + \sum_{i=1}^{3} u_i(x,t;\hbar). \tag{14.113}$$

By substituting $\hbar = -1$, the solution of the fractional heat-like equation is obtained as:

$$u(x,t) = x^2\left(1 + \frac{t^\alpha}{\Gamma(1+\alpha)} + \frac{t^{2\alpha}}{\Gamma(1+2\alpha)} + \frac{t^{3\alpha}}{\Gamma(1+3\alpha)} + \dots\right) = x^2 E(t^\alpha), \tag{14.114}$$

when $\alpha = 1$, Eq. (14.114) reduces to $u(x, t) = x^2 e^t$, which is same as the solution of Sadighi et al. (2008).

Further, applying Aboodh transform on both sides of Eq. (14.49) and using Eq. (14.51), we obtain

$$A[u(x,t)] = -\frac{x}{s^2} - s^{-\alpha} A\left[u \frac{\partial u(x,t)}{\partial x}\right].$$ (14.115)

From Eq. (14.49), the nonlinear operator is written as:

$$N[\phi(x,t;p)] = A[\phi(x,t;p)] + \frac{x}{s^2}(1-\chi_m) + s^{-\alpha} A\left[\phi(x,t;p)\frac{\partial \phi(x,t;p)}{\partial x}\right].$$ (14.116)

The mth-order deformation equation when $H(x, t) = 1$ is

$$A[u_m(x,t) - \chi_m u_{m-1}(x,t)] = \hbar R_m\left(\overrightarrow{u_{m-1}}(x,t)\right),$$ (14.117)

where

$$R_m\left(\overrightarrow{u_{m-1}}(x,t)\right) = A[u_{m-1}(x,t)] + \frac{x}{s^2}(1-\chi_m) + s^{-\alpha} A\left[\sum_{k=0}^{m-1} u_k(x,t)(u_{m-1-k}(x,t))_x\right].$$ (14.118)

Taking inverse Aboodh transform on both sides of Eq. (14.117), we get

$$u_m(x,t) = \chi_m u_{m-1}(x,t) + A^{-1}\left[\hbar R_m\left(\overrightarrow{u_{m-1}}(x,t)\right)\right].$$ (14.119)

Now, from Eqs. (14.51) and (14.119), the expressions are obtained as follows:

$$u_0(x,t) = -x,$$ (14.120)

$$u_1(x,t) = \frac{\hbar x t^\alpha}{\Gamma(\alpha+1)},$$ (14.121)

$$u_2(x,t) = \frac{\hbar x t^\alpha}{\Gamma(\alpha+1)} + \hbar\left(\frac{\hbar x t^\alpha}{\Gamma(\alpha+1)} - \frac{2\hbar x t^{2\alpha}}{\Gamma(2\alpha+1)}\right) = \frac{x t^\alpha}{\Gamma(\alpha+1)}\hbar(\hbar+1) - \frac{2\hbar^2 x t^{2\alpha}}{\Gamma(2\alpha+1)},$$ (14.122)

$$u_3(x,t) = \frac{x t^\alpha}{\Gamma(\alpha+1)}\hbar(\hbar+1)^2 - \frac{4x t^{2\alpha}}{\Gamma(2\alpha+1)}\hbar^2(\hbar+1) + \frac{x\hbar^3\Gamma(2\alpha+1)t^{3\alpha}}{(\Gamma(\alpha+1))^2\Gamma(3\alpha+1)} + \frac{4\hbar^3 x t^{3\alpha}}{\Gamma(3\alpha+1)},$$ (14.123)

and so on. The fourth-order series solution of Eq. (14.49) by *HAATM* can be written in the form:

$$u_{HAATM}(x,t;\hbar) = u_0(x,t;\hbar) + \sum_{i=1}^{3} u_i(x,t;\hbar).$$ (14.124)

or, especially when $\hbar = -1$,

$$u_{HAATM}(x,t) = -x - x\frac{t^\alpha}{\Gamma(1+\alpha)} - 2x\frac{t^{2\alpha}}{\Gamma(1+2\alpha)} - 4x\frac{t^{3\alpha}}{\Gamma(1+3\alpha)} - \frac{x\,\Gamma(1+2\alpha)}{(\Gamma(1+\alpha))^2}\frac{t^{3\alpha}}{\Gamma(1+3\alpha)} - \dots$$ (14.125)

Particularly at $\alpha = 1$, Eq. (14.125) reduces to a solution $u(x,t) = \frac{x}{t-1}$, which is same as the solution of Wazwaz (2007).

References

Aboodh, K.S. (2013). The new integral transform "Aboodh Transform". *Global Journal of Pure and Applied Mathematics* 9 (1): 35–43.

Aboodh, K.S., Idris, A., and Nuruddeen, R.I. (2017). On the Aboodh transform connections with some famous integral transforms. *International Journal of Engineering and Information Systems* 1 (9): 143–151.

Baleanu, D. and Jassim, H.K. (2019). A modification fractional homotopy perturbation method for solving helmholtz and coupled helmholtz equations on cantor sets. *Fractal Fract* 3 (2): 30.

Jena, R.M. and Chakraverty, S. (2019a). Analytical solution of Bagley–Torvik equations using Sumudu transformation method. *SN Applied Sciences* 1 (3): 246.

Jena, R.M. and Chakraverty, S. (2019b). Solving time-fractional Navier–Stokes equations using homotopy perturbation Elzaki transform. *SN Applied Sciences* 1 (1): 16.

Liao, S.J. (2003). *Beyond Perturbation: Introduction to the Homotopy Analysis Method*. Boca Raton: Chapman and Hall/CRC Press.

Liao, S.J. (2004). On the homotopy analysis method for nonlinear problems. *Applied Mathematics and Computation* 147: 499–513.

Maitama, S. and Zhao, W. (2020). New homotopy analysis transform method for solving multidimensional fractional diffusion equations. *Arab Journal of Basic and Applied Sciences* 27 (1): 27–44.

Mohamed, M.S., Al-Malki, F., and Al-humyani, M. (2014). Homotopy analysis transform method for time-space fractional gas dynamics equation. *General Mathematics Notes* 24: 1–16.

Özis, T. and Agırseven, D. (2008). He's homotopy perturbation method for solving heat-like and wave-like equations with variable coefficients. *Physics Letters A* 372: 5944–5950.

Saad, K.M. and Al-Shomrani, A.A. (2016). An application of homotopy analysis transform method for Riccati differential equation of fractional order. *Journal of Fractional Calculus and Applications* 7 (1): 61–72.

Sadighi, A., Ganji, D.D., Gorji, M., and Tolou, N. (2008). Numerical simulation of heat-like models with variable coefficients by the variational iteration method. *Journal of Physics Conference Series* 96: 012083.

Saratha, S.R., Bagyalakshmi, M., and Krishnan, G.S.S. (2020). Fractional generalised homotopy analysismethod for solving nonlinear fractional differential equations. *Computational and Applied Mathematics* 39: 1–32.

Wazwaz, A.M. (2007). A comparison between the variational iteration method and adomian decomposition method. *Journal of Computational and Applied Mathematics* 207: 129–136.

Ziane, D. and Cherif, M.H. (2017). Modified homotopy analysis method for nonlinear fractional partial differential equations. *International Journal of Analysis and Applications* 14 (1): 77–87.

15

q-Homotopy Analysis Method

15.1 Introduction

Chapter 13 has already discussed the homotopy analysis method (HAM), which is based on the coupling of the standard perturbation approach and homotopy in topology. This approach uses the auxiliary parameter $\hbar$, which gives us some versatility and great freedom to monitor and adjust the convergence region, including the convergence rate of the series solution. Subsequently, El-Tawil and Huseen in 2012 (El-Tawil and Huseen 2012) introduced an improvement to the HAM, which is called the q-HAM. The q-HAM is later used by Iyiola et al. (2013) to obtain a solution for the time-fractional foams drainage equation. Iyiola (2013) constructed the numerical solution of the fifth-order time-fractional Ito and Sawada–Kotera equations using q-HAM. The convergence of q-HAM was considered in El-Tawil and Huseen (2013). This method includes two auxiliary parameters n and $\hbar$ which helps us to adjust and control the convergence of the solution. It may be noted that the standard HAM is obtained by replacing $n = 1$ in the q-HAM. It is already mentioned that q-HAM is an improved HAM scheme and does not require discretization, perturbation, or linearization. The introduction of the additional parameter n in the q-HAM provides greater flexibility than the HAM in adjusting and controlling the convergence region and the convergence rate of the series solution. Several authors have recently used q-HAM to solve (integer and non-integer) linear and nonlinear differential equations due to its accuracy and usefulness (Soh et al. 2014; Huseen 2015; Iyiola 2015; Huseen 2016; Iyiola 2016). Compared to other approaches, this method can retain great accuracy while minimizing computational time.

15.2 Theory of q-HAM

To demonstrate the elementary idea of the q-HAM, let us consider the nonlinear fractional differential equation as follows (Akinyemi 2019):

$$\aleph\left(D_t^\alpha u(x,t)\right) - f(x,t) = 0, \quad (x,t) \in \Omega, \tag{15.1}$$

where D_t^α is the Caputo fractional derivative, $\aleph$ denotes nonlinear operator, $u(x,t)$ is the unknown function in the domain Ω, and $f(x,t)$ is the given function. Generalizing the traditional homotopy method (Liao 2003), the zero-order deformation equation is constructed and can be written as (Akinyemi 2019):

$$(1 - np)L[\phi(x,t;p) - u_0(x,t)] = p\hbar H(x,t)\left\{\aleph\left(D_t^\alpha \phi(x,t;p)\right) - f(x,t)\right\}, \tag{15.2}$$

for $0 \le p \le \dfrac{1}{n}, n \ge 1,$

where p is the embedding parameter, $\hbar \ne 0$ is a nonzero auxiliary parameter, $H(x,t) \ne 0$ is a nonzero auxiliary function, $u_0(x,t)$ is the initial condition of $u(x,t)$, $\phi(x,t;p)$ is the unknown function, and L denotes an auxiliary linear operator. By substituting $p = 0$ and $p = \dfrac{1}{n}$ in Eq. (15.2), we obtain

$$\phi(x,t;0) = u_0(x,t), \quad \text{and} \quad \phi\left(x,t;\frac{1}{n}\right) = u(x,t). \tag{15.3}$$

respectively. Thus, as p rises from 0 to $1/n$, the solution $\phi(x,t;p)$ ranges from initial guess $u_0(x,t)$ to the solution $u(x,t)$. If L, $u_0(x,t)$, $\hbar$, and $H(x,t) \ne 0$ are appropriately chosen and then the solution $\phi(x,t;p)$ in Eq. (15.2) is valid as long as $0 \le p \le \dfrac{1}{n}$. Expanding $\phi(x,t;p)$ using Taylor's series expansion with respect to p, we have

Computational Fractional Dynamical Systems: Fractional Differential Equations and Applications, First Edition.
Snehashish Chakraverty, Rajarama Mohan Jena, and Subrat Kumar Jena.
© 2023 John Wiley & Sons, Inc. Published 2023 by John Wiley & Sons, Inc.

$$\phi(x,t;p) = u_0(x,t) + \sum_{m=1}^{\infty} u_m(x,t)p^m, \tag{15.4}$$

where

$$u_m(x,t) = \frac{1}{m!} \frac{\partial^m \varphi(x,t;p)}{\partial p^m}\bigg|_{p=0}. \tag{15.5}$$

If we chose L, $u_0(x,t)$, $\hbar$, and $H(x,t)$ correctly so that Eq. (15.4) converges at $p = \frac{1}{n}$. Now from Eq. (15.3), we get

$$u(x,t) = u_0(x,t) + \sum_{m=1}^{\infty} u_m(x,t)\left(\frac{1}{n}\right)^m. \tag{15.6}$$

Now let us define the vector as:

$$\vec{u}_\omega = \{u_0(x,t), u_1(x,t), \dots, u_\omega(x,t)\}. \tag{15.7}$$

Differentiating Eq. (15.2) m times with respect to p, putting $p = 0$ and then dividing them by m!, we obtain the mth-order deformation equation with the assumption $H(x,t) = 1$ as follows (Liao 2003; Akinyemi 2019):

$$L[u_m(x,t) - \chi_m u_{m-1}(x,t)] = \hbar R_m\left(\vec{u}_{m-1}(x,t)\right), \tag{15.8}$$

subject to the following initial conditions:

$$u_m^{(k)}(x,0) = 0, k = 0,1,2,\dots,m-1, \tag{15.9}$$

where

$$R_m\left(\overrightarrow{u_{m-1}(x,t)}\right) = \frac{1}{(m-1)!} \frac{\partial^{m-1}\{\aleph\left(D_t^\alpha \phi(x,t;p)\right) - f(x,t)\}}{\partial p^{m-1}}\bigg|_{p=0}, \tag{15.10}$$

and

$$\chi_m = \begin{cases} 0, & m \le 1 \\ n, & m > 1 \end{cases} \tag{15.11}$$

Applying an integral operator L^{-1} on both sides of Eq. (15.8), we find

$$L^{-1}L[u_m(x,t) - \chi_m u_{m-1}(x,t)] = \hbar L^{-1}R_m\left(\overrightarrow{u_{m-1}(x,t)}\right),$$
$$u_m(x,t) = \chi_m u_{m-1}(x,t) + \hbar L^{-1}R_m\left(\overrightarrow{u_{m-1}(x,t)}\right). \tag{15.12}$$

The mth-order approximate solution is then written as:

$$u(x,t) = u_0(x,t) + \sum_{k=1}^{m-1} u_k(x,t)\left(\frac{1}{n}\right)^m. \tag{15.13}$$

Theorem 15.1 Amit and Hardish (2017) If we can find $0 < c < 1$ where c is a constant such that $\|u_{m+1}(x,t)\| \le c\|u_m(x,t)\|$ ∀ m. Furthermore, if the series $\sum_{m=1}^{r} u_m(x,t)\left(\frac{1}{n}\right)^m$ is assumed as an approximate solution of $u(x,t)$, then the maximum absolute truncated error is written as:

$$\left\|u(x,t) - \sum_{m=0}^{r} u_m(x,t)\left(\frac{1}{n}\right)^m\right\| \le \frac{c^{r+1}}{n^r(n-c)}u_0(x,t). \tag{15.14}$$

Proof: One may see the proof of this theorem in Amit and Hardish (2017).

15.3 Illustrative Examples

Here, we apply the present method to solve a time-fractional one-dimensional heat-like equation in Example 15.1 and a fractional nonlinear advection equation in Example 15.2.

Example 15.1 Let us consider the one-dimensional heat-like model (Özis and Agırseven 2008; Sadighi et al. 2008) as:

$$D_t^\alpha u(x,t) = \frac{1}{2}x^2 u_{xx}(x,t), 0 < x < 1, \quad t > 0, \quad 0 < \alpha \le 1, \tag{15.15}$$

with the boundary conditions (BCs):

$$u(0,t) = 0, u(1,t) = e^t, \tag{15.16}$$

and initial condition:

$$u(x,0) = x^2. \tag{15.17}$$

Solution

To solve Eqs. (15.15)–(15.17) using q-HAM, we chose the initial approximation as:

$$u_0(x,t) = u(x,0) = x^2. \tag{15.18}$$

Equation (15.15) suggests the nonlinear operator as:

$$\aleph[\phi(x,t;p)] = D_t^\alpha \phi(x,t;p) - \frac{1}{2}x^2 \phi_{xx}(x,t;p), \tag{15.19}$$

and the linear operator:

$$L[\phi(x,t;p)] = D_t^\alpha \phi(x,t;p), \tag{15.20}$$

with the property $L[c] = 0$ where c is the integration constant. Using the aforementioned definition with the assumption $H(x,t) = 1$, the zeroth-order deformation equation may be constructed as (Akinyemi 2019):

$$(1 - np)L[\phi(x,t;p) - u_0(x,t)] = p\hbar\aleph(\phi(x,t;p)). \tag{15.21}$$

Obviously, at $p = 0$ and $p = \frac{1}{n}$ in Eq. (15.21) gives

$$\phi(x,t;0) = u_0(x,t), \quad \text{and} \quad \phi\left(x,t;\frac{1}{n}\right) = u(x,t), \tag{15.22}$$

respectively. So, as p increases from 0 to $\frac{1}{n}$, the solution $\phi(x,t;p)$ varies from the initial guess $u_0(x,t)$ to the solution $u(x,t)$. Now, the mth-order deformation equation can be written as:

$$L[u_m(x,t) - \chi_m u_{m-1}(x,t)] = \hbar R_m\left(\overrightarrow{u_{m-1}}(x,t)\right), \tag{15.23}$$

where

$$R_m\left(\overrightarrow{u_{m-1}}(x,t)\right) = D_t^\alpha u_{m-1}(x,t) - \frac{1}{2}x^2(u_{m-1}(x,t))_{xx}. \tag{15.24}$$

The solution of mth-order deformation Eq. (15.23) for $m \ge 1$ becomes

$$u_m(x,t) = \chi_m u_{m-1}(x,t) + \hbar J_t^\alpha R_m\left(\overrightarrow{u_{m-1}}(x,t)\right). \tag{15.25}$$

From Eqs. (15.18) and (15.25), we have

$$u_0(x,t) = x^2, \tag{15.26}$$

$$u_1(x,t) = \chi_1 u_0(x,t) + \hbar J_t^\alpha R_1\left(\overrightarrow{u_0(x,t)}\right) = \hbar J_t^\alpha\left[D_t^\alpha u_0(x,t) - \frac{1}{2}x^2(u_0(x,t))_{xx}\right]$$

$$= \hbar J_t^\alpha\left(0-x^2\right) = \frac{-\hbar x^2 t^\alpha}{\Gamma(\alpha+1)}, \tag{15.27}$$

$$u_2(x,t) = \chi_2 u_1(x,t) + \hbar J_t^\alpha R_2\left(\overrightarrow{u_1(x,t)}\right) = \frac{-n\hbar x^2 t^\alpha}{\Gamma(\alpha+1)} + \hbar J_t^\alpha\left[D_t^\alpha u_1(x,t) - \frac{1}{2}x^2(u_1(x,t))_{xx}\right]$$

$$= \frac{-n\hbar x^2 t^\alpha}{\Gamma(\alpha+1)} + \hbar J_t^\alpha\left(-\hbar x^2 + \frac{\hbar x^2 t^\alpha}{\Gamma(\alpha+1)}\right) = \frac{-n\hbar x^2 t^\alpha}{\Gamma(\alpha+1)} + \hbar\left(\frac{-\hbar x^2 t^\alpha}{\Gamma(\alpha+1)} + \frac{\hbar x^2 t^{2\alpha}}{\Gamma(2\alpha+1)}\right) \tag{15.28}$$

$$= \frac{-x^2 t^\alpha}{\Gamma(\alpha+1)}\hbar(\hbar+n) + \frac{\hbar^2 x^2 t^{2\alpha}}{\Gamma(2\alpha+1)},$$

$$u_3(x,t) = \chi_3 u_2(x,t) + \hbar J_t^\alpha R_3\left(\overrightarrow{u_2(x,t)}\right) = \frac{-x^2 t^\alpha}{\Gamma(\alpha+1)}\hbar(\hbar+n)^2 + \frac{x^2 t^{2\alpha}}{\Gamma(1+2\alpha)}2\hbar^2(n+\hbar)$$

$$-\frac{\hbar^3 x^2 t^{3\alpha}}{\Gamma(3\alpha+1)} \tag{15.29}$$

and so on. Therefore, the four-term approximate solution of Eq. (15.15) is given by:

$$u_{q-HAM}(x,t;n,\hbar) = u_0(x,t;n,\hbar) + \sum_{i=1}^{3} u_i(x,t;n,\hbar)\left(\frac{1}{n}\right)^i. \tag{15.30}$$

By substituting $\hbar = -1$ and $n = 1$, the solution of the fractional heat-like equation may be obtained as:

$$u(x,t) = x^2 + x^2\frac{t^\alpha}{\Gamma(1+\alpha)} + x^2\frac{t^{2\alpha}}{\Gamma(1+2\alpha)} + x^2\frac{t^{3\alpha}}{\Gamma(1+3\alpha)} + \ldots,$$

$$= x^2\left(1 + \frac{t^\alpha}{\Gamma(1+\alpha)} + \frac{t^{2\alpha}}{\Gamma(1+2\alpha)} + \frac{t^{3\alpha}}{\Gamma(1+3\alpha)} + \ldots\right) = x^2 E(t^\alpha), \tag{15.31}$$

where $E(t^\alpha)$ is called the Mittag-Leffler function. In particular, at $\alpha = 1$, Eq. (15.31) reduces to $u(x,t) = x^2 e^t$, which is same as the solution of Sadighi et al. (Sadighi et al. 2008). The third-order approximate solution plot is depicted in Figure 15.1 for $\hbar = -0.3, -0.6, -0.9, -1.0$ at $\alpha = 1, n = 1$ and $x = 1$. In this figure, one may see the comparison of the present results with the exact solution. From the results presented in Figure 15.1, we observe that at $\hbar = -1.0$ and $n = 1$, the q-HAM solution is the same as the exact solution.

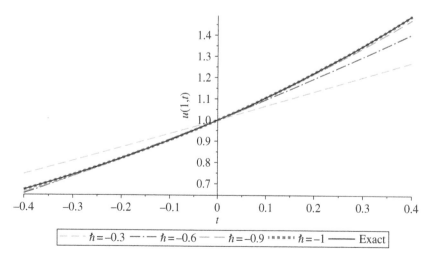

Figure 15.1 Solution of Eq. (15.15) for different values of $\hbar$.

Example 15.2 Consider the following nonlinear advection equation (Wazwaz 2007):

$$\frac{\partial^\alpha u}{\partial t^\alpha} + u\frac{\partial u}{\partial x} = 0, \quad 0 < \alpha \le 1, \tag{15.32}$$

with initial condition:

$$u(x, 0) = -x. \tag{15.33}$$

Solution

Here, we chose the initial approximation as:

$$u_0(x, t) = u(x, 0) = -x. \tag{15.34}$$

From Eq. (15.32), the nonlinear operator is written as:

$$\aleph[\phi(x, t; p)] = D_t^\alpha \phi(x, t; p) + \phi(x, t; p)\phi_x(x, t; p), \tag{15.35}$$

and the linear operator as:

$$L[\phi(x, t; p)] = D_t^\alpha \phi(x, t; p), \tag{15.36}$$

Using the aforementioned definition with an assumption $H(x, t) = 1$, the zeroth-order deformation equation can be constructed as (Akinyemi 2019):

$$(1 - np)L[\phi(x, t; p) - u_0(x, t)] = p\hbar\aleph(\phi(x, t; p)). \tag{15.37}$$

The mth-order deformation equation is

$$L[u_m(x, t) - \chi_m u_{m-1}(x, t)] = \hbar R_m\left(\overrightarrow{u_{m-1}}(x, t)\right), \tag{15.38}$$

where

$$R_m\left(\overrightarrow{u_{m-1}}(x, t)\right) = D_t^\alpha u_{m-1}(x, t) + \sum_{k=0}^{m-1} u_k(x, t)(u_{m-1-k}(x, t))_x. \tag{15.39}$$

The solution of mth-order deformation Eq. (15.38) for $m \ge 1$ becomes

$$u_m(x, t) = \chi_m u_{m-1}(x, t) + \hbar J_t^\alpha R_m\left(\overrightarrow{u_{m-1}}(x, t)\right). \tag{15.40}$$

From Eqs. (15.34) and (15.40), we obtain

$$u_0(x, t) = -x, \tag{15.41}$$

$$u_1(x, t) = \chi_1 u_0(x, t) + \hbar J_t^\alpha R_1\left(\overrightarrow{u_0}(x, t)\right) = \hbar J_t^\alpha\left[D_t^\alpha u_0(x, t) + u_0(x, t)(u_0(x, t))_x\right]$$
$$= \hbar J_t^\alpha(0 + (-x)(-1)) = \frac{\hbar x t^\alpha}{\Gamma(\alpha + 1)}, \tag{15.42}$$

$$u_2(x, t) = \chi_2 u_1(x, t) + \hbar J_t^\alpha R_2\left(\overrightarrow{u_1}(x, t)\right) = \frac{n\hbar x t^\alpha}{\Gamma(\alpha + 1)} + \hbar J_t^\alpha\begin{bmatrix} D_t^\alpha u_1(x, t) + u_0(x, t)(u_1(x, t))_x + \\ u_1(x, t)(u_0(x, t))_x \end{bmatrix}$$

$$= \frac{n\hbar x t^\alpha}{\Gamma(\alpha + 1)} + \hbar J_t^\alpha\left(\hbar x - \frac{\hbar x t^\alpha}{\Gamma(\alpha + 1)} - \frac{\hbar x t^\alpha}{\Gamma(\alpha + 1)}\right) = \frac{n\hbar x t^\alpha}{\Gamma(\alpha + 1)} + \hbar J_t^\alpha\left(\hbar x - \frac{2\hbar x t^\alpha}{\Gamma(\alpha + 1)}\right) \tag{15.43}$$

$$= \frac{n\hbar x t^\alpha}{\Gamma(\alpha + 1)} + \hbar\left(\frac{\hbar x t^\alpha}{\Gamma(\alpha + 1)} - \frac{2\hbar x t^{2\alpha}}{\Gamma(2\alpha + 1)}\right) = \frac{x t^\alpha}{\Gamma(\alpha + 1)}\hbar(\hbar + n) - \frac{2\hbar^2 x t^{2\alpha}}{\Gamma(2\alpha + 1)},$$

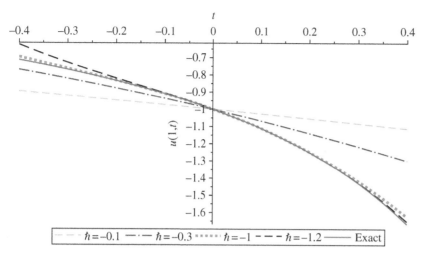

Figure 15.2 Solution of Eq. (15.32) for different values of $\hbar$.

$$u_3(x,t) = \chi_3 u_2(x,t) + \hbar J_t^\alpha R_3\left(\overrightarrow{u_2(x,t)}\right) = \frac{xt^\alpha}{\Gamma(\alpha+1)} n\left(\hbar^2 + \hbar n\right) - \frac{2\hbar^2 xnt^{2\alpha}}{\Gamma(2\alpha+1)} + \hbar J_t^\alpha$$

$$\begin{bmatrix} D_t^\alpha u_2(x,t) + u_0(x,t)(u_2(x,t))_x + \\ u_1(x,t)(u_1(x,t))_x + u_2(x,t)(u_0(x,t))_x \end{bmatrix} = \frac{xt^\alpha}{\Gamma(\alpha+1)} \hbar(\hbar+n)^2 - \frac{4xt^{2\alpha}}{\Gamma(2\alpha+1)} \hbar^2(\hbar+n) \tag{15.44}$$

$$+ \frac{x\hbar^3 \Gamma(2\alpha+1)t^{3\alpha}}{(\Gamma(\alpha+1))^2 \Gamma(3\alpha+1)} + \frac{4\hbar^3 xt^{3\alpha}}{\Gamma(3\alpha+1)},$$

and so on. The fourth-order series solution of Eq. (15.32) by q-HAM can be written in the form:

$$u_{q-HAM}(x,t;n,\hbar) = u_0(x,t;n,\hbar) + \sum_{i=1}^{3} u_i(x,t;n,\hbar)\left(\frac{1}{n}\right)^i.$$

or, especially when $\hbar = -1$ and $n = 1$,

$$u_{q-HAM}(x,t) = -x - x\frac{t^\alpha}{\Gamma(1+\alpha)} - 2x\frac{t^{2\alpha}}{\Gamma(1+2\alpha)} - 4x\frac{t^{3\alpha}}{\Gamma(1+3\alpha)} - \frac{x\ \Gamma(1+2\alpha)}{(\Gamma(1+\alpha))^2}\frac{t^{3\alpha}}{\Gamma(1+3\alpha)} - \cdots. \tag{15.45}$$

Particularly at $\alpha = 1$, Eq. (15.45) reduces to a solution $u(x,t) = \dfrac{x}{t-1}$, which is same as the solution of Wazwaz (2007). Figure 15.2 shows that the q-HAM solution converges to the exact solution at $\hbar = -1.0$ for $\alpha = 1$, $n = 1$ and $x = 1$. It may be noted that it is not always true to achieve a better result at $\hbar = -1.0$, $n = 1$. In this case, one may obtain a closed-form solution at $\hbar = -1.2$, $n = 1$. It can be concluded from the aforementioned discussion that the proper values of the control parameter $\hbar$ should be selected to provide better results with less errors. It is worth mentioning that the q-HAM produces not only the approximate convergent series solution but also an exact solution depending on the considered problem with the proper $\hbar$ value.

References

Akinyemi, L. (2019). q-Homotopy analysis method for solving the seventh-order time-fractional Lax's Korteweg–de Vries and Sawada–Kotera equations. *Comput. Appl. Math.* 38: 191–212.

Amit, P. and Hardish, K. (2017). Numerical solution for fractional model of Fokker Planck equation by using q-HATM. *Chaos, Solitons Fractals* 105: 99–110.

El-Tawil, M.A. and Huseen, S.N. (2012). The Q-homotopy analysis method (Q-HAM). *Int. J. Appl. Math. Mech.* 8 (15): 51–75.

El-Tawil, M.A. and Huseen, S.N. (2013). On convergence of the q -homotopy analysis method. *Int. J. Contemp. Math. Sci.* 8: 481–497.

Huseen, S.N. (2015). Solving the K(2,2) equation by means of the q-homotopy analysis method (q-HAM). *Int. J. Innov. Sci. Eng. Technol.* 2 (8): 805–817.

Huseen, S.N. (2016). Series solutions of fractional initial-value problems by q-homotopy analysis method. *Int. J. Innov. Sci. Eng. Technol.* 3 (1): 27–41.

Iyiola, O.S. (2013). A numerical study of ito equation and Sawada-Kotera equation both of time-fractional type. *Adv. Math. Sci. J.* 2 (2): 71–79.

Iyiola, O.S. (2015). On the solutions of non-linear time-fractional gas dynamic equations: an analytical approach. *Int. J. Pure Appl. Math.* 98 (4): 491–502.

Iyiola, O.S. (2016). Exact and approximate solutions of fractional diffusion equations with fractional reaction terms. *Prog. Fract. Differ. Appl.* 2 (1): 21–30.

Iyiola, O.S., Soh, M.E., and Enyi, C.D. (2013). Generalised homotopy analysis method (q-HAM) for solving foam drainage equation of time fractional type. *Math. Eng. Sci. Aerosp.* 4 (4): 105.

Liao, S.J. (2003). *Beyond Perturbation: Introduction to the Homotopy Analysis Method*. Boca Raton: Chapman and Hall/CRC Press.

Özis, T. and Agırseven, D. (2008). He's homotopy perturbation method for solving heat-like and wave-like equations with variable coefficients. *Phys. Lett. A* 372: 5944–5950.

Sadighi, A., Ganji, D.D., Gorji, M., and Tolou, N. (2008). Numerical simulation of heat-like models with variable coefficients by the variational iteration method. *J. Phys. Conf. Ser.* 96: 012083.

Soh, M.E., Enyi, C.D., Iyiola, O.S., and Audu, J.D. (2014). Approximate analytical solutions of strongly nonlinear fractional BBM-Burger's equations with dissipative term. *Appl. Math. Sci.* 8 (155): 7715–7726.

Wazwaz, A.M. (2007). A comparison between the variational iteration method and adomian decomposition method. *J. Comput. Appl. Math.* 207: 129–136.

16

q-Homotopy Analysis Transform Method

16.1 Introduction

A Chinese mathematician, Liao, proposed the homotopy analysis method (HAM) (Liao 2003, 2004) by employing the fundamental concept of differential geometry and topology. HAM has recently been used effectively to obtain solutions to problems in various fields of science and technology. In accordance with this, the q-homotopy analysis transform method (q-HATM) was developed, which is a combination of q-HAM and the different transform methods. This method monitors and manipulates the series solution, which converges to the exact solution. As a result, several authors have recently studied different phenomena using q-HATM (Srivastava et al. 2017; Jena and Chakraverty 2019; Veeresha et al. 2019; Jena et al. 2020; Veeresha and Prakasha 2020). The HAM takes longer for computing and large computer memory. There has been a need to integrate this technique with transformation techniques to address these limitations. The present method has many strong properties, including a nonlocal effect, a simple solution procedure, a broad convergence region free from assumptions, discretization, and perturbation. It is worth mentioning that the transform methods with semi-analytical techniques require less CPU time to evaluate the solutions for nonlinear fractional complex models. Again, the q-HATM solution involves two auxiliary parameters, n and $\hbar$, which help us to adjust and control the convergence of the solution.

16.2 Transform Methods for the Caputo Sense Derivative

Definition 16.1 The Laplace transform of the Caputo fractional derivative is defined as (Baleanu and Jassim 2019):

$$L\left[D_t^\alpha u(x,t)\right] = s^\alpha L[u(x,t)] - \sum_{k=0}^{n-1} s^{(\alpha-k-1)} u^{(k)}(x,0), \quad n-1 < \alpha \le 1, \quad n \in \mathrm{N}. \tag{16.1}$$

Definition 16.2 The Sumudu transform of the Caputo fractional derivative is defined as (Jena and Chakraverty 2019):

$$S\left[D_t^\alpha u(x,t)\right] = s^{-\alpha} S[u(x,t)] - \sum_{k=0}^{n-1} s^{-\alpha+k} u^{(k)}(x,0), \quad n-1 < \alpha \le 1, \quad n \in \mathrm{N}. \tag{16.2}$$

Definition 16.3 The Elzaki transform of the Caputo fractional derivative is defined as (Jena and Chakraverty 2019):

$$E\left[D_t^\alpha u(x,t)\right] = \frac{E[u(x,t)]}{s^\alpha} - \sum_{k=0}^{n-1} s^{k-\alpha+2} u^{(k)}(x,0), \quad n-1 < \alpha \le n, \quad n \in \mathrm{N}. \tag{16.3}$$

Definition 16.4 The Aboodh transform of the Caputo fractional derivative is defined as (Aboodh 2013; Aboodh et al. 2017):

$$A\left[D_t^\alpha u(x,t)\right] = s^\alpha A[u(x,t)] - \sum_{k=0}^{n-1} s^{-k+\alpha-2} u^{(k)}(x,0), \quad n-1 < \alpha \le n, \quad n \in \mathrm{N}. \tag{16.4}$$

Table 16.1   Transforms of some essential functions.

Functions	Laplace transform	Sumudu transform	Elzaki transform	Aboodh transform
Definitions	$L[f(t)] = f(s)$	$S[g(t)] = g(s)$	$E[h(t)] = h(s)$	$A[p(t)] = p(s)$
	$= \int_0^\infty e^{-st} f(t) dt$	$= \int_0^\infty e^{-t} f(st) dt$	$= s \int_0^\infty e^{\frac{-t}{s}} f(t) dt$	$= \frac{1}{s} \int_0^\infty e^{-st} f(t) dt$
		$= \frac{1}{s} f\left(\frac{1}{s}\right)$	$= s f\left(\frac{1}{s}\right)$	$= \frac{1}{s} f(s)$
1	$\dfrac{1}{s}$	1	s^2	$\dfrac{1}{s^2}$
t^α	$\dfrac{\Gamma(1+\alpha)}{s^{\alpha+1}}$	$s^\alpha \Gamma(1+\alpha)$	$s^{\alpha+2} \Gamma(1+\alpha)$	$\dfrac{\Gamma(1+\alpha)}{s^{\alpha+2}}$
e^{at}	$\dfrac{1}{s-a}$	$\dfrac{1}{1-as}$	$\dfrac{s^2}{1-as}$	$\dfrac{1}{s(1-s)}$
$\sin(at)$	$\dfrac{a}{s^2+a^2}$	$\dfrac{as}{1+a^2 s^2}$	$\dfrac{as^3}{1+a^2 s^2}$	$\dfrac{a}{s(s^2+a^2)}$
$\cos(at)$	$\dfrac{s}{s^2+a^2}$	$\dfrac{1}{1+s^2 a^2}$	$\dfrac{s}{1+s^2 a^2}$	$\dfrac{1}{s^2+a^2}$

In view of the above, Table 16.1 is included, which shows the transforms of some standard functions with respect to the aforementioned four transform methods and their definitions.

Following section deals with the systematic study of four hybrid methods, namely q-homotopy analysis Laplace transform method (q-HALTM), q-homotopy analysis Sumudu transform method (q-HASTM), q-homotopy analysis Elzaki transform method (q-HAETM), and q-homotopy analysis Aboodh transform method (q-HAATM), one after another.

16.3   q-Homotopy Analysis Laplace Transform Method (q-HALTM)

To illustrate the basic concept of q-HALTM, we consider the nonlinear nonhomogenous fractional partial differential equation (PDE) as follows (Jena et al. 2020):

$$D_t^\alpha u(x,t) + Ru(x,t) + Nu(x,t) = h(x,t), \quad n-1 < \alpha \le n, \tag{16.5}$$

where D_t^α is the Caputo fractional derivative, R and N are the linear and nonlinear differential operator, respectively, and $h(x, t)$ is the source term. Applying Laplace transform on both sides of Eq. (16.5) and using Eq. (16.1), we have

$$L[u(x,t)] = \sum_{k=0}^{n-1} s^{-k-1} u^{(k)}(x,0) + s^{-\alpha} L[h(x,t) - Ru(x,t) - Nu(x,t)], \tag{16.6}$$

Define a nonlinear operator as:

$$N[\varphi(x,t;p)] = L[\varphi(x,t;p)] - \sum_{k=0}^{n-1} s^{-k-1} \varphi^{(k)}(x,t;p)(0^+) + s^{-\alpha} L\left[\begin{array}{c} R\varphi(x,t;p) + N\varphi(x,t;p) \\ -h(x,t) \end{array}\right], \tag{16.7}$$

where $p \in \left[0, \dfrac{1}{n}\right]$ is the embedding parameter, and $\varphi(x, t; p)$ is the unknown function. Constructing a homotopy, we get

$$(1-np)L[\varphi(x,t;p) - u_0(x,t)] = \hbar p H(x,t) N[\varphi(x,t;p)], \tag{16.8}$$

where $H(x, t)$ represent the nonzero auxiliary function, $\hbar \ne 0$ is an auxiliary parameter, and $u_0(x, t)$ is the initial value of $u(x, t)$.

If $p = 0$ and $p = \dfrac{1}{n}$ then we obtain

$$\varphi(x, t; 0) = u_0(x, t), \quad \text{and} \quad \varphi\left(x, t; \frac{1}{n}\right) = u(x, t), \tag{16.9}$$

respectively. As p increases from 0 to $\frac{1}{n}$ then $\varphi(x, t; p)$ varies from $u_0(x, t)$ to the solution of Eq. (16.5). Using Taylor's series expansion on $\varphi(x, t; p)$, one has

$$\varphi(x, t; p) = u_0(x, t) + \sum_{m=1}^{\infty} u_m(x, t) p^m, \tag{16.10}$$

where

$$u_m(x, t) = \frac{1}{m!} \frac{\partial^m u(x, t; p)}{\partial p^m}\bigg|_{p=0}, \tag{16.11}$$

If R, $u_0(x, t)$, $H(x, t)$, and $\hbar$ are properly chosen, then Eq. (16.10) converges at $p = \frac{1}{n}$. So we have

$$u(x, t) = u_0(x, t) + \sum_{m=1}^{\infty} u_m(x, t) \left(\frac{1}{n}\right)^m. \tag{16.12}$$

Let us define the vector:

$$\vec{u}_m = \{u_0, u_1, \dots, \dots u_m\}. \tag{16.13}$$

Differentiating Eq. (16.8) m-times with respect to p, setting $p = 0$ and then dividing by m!, we have the mth-order deformation equation (Liao 2003; Liao 2004) as follows:

$$L[u_m(x, t) - \chi_m u_{m-1}(x, t)] = \hbar H(x, t) \Re_m\left(\overrightarrow{u_{m-1}(x, t)}\right), \tag{16.14}$$

By applying inverse Laplace transform on both sides of Eq. (16.14), we obtain

$$u_m(x, t) = \chi_m u_{m-1}(x, t) + L^{-1}\left[\hbar H(x, t) \Re_m\left(\overrightarrow{u_{m-1}(x, t)}\right)\right], \tag{16.15}$$

where

$$\Re_m\left(\vec{u}_{m-1}(x, t)\right) = \frac{1}{(m-1)!} \frac{\partial^{m-1} N(\varphi(x, t; p))}{\partial p^{m-1}}\bigg|_{p=0}, \tag{16.16}$$

and

$$\chi_m = \begin{cases} 0, & m \leq 1, \\ n, & m > 1. \end{cases} \tag{16.17}$$

16.4 q-Homotopy Analysis Sumudu Transform Method (q-HASTM)

Taking Sumudu transform on both sides of Eq. (16.5) and using Eq. (16.2), we have

$$S[u(x, t)] = \sum_{k=0}^{n-1} s^k u^{(k)}(x, 0) + s^\alpha S[h(x, t) - Ru(x, t) - Nu(x, t)]. \tag{16.18}$$

The nonlinear operator is as:

$$N[\varphi(x, t; p)] = S[\varphi(x, t; p)] - \sum_{k=0}^{n-1} s^k \varphi^{(k)}(x, t; p)(0^+) + s^\alpha S\left[\begin{matrix} R\varphi(x, t; p) + N\varphi(x, t; p) \\ -h(x, t) \end{matrix}\right]. \tag{16.19}$$

Constructing the homotopy, we get

$$(1-np)S[\varphi(x,t;p) - u_0(x,t)] = \hbar pH(x,t)N[\varphi(x,t;p)]. \tag{16.20}$$

Substituting $p = 0$ and $p = \dfrac{1}{n}$ in Eq. (16.20), we obtain Eq. (16.9). Using Taylor's series expansion on $\varphi(x,t;p)$, we obtain Eqs. (16.10) and (16.11). If R, $u_0(x,t)$, $H(x,t)$, and $\hbar$ are correctly chosen, then Eq. (16.10) converges at $p = \dfrac{1}{n}$.

Let us define the vector Eq. (16.13). Differentiating Eq. (16.20) m-times with respect to p, setting $p = 0$ and then dividing by m!, we have the mth-order deformation equation as follows:

$$S[u_m(x,t) - \chi_m u_{m-1}(x,t)] = \hbar H(x,t)\mathfrak{R}_m\left(\overrightarrow{u_{m-1}(x,t)}\right). \tag{16.21}$$

By applying inverse Sumudu transform on both sides of Eq. (16.21), we obtain

$$u_m(x,t) = \chi_m u_{m-1}(x,t) + S^{-1}\left[\hbar H(x,t)\mathfrak{R}_m\left(\overrightarrow{u_{m-1}(x,t)}\right)\right], \tag{16.22}$$

with Eqs. (16.16) and (16.17).

16.5 q-Homotopy Analysis Elzaki Transform Method (q-HAETM)

Taking Elzaki transform on both sides of Eq. (16.5) and with the help of Eq. (16.3), we have

$$E[u(x,t)] = \sum_{k=0}^{n-1} s^{k+2}u^{(k)}(x,0) + s^{\alpha}E[h(x,t) - Ru(x,t) - Nu(x,t)]. \tag{16.23}$$

Let us define the nonlinear operator as:

$$N[\varphi(x,t;p)] = E[\varphi(x,t;p)] - \sum_{k=0}^{n-1} s^{k+2}\varphi^{(k)}(x,t;p)(0^+) + s^{\alpha}E\left[\begin{array}{c} R\varphi(x,t;p) + N\varphi(x,t;p) \\ -h(x,t) \end{array}\right]. \tag{16.24}$$

The homotopy may be constructed as:

$$(1-np)E[\varphi(x,t;p) - u_0(x,t)] = \hbar pH(x,t)N[\varphi(x,t;p)]. \tag{16.25}$$

Let us define the vector Eq. (16.13). Differentiating Eq. (16.25) m-times with respect to p, setting $p = 0$ and then dividing by m!, the mth-order deformation equation is obtained as follows:

$$E[u_m(x,t) - \chi_m u_{m-1}(x,t)] = \hbar H(x,t)\mathfrak{R}_m\left(\overrightarrow{u_{m-1}(x,t)}\right). \tag{16.26}$$

By applying inverse Elzaki transform on both sides of Eq. (16.26), we find

$$u_m(x,t) = \chi_m u_{m-1}(x,t) + E^{-1}\left[\hbar H(x,t)\mathfrak{R}_m\left(\overrightarrow{u_{m-1}(x,t)}\right)\right], \tag{16.27}$$

with Eqs. (16.16) and (16.17).

16.6 q-Homotopy Analysis Aboodh Transform Method (q-HAATM)

Applying Aboodh transform on both sides of Eq. (16.5) and using Eq. (16.4), we have

$$A[u(x,t)] = \sum_{k=0}^{n-1} s^{-k-2}u^{(k)}(x,0) + s^{-\alpha}A[h(x,t) - Ru(x,t) - Nu(x,t)], \tag{16.28}$$

Define a nonlinear operator as:

$$N[\varphi(x,t;p)] = A[\varphi(x,t;p)] - \sum_{k=0}^{n-1} s^{-k-2}\varphi^{(k)}(x,t;p)(0^+) + s^{-\alpha}A\begin{bmatrix} R\varphi(x,t;p) + N\varphi(x,t;p) \\ -h(x,t) \end{bmatrix}, \tag{16.29}$$

where $p \in \left[0, \dfrac{1}{n}\right]$, and $\varphi(x, t; p)$ is the unknown function.

Constructing a homotopy, we get

$$(1-np)A[\varphi(x,t;p) - u_0(x,t)] = \hbar pH(x,t)N[\varphi(x,t;p)], \tag{16.30}$$

where $H(x, t)$ represent the nonzero auxiliary function, $\hbar \neq 0$ is an auxiliary parameter, and $u_0(x, t)$ is the initial value. If $p = 0$ and $p = \dfrac{1}{n}$ then, we obtain Eq. (16.9). As p increases from 0 to $\dfrac{1}{n}$, then $\varphi(x, t; p)$ varies from $u_0(x, t)$ to the solution of Eq. (16.5). Using Taylor's series expansion on $\varphi(x, t; p)$, one has get Eq. (16.10) with Eq. (16.11). If $R, u_0(x, t), H(x, t)$, and $\hbar$ are properly chosen, then Eq. (16.10) converges at $p = \dfrac{1}{n}$. So we get Eq. (16.12).

Let us define the vector Eq. (16.13). Differentiating Eq. (16.30) m-times with respect to p, setting $p = 0$ and then dividing by m!, we have the mth-order deformation equation as:

$$A[u_m(x,t) - \chi_m u_{m-1}(x,t)] = \hbar H(x,t)\Re_m\left(\overrightarrow{u_{m-1}}(x,t)\right). \tag{16.31}$$

Applying inverse Aboodh transform on both sides of Eq. (16.31), we obtain

$$u_m(x,t) = \chi_m u_{m-1}(x,t) + A^{-1}\left[\hbar H(x,t)\Re_m\left(\overrightarrow{u_{m-1}}(x,t)\right)\right], \tag{16.32}$$

with Eqs. (16.16) and (16.17).

Next, we solve two test problems to demonstrate the present methods.

16.7 Test Problems

16.7.1 Implementation of q-HALTM

Example 16.1 Let us consider the one-dimensional heat-like model (Özis and Agırseven 2008; Sadighi et al. 2008) as:

$$D_t^\alpha u(x,t) = \frac{1}{2}x^2 u_{xx}(x,t), \quad 0 < x < 1, \quad t > 0, \quad 0 < \alpha \leq 1, \tag{16.33}$$

with the boundary conditions:

$$u(0,t) = 0, u(1,t) = e^t, \tag{16.34}$$

and initial condition:

$$u(x,0) = x^2. \tag{16.35}$$

Solution

To solve Eqs. (16.33)–(16.35) using q-HALTM, we chose the initial approximation as:

$$u_0(x,t) = u(x,0) = x^2. \tag{16.36}$$

Applying Laplace transform on Eq. (16.33) and simplifying, we obtain

$$L[u(x,t)] = \frac{x^2}{s} + \frac{s^{-\alpha}}{2}L\left[x^2\frac{\partial^2 u(x,t)}{\partial x^2}\right]. \tag{16.37}$$

The nonlinear operator may be written as:

$$N[\varphi(x,t;p)] = L[\varphi(x,t;p)] - \frac{x^2}{s}\left(1 - \frac{\chi_m}{n}\right) - \frac{s^{-\alpha}}{2}L\left[x^2\frac{\partial^2 \varphi(x,t;p)}{\partial x^2}\right]. \tag{16.38}$$

Now, the mth-order deformation equation with the assumption $H(x, t) = 1$ can be written as:

$$L[u_m(x,t) - \chi_m u_{m-1}(x,t)] = \hbar R_m\left(\overrightarrow{u_{m-1}(x,t)}\right),\tag{16.39}$$

where

$$R_m\left(\overrightarrow{u_{m-1}(x,t)}\right) = L[u_{m-1}(x,t)] - \frac{x^2}{s}\left(1 - \frac{\chi_m}{n}\right) - \frac{s^{-\alpha}}{2}L\left[x^2\frac{\partial^2 u_{m-1}(x,t)}{\partial x^2}\right].\tag{16.40}$$

Applying inverse Laplace transform in Eq. (16.39), we have

$$u_m(x,t) = \chi_m u_{m-1}(x,t) + L^{-1}\left\{\hbar \Re_m\left(\overrightarrow{u_{m-1}(x,t)}\right)\right\}.\tag{16.41}$$

From Eqs. (16.36) and (16.41), we have

$$u_0(x,t) = x^2,\tag{16.42}$$

$$u_1(x,t) = \chi_1 u_0(x,t) + L^{-1}\left[\hbar R_1\left(\overrightarrow{u_0(x,t)}\right)\right] = \frac{-\hbar x^2 t^\alpha}{\Gamma(\alpha+1)},\tag{16.43}$$

$$u_2(x,t) = \chi_2 u_1(x,t) + L^{-1}\left[\hbar R_2\left(\overrightarrow{u_1(x,t)}\right)\right] = \frac{-n\hbar x^2 t^\alpha}{\Gamma(\alpha+1)} + \hbar\left(\frac{-\hbar x^2 t^\alpha}{\Gamma(\alpha+1)} + \frac{\hbar x^2 t^{2\alpha}}{\Gamma(2\alpha+1)}\right)$$
$$= \frac{-x^2 t^\alpha}{\Gamma(\alpha+1)}\hbar(\hbar+n) + \frac{\hbar^2 x^2 t^{2\alpha}}{\Gamma(2\alpha+1)},\tag{16.44}$$

$$u_3(x,t) = \chi_3 u_2(x,t) + L^{-1}\left[\hbar R_3\left(\overrightarrow{u_2(x,t)}\right)\right] = \frac{-x^2 t^\alpha}{\Gamma(\alpha+1)}\hbar(\hbar+n)^2 + \frac{x^2 t^{2\alpha}}{\Gamma(1+2\alpha)}2\hbar^2(n+\hbar)$$
$$- \frac{\hbar^3 x^2 t^{3\alpha}}{\Gamma(3\alpha+1)},\tag{16.45}$$

and so on. Therefore, the fourth-order approximate solution of Eq. (16.33) is given by:

$$u_{q-HALTM}(x,t;n,\hbar) = u_0(x,t;n,\hbar) + \sum_{i=1}^{3} u_i(x,t;n,\hbar)\left(\frac{1}{n}\right)^i.\tag{16.46}$$

By substituting $\hbar = -1$ and $n = 1$, the solution of the fractional heat-like equation may be obtained as:

$$u(x,t) = x^2 + x^2\frac{t^\alpha}{\Gamma(1+\alpha)} + x^2\frac{t^{2\alpha}}{\Gamma(1+2\alpha)} + x^2\frac{t^{3\alpha}}{\Gamma(1+3\alpha)} + \cdots,$$
$$= x^2\left(1 + \frac{t^\alpha}{\Gamma(1+\alpha)} + \frac{t^{2\alpha}}{\Gamma(1+2\alpha)} + \frac{t^{3\alpha}}{\Gamma(1+3\alpha)} + \cdots\right) = x^2 E(t^\alpha),\tag{16.47}$$

where $E(t^\alpha)$ is called the Mittag-Leffler function. In particular, at $\alpha = 1$, Eq. (16.47) reduces to $u(x,t) = x^2 e^t$, which is same as the solution of Sadighi et al. (2008).

Example 16.2 Consider the following nonlinear advection equation (Wazwaz 2007):

$$\frac{\partial^\alpha u}{\partial t^\alpha} + u\frac{\partial u}{\partial x} = 0, \quad 0 < \alpha \le 1,\tag{16.48}$$

with the initial condition:

$$u(x,0) = -x.\tag{16.49}$$

Solution

Here, we chose the initial approximation as:

$$u_0(x,t) = u(x,0) = -x.\tag{16.50}$$

Applying Laplace transform on both sides of Eq. (16.48) and simplifying, we obtain

$$L[u(x,t)] = \frac{-x}{s} - \frac{1}{s^\alpha}L\left[u\frac{\partial u(x,t)}{\partial x}\right]. \tag{16.51}$$

From Eq. (16.48), the nonlinear operator is written as:

$$N[\phi(x,t;p)] = L[\phi(x,t;p)] + \frac{x}{s}\left(1-\frac{\chi_m}{n}\right) + \frac{1}{s^\alpha}L\left[\phi(x,t;p)\frac{\partial\phi(x,t;p)}{\partial x}\right]. \tag{16.52}$$

The mth-order deformation equation when $H(x,t) = 1$ is

$$L[u_m(x,t) - \chi_m u_{m-1}(x,t)] = \hbar R_m\left(\overrightarrow{u_{m-1}}(x,t)\right), \tag{16.53}$$

where

$$R_m\left(\overrightarrow{u_{m-1}}(x,t)\right) = L[u_{m-1}(x,t)] + \frac{x}{s}\left(1-\frac{\chi_m}{n}\right) + \frac{1}{s^\alpha}L\left[\sum_{k=0}^{m-1}u_k(x,t)(u_{m-1-k}(x,t))_x\right]. \tag{16.54}$$

Taking inverse Laplace transform on Eq. (16.53), we get

$$u_m(x,t) = \chi_m u_{m-1}(x,t) + L^{-1}\left[\hbar R_m\left(\overrightarrow{u_{m-1}}(x,t)\right)\right]. \tag{16.55}$$

Now, from Eq. (16.55) for $m \geq 1$, we find

$$u_1(x,t) = \chi_1 u_0(x,t) + L^{-1}\left[\hbar R_1\left(\overrightarrow{u_0}(x,t)\right)\right] = \frac{\hbar x t^\alpha}{\Gamma(\alpha+1)}, \tag{16.56}$$

$$u_2(x,t) = \chi_2 u_1(x,t) + L^{-1}\left[\hbar R_2\left(\overrightarrow{u_1}(x,t)\right)\right] = \frac{n\hbar x t^\alpha}{\Gamma(\alpha+1)} + \hbar\left(\frac{\hbar x t^\alpha}{\Gamma(\alpha+1)} - \frac{2\hbar x t^{2\alpha}}{\Gamma(2\alpha+1)}\right)$$
$$= \frac{x t^\alpha}{\Gamma(\alpha+1)}\hbar(\hbar+n) - \frac{2\hbar^2 x t^{2\alpha}}{\Gamma(2\alpha+1)}, \tag{16.57}$$

$$u_3(x,t) = \chi_3 u_2(x,t) + L^{-1}\left[\hbar R_3\left(\overrightarrow{u_2}(x,t)\right)\right] = \frac{x t^\alpha}{\Gamma(\alpha+1)}\hbar(\hbar+n)^2 - \frac{4x t^{2\alpha}}{\Gamma(2\alpha+1)}\hbar^2(\hbar+n)$$
$$+ \frac{x\hbar^3\Gamma(2\alpha+1)t^{3\alpha}}{(\Gamma(\alpha+1))^2\Gamma(3\alpha+1)} + \frac{4\hbar^3 x t^{3\alpha}}{\Gamma(3\alpha+1)}, \tag{16.58}$$

and so on. The fourth-order series solution of Eq. (16.48) by q-HALTM can be written in the form:

$$u_{q-HALTM}(x,t;n,\hbar) = u_0(x,t;n,\hbar) + \sum_{i=1}^{3}u_i(x,t;n,\hbar)\left(\frac{1}{n}\right)^i.$$

or, especially when $\hbar = -1$ and $n = 1$,

$$u_{q-HALTM}(x,t) = -x - x\frac{t^\alpha}{\Gamma(1+\alpha)} - 2x\frac{t^{2\alpha}}{\Gamma(1+2\alpha)} - 4x\frac{t^{3\alpha}}{\Gamma(1+3\alpha)} - \frac{x\Gamma(1+2\alpha)}{(\Gamma(1+\alpha))^2}\frac{t^{3\alpha}}{\Gamma(1+3\alpha)} - \cdots. \tag{16.59}$$

Particularly at $\alpha = 1$, Eq. (16.59) reduces to a solution $u(x,t) = \frac{x}{t-1}$, which is same as the solution of Wazwaz (2007).

16.7.2 Implementation of q-HASTM

In order to solve Eqs. (16.33)–(16.35) using q-HASTM, let us choose the initial approximation as Eq. (16.36). Applying Sumudu transform on both sides of Eq. (16.33) and substituting the initial condition, we obtain

$$S[u(x,t)] = x^2 + \frac{s^\alpha}{2}S\left[x^2\frac{\partial^2 u(x,t)}{\partial x^2}\right]. \tag{16.60}$$

The nonlinear operator is

$$N[\varphi(x,t;p)] = S[\varphi(x,t;p)] - x^2\left(1 - \frac{\chi_m}{n}\right) - \frac{s^\alpha}{2}S\left[x^2\frac{\partial^2\varphi(x,t;p)}{\partial x^2}\right]. \tag{16.61}$$

Now, the mth-order deformation equation when $H(x,t) = 1$ can be written as:

$$S[u_m(x,t) - \chi_m u_{m-1}(x,t)] = \hbar R_m\left(\overrightarrow{u_{m-1}}(x,t)\right), \tag{16.62}$$

where

$$R_m\left(\overrightarrow{u_{m-1}}(x,t)\right) = S[u_{m-1}(x,t)] - x^2\left(1 - \frac{\chi_m}{n}\right) - \frac{s^\alpha}{2}S\left[x^2\frac{\partial^2 u_{m-1}(x,t)}{\partial x^2}\right]. \tag{16.63}$$

Applying inverse Sumudu transform on Eq. (16.62), it gives

$$u_m(x,t) = \chi_m u_{m-1}(x,t) + S^{-1}\left\{\hbar\mathfrak{R}_m\left(\overrightarrow{u_{m-1}}(x,t)\right)\right\}. \tag{16.64}$$

From Eqs. (16.36) and (16.64), we have

$$u_0(x,t) = x^2, \tag{16.65}$$

$$u_1(x,t) = \chi_1 u_0(x,t) + S^{-1}\left[\hbar R_1\left(\overrightarrow{u_0}(x,t)\right)\right] = \frac{-\hbar x^2 t^\alpha}{\Gamma(\alpha+1)}, \tag{16.66}$$

$$\begin{aligned} u_2(x,t) = \chi_2 u_1(x,t) + S^{-1}\left[\hbar R_2\left(\overrightarrow{u_1}(x,t)\right)\right] &= \frac{-n\hbar x^2 t^\alpha}{\Gamma(\alpha+1)} + \hbar\left(\frac{-\hbar x^2 t^\alpha}{\Gamma(\alpha+1)} + \frac{\hbar x^2 t^{2\alpha}}{\Gamma(2\alpha+1)}\right) \\ &= \frac{-x^2 t^\alpha}{\Gamma(\alpha+1)}\hbar(\hbar+n) + \frac{\hbar^2 x^2 t^{2\alpha}}{\Gamma(2\alpha+1)}, \end{aligned} \tag{16.67}$$

$$\begin{aligned} u_3(x,t) = \chi_3 u_2(x,t) + S^{-1}\left[\hbar R_3\left(\overrightarrow{u_2}(x,t)\right)\right] &= \frac{-x^2 t^\alpha}{\Gamma(\alpha+1)}\hbar(\hbar+n)^2 + \frac{x^2 t^{2\alpha}}{\Gamma(1+2\alpha)}2\hbar^2(n+\hbar) \\ &\quad - \frac{\hbar^3 x^2 t^{3\alpha}}{\Gamma(3\alpha+1)} \end{aligned} \tag{16.68}$$

and so on. Therefore, the fourth-order approximate solution of Eq. (16.33) is given by:

$$u_{q-HASTM}(x,t;n,\hbar) = u_0(x,t;n,\hbar) + \sum_{i=1}^{3} u_i(x,t;n,\hbar)\left(\frac{1}{n}\right)^i. \tag{16.69}$$

By substituting $\hbar = -1$ and $n = 1$, the solution of Eq. (16.33) may be written as:

$$u(x,t) = x^2\left(1 + \frac{t^\alpha}{\Gamma(1+\alpha)} + \frac{t^{2\alpha}}{\Gamma(1+2\alpha)} + \frac{t^{3\alpha}}{\Gamma(1+3\alpha)} + \cdots\right) = x^2 E(t^\alpha). \tag{16.70}$$

where $E(t^\alpha)$ is called the Mittag-Leffler function. In particular, at $\alpha = 1$, Eq. (16.70) reduces to $u(x,t) = x^2 e^t$, which is same as the solution of Sadighi et al. (2008).

Further, taking Sumudu transform on both sides of Eq. (16.48) and simplifying, we have

$$S[u(x,t)] = -x - s^\alpha S\left[u\frac{\partial u(x,t)}{\partial x}\right]. \tag{16.71}$$

From Eq. (16.48), the nonlinear operator is written as:

$$N[\phi(x,t;p)] = S[\phi(x,t;p)] + x\left(1 - \frac{\chi_m}{n}\right) + s^\alpha S\left[\phi(x,t;p)\frac{\partial\phi(x,t;p)}{\partial x}\right]. \tag{16.72}$$

The mth-order deformation equation with the assumption $H(x,t) = 1$ is Eq. (16.62), where

$$R_m\left(\overrightarrow{u_{m-1}}(x,t)\right) = S[u_{m-1}(x,t)] + x\left(1 - \frac{\chi_m}{n}\right) + s^\alpha S\left[\sum_{k=0}^{m-1} u_k(x,t)(u_{m-1-k}(x,t))_x\right]. \tag{16.73}$$

Taking inverse Sumudu transform on Eq. (16.62), it gives Eq. (16.64).

Now, from Eqs. (16.50) and (16.64) for $m \geq 1$, we find

$$u_0(x,t) = -x, \tag{16.74}$$

$$u_1(x,t) = \frac{\hbar x t^\alpha}{\Gamma(\alpha+1)}, \tag{16.75}$$

$$u_2(x,t) = \frac{n\hbar x t^\alpha}{\Gamma(\alpha+1)} + \hbar\left(\frac{\hbar x t^\alpha}{\Gamma(\alpha+1)} - \frac{2\hbar x t^{2\alpha}}{\Gamma(2\alpha+1)}\right) = \frac{x t^\alpha}{\Gamma(\alpha+1)}\hbar(\hbar+n) - \frac{2\hbar^2 x t^{2\alpha}}{\Gamma(2\alpha+1)}, \tag{16.76}$$

$$u_3(x,t) = \frac{x t^\alpha}{\Gamma(\alpha+1)}\hbar(\hbar+n)^2 - \frac{4x t^{2\alpha}}{\Gamma(2\alpha+1)}\hbar^2(\hbar+n) + \frac{x\hbar^3\Gamma(2\alpha+1)t^{3\alpha}}{(\Gamma(\alpha+1))^2\Gamma(3\alpha+1)} + \frac{4\hbar^3 x t^{3\alpha}}{\Gamma(3\alpha+1)}, \tag{16.77}$$

and so on. The fourth-order series solution is written in the form:

$$u_{q-HASTM}(x,t;n,\hbar) = u_0(x,t;n,\hbar) + \sum_{i=1}^{3} u_i(x,t;n,\hbar)\left(\frac{1}{n}\right)^i. \tag{16.78}$$

or, especially when $\hbar = -1$ and $n = 1$,

$$u_{q-HASTM}(x,t) = -x - x\frac{t^\alpha}{\Gamma(1+\alpha)} - 2x\frac{t^{2\alpha}}{\Gamma(1+2\alpha)} - 4x\frac{t^{3\alpha}}{\Gamma(1+3\alpha)} - \frac{x\,\Gamma(1+2\alpha)}{(\Gamma(1+\alpha))^2}\frac{t^{3\alpha}}{\Gamma(1+3\alpha)} - .... \tag{16.79}$$

Particularly at $\alpha = 1$, Eq. (16.79) reduces to a solution $u(x,t) = \frac{x}{t-1}$, which is same as the solution of Wazwaz (2007).

16.7.3 Implementation of q-HAETM

To solve Eqs. (16.33)–(16.35) using q-HAETM, we take the initial approximation as:

$$u_0(x,t) = u(x,0) = x^2. \tag{16.80}$$

Applying Elzaki transform on Eq. (16.33) and using Eq. (16.80), we get

$$E[u(x,t)] = x^2 s^2 + \frac{s^\alpha}{2}E\left[x^2\frac{\partial^2 u(x,t)}{\partial x^2}\right]. \tag{16.81}$$

The nonlinear operator may be written as:

$$N[\varphi(x,t;p)] = E[\varphi(x,t;p)] - x^2 s^2\left(1 - \frac{\chi_m}{n}\right) - \frac{s^\alpha}{2}E\left[x^2\frac{\partial^2 \varphi(x,t;p)}{\partial x^2}\right]. \tag{16.82}$$

Now, the mth-order deformation equation with the assumption $H(x,t) = 1$ may be written as:

$$E[u_m(x,t) - \chi_m u_{m-1}(x,t)] = \hbar R_m\left(\overrightarrow{u_{m-1}}(x,t)\right), \tag{16.83}$$

where

$$R_m\left(\overrightarrow{u_{m-1}}(x,t)\right) = E[u_{m-1}(x,t)] - x^2 s^2\left(1 - \frac{\chi_m}{n}\right) - \frac{s^\alpha}{2}E\left[x^2\frac{\partial^2 u_{m-1}(x,t)}{\partial x^2}\right]. \tag{16.84}$$

Applying inverse Elzaki transform in Eq. (16.83), we have

$$u_m(x,t) = \chi_m u_{m-1}(x,t) + E^{-1}\left\{\hbar\mathfrak{R}_m\left(\overrightarrow{u_{m-1}}(x,t)\right)\right\}. \tag{16.85}$$

From Eqs. (16.80) and (16.85), we have

$$u_0(x,t) = x^2, \tag{16.86}$$

$$u_1(x,t) = \chi_1 u_0(x,t) + E^{-1}\left[\hbar R_1\left(\overrightarrow{u_0}(x,t)\right)\right] = \frac{-\hbar x^2 t^\alpha}{\Gamma(\alpha+1)}, \tag{16.87}$$

$$u_2(x,t) = \chi_2 u_1(x,t) + E^{-1}\left[\hbar R_2\left(\overrightarrow{u_1(x,t)}\right)\right] = \frac{-n\hbar x^2 t^\alpha}{\Gamma(\alpha+1)} + \hbar\left(\frac{-\hbar x^2 t^\alpha}{\Gamma(\alpha+1)} + \frac{\hbar x^2 t^{2\alpha}}{\Gamma(2\alpha+1)}\right)$$
$$= \frac{-x^2 t^\alpha}{\Gamma(\alpha+1)}\hbar(\hbar+n) + \frac{\hbar^2 x^2 t^{2\alpha}}{\Gamma(2\alpha+1)}, \tag{16.88}$$

$$u_3(x,t) = \chi_3 u_2(x,t) + E^{-1}\left[\hbar R_3\left(\overrightarrow{u_2(x,t)}\right)\right] = \frac{-x^2 t^\alpha}{\Gamma(\alpha+1)}\hbar(\hbar+n)^2 + \frac{x^2 t^{2\alpha}}{\Gamma(1+2\alpha)}2\hbar^2(n+\hbar)$$
$$- \frac{\hbar^3 x^2 t^{3\alpha}}{\Gamma(3\alpha+1)} \tag{16.89}$$

and so on. Therefore, the fourth-order approximate solution of Eq. (16.33) is given by:

$$u_{q-HAETM}(x,t;n,\hbar) = u_0(x,t;n,\hbar) + \sum_{i=1}^{3} u_i(x,t;n,\hbar)\left(\frac{1}{n}\right)^i. \tag{16.90}$$

By substituting $\hbar = -1$ and $n = 1$, the solution of the fractional heat-like equation may be obtained as:

$$u(x,t) = x^2\left(1 + \frac{t^\alpha}{\Gamma(1+\alpha)} + \frac{t^{2\alpha}}{\Gamma(1+2\alpha)} + \frac{t^{3\alpha}}{\Gamma(1+3\alpha)} + ...\right) = x^2 E(t^\alpha). \tag{16.91}$$

When $\alpha = 1$ Eq. (16.91) reduces to $u(x,t) = x^2 e^t$, which is same as the solution of Sadighi et al. (2008). Again, applying Elzaki transform on both sides of Eq. (16.48) and using Eq. (16.50), we obtain

$$E[u(x,t)] = -xs^2 - s^\alpha E\left[u\frac{\partial u(x,t)}{\partial x}\right]. \tag{16.92}$$

From Eq. (16.48), the nonlinear operator is written as:

$$N[\phi(x,t;p)] = E[\phi(x,t;p)] + xs^2\left(1 - \frac{\chi_m}{n}\right) + s^\alpha E\left[\phi(x,t;p)\frac{\partial\phi(x,t;p)}{\partial x}\right]. \tag{16.93}$$

The mth-order deformation equation when $H(x,t) = 1$ is

$$E[u_m(x,t) - \chi_m u_{m-1}(x,t)] = \hbar R_m\left(\overrightarrow{u_{m-1}(x,t)}\right), \tag{16.94}$$

where

$$R_m\left(\overrightarrow{u_{m-1}(x,t)}\right) = E[u_{m-1}(x,t)] + xs^2\left(1 - \frac{\chi_m}{n}\right) + s^\alpha E\left[\sum_{k=0}^{m-1} u_k(x,t)(u_{m-1-k}(x,t))_x\right]. \tag{16.95}$$

Taking inverse Elzaki transform on Eq. (16.94), we get

$$u_m(x,t) = \chi_m u_{m-1}(x,t) + E^{-1}\left[\hbar R_m\left(\overrightarrow{u_{m-1}(x,t)}\right)\right]. \tag{16.96}$$

Now, from Eqs. (16.96) and (16.50) for $m \geq 1$, the following expressions are obtained as:

$$u_0(x,t) = -x, \tag{16.97}$$

$$u_1(x,t) = \frac{\hbar x t^\alpha}{\Gamma(\alpha+1)}, \tag{16.98}$$

$$u_2(x,t) = \frac{n\hbar x t^\alpha}{\Gamma(\alpha+1)} + \hbar\left(\frac{\hbar x t^\alpha}{\Gamma(\alpha+1)} - \frac{2\hbar x t^{2\alpha}}{\Gamma(2\alpha+1)}\right) = \frac{x t^\alpha}{\Gamma(\alpha+1)}\hbar(\hbar+n) - \frac{2\hbar^2 x t^{2\alpha}}{\Gamma(2\alpha+1)}, \tag{16.99}$$

$$u_3(x,t) = \frac{x t^\alpha}{\Gamma(\alpha+1)}\hbar(\hbar+n)^2 - \frac{4x t^{2\alpha}}{\Gamma(2\alpha+1)}\hbar^2(\hbar+n) + \frac{x\hbar^3\Gamma(2\alpha+1)t^{3\alpha}}{(\Gamma(\alpha+1))^2\Gamma(3\alpha+1)} + \frac{4\hbar^3 x t^{3\alpha}}{\Gamma(3\alpha+1)}, \tag{16.100}$$

and so on. The fourth-order series solution of Eq. (16.48) by q-HAETM can be written in the form:

$$u_{q-HAETM}(x,t;n,\hbar) = u_0(x,t;n,\hbar) + \sum_{i=1}^{3} u_i(x,t;n,\hbar)\left(\frac{1}{n}\right)^i. \tag{16.101}$$

or, especially when $\hbar = -1$ and $n = 1$,

$$u_{q-HAETM}(x,t) = -x - x\frac{t^\alpha}{\Gamma(1+\alpha)} - 2x\frac{t^{2\alpha}}{\Gamma(1+2\alpha)} - 4x\frac{t^{3\alpha}}{\Gamma(1+3\alpha)} - \frac{x\Gamma(1+2\alpha)}{(\Gamma(1+\alpha))^2}\frac{t^{3\alpha}}{\Gamma(1+3\alpha)} - \dots, \qquad (16.102)$$

Particularly at $\alpha = 1$, Eq. (16.102) reduces to a solution $u(x,t) = \dfrac{x}{t-1}$, which is same as the solution of Wazwaz (2007).

16.7.4 Implementation of q-HAATM

To solve Eqs. (16.33)–(16.35) using q-HAATM, we have taken the initial approximation as:

$$u_0(x,t) = u(x,0) = x^2. \qquad (16.103)$$

Applying Aboodh transform on Eq. (16.33) and using initial approximation Eq. (16.103), we get

$$A[u(x,t)] = \frac{x^2}{s^2} + \frac{s^{-\alpha}}{2}A\left[x^2\frac{\partial^2 u(x,t)}{\partial x^2}\right]. \qquad (16.104)$$

The nonlinear operator is obtained as:

$$N[\varphi(x,t;p)] = A[\varphi(x,t;p)] - \frac{x^2}{s^2}\left(1-\frac{\chi_m}{n}\right) - \frac{s^{-\alpha}}{2}A\left[x^2\frac{\partial^2\varphi(x,t;p)}{\partial x^2}\right]. \qquad (16.105)$$

Now, the mth-order deformation equation for $H(x,t) = 1$ may be written as:

$$A[u_m(x,t) - \chi_m u_{m-1}(x,t)] = \hbar R_m\left(\overrightarrow{u_{m-1}}(x,t)\right), \qquad (16.106)$$

where

$$R_m\left(\overrightarrow{u_{m-1}}(x,t)\right) = A[u_{m-1}(x,t)] - \frac{x^2}{s^2}\left(1-\frac{\chi_m}{n}\right) - \frac{s^{-\alpha}}{2}A\left[x^2\frac{\partial^2 u_{m-1}(x,t)}{\partial x^2}\right]. \qquad (16.107)$$

Applying inverse Aboodh transform, Eq. (16.106) reduces to

$$u_m(x,t) = \chi_m u_{m-1}(x,t) + A^{-1}\left\{\hbar\Re_m\left(\overrightarrow{u_{m-1}}(x,t)\right)\right\}. \qquad (16.108)$$

From Eqs. (16.108) for $m \geq 1$, the following expressions are obtained:

$$u_1(x,t) = \chi_1 u_0(x,t) + A^{-1}\left[\hbar R_1\left(\overrightarrow{u_0}(x,t)\right)\right] = \frac{-\hbar x^2 t^\alpha}{\Gamma(\alpha+1)}, \qquad (16.109)$$

$$u_2(x,t) = \chi_2 u_1(x,t) + A^{-1}\left[\hbar R_2\left(\overrightarrow{u_1}(x,t)\right)\right] = \frac{-n\hbar x^2 t^\alpha}{\Gamma(\alpha+1)} + \hbar\left(\frac{-\hbar x^2 t^\alpha}{\Gamma(\alpha+1)} + \frac{\hbar x^2 t^{2\alpha}}{\Gamma(2\alpha+1)}\right)$$
$$= \frac{-x^2 t^\alpha}{\Gamma(\alpha+1)}\hbar(\hbar+n) + \frac{\hbar^2 x^2 t^{2\alpha}}{\Gamma(2\alpha+1)}, \qquad (16.110)$$

$$u_3(x,t) = \chi_3 u_2(x,t) + A^{-1}\left[\hbar R_3\left(\overrightarrow{u_2}(x,t)\right)\right] = \frac{-x^2 t^\alpha}{\Gamma(\alpha+1)}\hbar(\hbar+n)^2 + \frac{x^2 t^{2\alpha}}{\Gamma(1+2\alpha)}2\hbar^2(n+\hbar)$$
$$- \frac{\hbar^3 x^2 t^{3\alpha}}{\Gamma(3\alpha+1)} \qquad (16.111)$$

and so on. Hence, the fourth-order approximate solution of Eq. (16.33) is given by:

$$u_{q-HAATM}(x,t;n,\hbar) = u_0(x,t;n,\hbar) + \sum_{i=1}^{3}u_i(x,t;n,\hbar)\left(\frac{1}{n}\right)^i. \qquad (16.112)$$

By substituting $\hbar = -1$ and $n = 1$, the solution of the fractional heat-like equation is obtained as:

$$u(x,t) = x^2\left(1 + \frac{t^\alpha}{\Gamma(1+\alpha)} + \frac{t^{2\alpha}}{\Gamma(1+2\alpha)} + \frac{t^{3\alpha}}{\Gamma(1+3\alpha)} + \dots\right) = x^2 E(t^\alpha). \qquad (16.113)$$

When $\alpha = 1$, Eq. (16.113) reduces to $u(x, t) = x^2 e^t$, which is same as the solution of Sadighi et al. (2008). Further, applying Aboodh transform on both sides of Eq. (16.48) and using Eq. (16.50), we obtain

$$A[u(x,t)] = -\frac{x}{s^2} - s^{-\alpha} A\left[u\frac{\partial u(x,t)}{\partial x}\right].$$ (16.114)

From Eq. (16.48), the nonlinear operator is written as:

$$N[\phi(x,t;p)] = A[\phi(x,t;p)] + \frac{x}{s^2}\left(1 - \frac{\chi_m}{n}\right) + s^{-\alpha} A\left[\phi(x,t;p)\frac{\partial\phi(x,t;p)}{\partial x}\right].$$ (16.115)

The mth-order deformation equation when $H(x, t) = 1$ is

$$A[u_m(x,t) - \chi_m u_{m-1}(x,t)] = \hbar R_m\left(\vec{u}_{m-1}(x,t)\right),$$ (16.116)

where

$$R_m\left(\vec{u}_{m-1}(x,t)\right) = A[u_{m-1}(x,t)] + \frac{x}{s^2}\left(1 - \frac{\chi_m}{n}\right) + s^{-\alpha} A\left[\sum_{k=0}^{m-1} u_k(x,t)(u_{m-1-k}(x,t))_x\right].$$ (16.117)

Taking inverse Aboodh transform on Eq. (16.116), we get

$$u_m(x,t) = \chi_m u_{m-1}(x,t) + A^{-1}\left[\hbar R_m\left(\vec{u_{m-1}}(x,t)\right)\right].$$ (16.118)

Now, from Eqs. (16.118) and (16.50), the following expressions are obtained as:

$$u_0(x,t) = -x,$$ (16.119)

$$u_1(x,t) = \frac{\hbar x t^\alpha}{\Gamma(\alpha+1)},$$ (16.120)

$$u_2(x,t) = \frac{n\hbar x t^\alpha}{\Gamma(\alpha+1)} + \hbar\left(\frac{\hbar x t^\alpha}{\Gamma(\alpha+1)} - \frac{2\hbar x t^{2\alpha}}{\Gamma(2\alpha+1)}\right) = \frac{x t^\alpha}{\Gamma(\alpha+1)}\hbar(\hbar+n) - \frac{2\hbar^2 x t^{2\alpha}}{\Gamma(2\alpha+1)},$$ (16.121)

$$u_3(x,t) = \frac{x t^\alpha}{\Gamma(\alpha+1)}\hbar(\hbar+n)^2 - \frac{4x t^{2\alpha}}{\Gamma(2\alpha+1)}\hbar^2(\hbar+n) + \frac{x\hbar^3\Gamma(2\alpha+1)t^{3\alpha}}{(\Gamma(\alpha+1))^2\Gamma(3\alpha+1)} + \frac{4\hbar^3 x t^{3\alpha}}{\Gamma(3\alpha+1)},$$ (16.122)

and so on. The fourth-order series solution of Eq. (16.48) by q-HAATM can be written in the form:

$$u_{q-HAATM}(x,t;n,\hbar) = u_0(x,t;n,\hbar) + \sum_{i=1}^{3} u_i(x,t;n,\hbar)\left(\frac{1}{n}\right)^i.$$ (16.123)

or, especially when $\hbar = -1$ and $n = 1$,

$$u_{q-HAATM}(x,t) = -x - x\frac{t^\alpha}{\Gamma(1+\alpha)} - 2x\frac{t^{2\alpha}}{\Gamma(1+2\alpha)} - 4x\frac{t^{3\alpha}}{\Gamma(1+3\alpha)} - \frac{x\,\Gamma(1+2\alpha)}{(\Gamma(1+\alpha))^2}\frac{t^{3\alpha}}{\Gamma(1+3\alpha)} - \cdots$$ (16.124)

Particularly at $\alpha = 1$, Eq. (16.124) reduces to a solution $u(x,t) = \frac{x}{t-1}$, which is same as the solution of Wazwaz (2007).

References

Aboodh, K.S. (2013). The new integral transform "AboodhTransform". *Global Journal of Pure and Applied Mathematics* 9 (1): 35–43.

Aboodh, K.S., Idris, A., and Nuruddeen, R.I. (2017). On the Aboodh transform connections with some famous integral transforms. *International Journal of Engineering and Information Systems* 1 (9): 143–151.

Baleanu, D. and Jassim, H.K. (2019). A modification fractional homotopy perturbation method for solving Helmholtz and Coupled Helmholtz equations on cantor sets. *Fractal and Fractional* 3 (2): 30.

Jena, R.M. and Chakraverty, S. (2019). Q-Homotopy analysis Aboodh transform method based solution of proportional delay time-fractional partial differential equations. *Journal of Interdisciplinary Mathematics* 22: 931–950.

Jena, R.M. and Chakraverty, S. (2019). Analytical solution of Bagley–Torvik equations using Sumudu transformation method. *SN Applied Sciences* 1 (3): 246.

Jena, R.M. and Chakraverty, S. (2019). Solving time-fractional Navier–Stokes equations using homotopy perturbation Elzaki transform. *SN Applied Sciences* 1 (1): 16.

Jena, R.M., Chakraverty, S., and Yavuz, M. (2020). Two-hybrid techniques coupled with an integral transformation for caputo time-fractional Navier–Stokes equations. *Progress in Fractional Differentiation and Applications* 6 (3): 201–213.

Liao, S.J. (2003). *Beyond Perturbation: Introduction to the Homotopy Analysis Method*. Boca Raton: Chapman and Hall/CRC Press.

Liao, S.J. (2004). On the homotopy analysis method for nonlinear problems. *Applied Mathematics and Computation* 147: 499–513.

Özis, T. and Agırseven, D. (2008). He's homotopy perturbation method for solving heat-like and wave-like equations with variable coefficients. *Physics Letter A* 372: 5944–5950.

Sadighi, A., Ganji, D.D., Gorji, M., and Tolou, N. (2008). Numerical simulation of heat-like models with variable coefficients by the variational iteration method. *Journal of Physics: Conference Series* 96: 012083.

Srivastava, H.M., Kumar, D., and Singh, J. (2017). An efficient analytical technique for fractional model of vibration equation. *Applied Mathematical Modelling* 45: 192–204.

Veeresha, P. and Prakasha, D.G. (2020). A reliable analytical technique for fractional Caudrey–Dodd–Gibbon equation with Mittag–Leffler kernel. *Nonlinear Engineering* 9: 319–328.

Veeresha, P., Prakasha, D.G., and Baleanu, D. (2019). An efficient numerical technique for the nonlinear fractional Kolmogorov–Petrovskii–Piskunov equation. *Mathematics* 7: 265.

Wazwaz, A.M. (2007). A comparison between the variational iteration method and adomian decomposition method. *Journal of Computational and Applied Mathematics* 207: 129–136.

17

(G'/G)-Expansion Method

17.1 Introduction

Several mathematical approaches (Diethelm et al. 2002; Erturk and Momani 2008; Behroozifar and Ahmadpour 2017; Jena et al. 2020a, 2020b) have been developed for solving fractional differential equations. However, finding exact solutions to nonlinear fractional differential equations (FDEs) was challenging until (Li and He 2010) introduced a fractional complex transform. Fractional complex transform turns FDEs into ordinary differential equations (ODEs), allowing all analytical methods used to solve ODEs. The (G'/G)-expansion method (Bin 2012; Gepreel and Omran 2012; Bekir and Guner 2013; Bekir and Guner 2014; Bekir et al. 2015; Khan et al. 2019) is a powerful method among them to solve FDEs. One advantage of this technique is that it allows us to find the exact solution of FDEs without initial or boundary conditions. Exact solutions can be achieved via this method by solving a set of linear or nonlinear algebraic equations.

17.2 Description of the (G'/G)-Expansion Method

This part will briefly describe the main steps of this method for solving fractional partial differential equations.

Step 1. In order to understand the (G'/G)-expansion method (Bin 2012; Gepreel and Omran 2012; Bekir et al. 2015), let us consider the following nonlinear fractional partial differential equation in two independent variables x and t of the type

$$Q\left(u, u_x, u_{xx}, ..., D_t^\alpha u, ...\right) = 0, \quad 0 < \alpha \le 1, \tag{17.1}$$

where $u = u(x, t)$ is an unknown function, and Q is a polynomial of u and its partial fractional derivatives, in which the highest order derivatives and the nonlinear terms are involved.

Step 2. The traveling wave variable is (Bekir et al. 2015)

$$u(x, t) = U(\xi), \tag{17.2}$$

$$\xi = kx - \frac{ct^\alpha}{\Gamma(1 + \alpha)}, \tag{17.3}$$

where c and k are nonzero constants to be determined later. By using the chain rule (Güner et al. 2015), we have

$$D_t^\alpha u = \Delta_t u_\zeta D_t^\alpha \zeta,$$
$$D_x^\alpha u = \Delta_x u_\zeta D_x^\alpha \zeta,$$

where Δ_t and Δ_x are the fractal indexes (Güner et al. 2015). Without loss of generality, we can consider $\Delta_t = \Delta_x = \kappa$, where κ is a constant. Using Eqs. (17.2) and (17.3), Eq. (17.1) can be rewritten in the following nonlinear ODE form

$$P\left(U, kU', k^2 U'', ..., -cU', ...\right) = 0, \tag{17.4}$$

where the prime denotes the derivative with respect to ξ.

Computational Fractional Dynamical Systems: Fractional Differential Equations and Applications, First Edition.
Snehashish Chakraverty, Rajarama Mohan Jena, and Subrat Kumar Jena.
© 2023 John Wiley & Sons, Inc. Published 2023 by John Wiley & Sons, Inc.

Step 3. We suppose that the (G'/G)-expansion approach is based on the assumption that the solution of the ordinary differential equation Eq. (17.4) may be represented in the following form (Bekir et al. 2015)

$$U(\xi) = \sum_{i=0}^{m} a_i \left(\frac{G'}{G}\right)^i, \qquad a_m \neq 0, \tag{17.5}$$

where a_i $(i = 0, 1, 2, ..., m)$ are constants, while $G = G(\xi)$ will satisfy the following second-order linear differential equation (Bekir et al. 2015)

$$G''(\xi) + \lambda G'(\xi) + \mu\, G(\xi) = 0, \tag{17.6}$$

where λ and μ are constants. Then, the positive integer m may be calculated by considering the homogenous balance between the highest order derivatives and the nonlinear terms arising in Eq. (17.4). More precisely, we define the degree of $U(\xi)$ as $\deg[U(\xi)] = m$, which leads to the degrees of the other expressions as follows (Gepreel and Omran 2012):

$$\deg\left[\frac{d^p U}{d\xi^p}\right] = m + p,$$
$$\deg\left[U^s\left(\frac{d^p U}{d\xi^p}\right)^q\right] = ms + q\,(p + m). \tag{17.7}$$

Therefore, we can obtain the value of m in Eq. (17.5).

Now, Eq. (17.6) can be rewritten as

$$\frac{d}{d\xi}\left(\frac{G'}{G}\right) = -\left(\frac{G'}{G}\right)^2 - \lambda\left(\frac{G'}{G}\right) - \mu. \tag{17.8}$$

By the generalized solutions of Eq. (17.6), we have (Shang and Zheng 2013)

$$\left(\frac{G'}{G}\right) = \begin{cases} -\dfrac{\lambda}{2} + \dfrac{\sqrt{\lambda^2 - 4\mu}}{2}\left(\dfrac{c_1 \sinh\left(\dfrac{\sqrt{\lambda^2 - 4\mu}}{2}\xi\right) + c_2 \cosh\left(\dfrac{\sqrt{\lambda^2 - 4\mu}}{2}\xi\right)}{c_1 \cosh\left(\dfrac{\sqrt{\lambda^2 - 4\mu}}{2}\xi\right) + c_2 \sinh\left(\dfrac{\sqrt{\lambda^2 - 4\mu}}{2}\xi\right)}\right), & \lambda^2 - 4\mu > 0, \\[4em] -\dfrac{\lambda}{2} + \dfrac{\sqrt{4\mu - \lambda^2}}{2}\left(\dfrac{-c_1 \sin\left(\dfrac{\sqrt{4\mu - \lambda^2}}{2}\xi\right) + c_2 \cos\left(\dfrac{\sqrt{4\mu - \lambda^2}}{2}\xi\right)}{c_1 \cos\left(\dfrac{\sqrt{4\mu - \lambda^2}}{2}\xi\right) + c_2 \sin\left(\dfrac{\sqrt{4\mu - \lambda^2}}{2}\xi\right)}\right), & \lambda^2 - 4\mu < 0, \\[4em] -\dfrac{\lambda}{2} + \dfrac{c_2}{c_1 + c_2\xi}, & \lambda^2 - 4\mu = 0, \end{cases} \tag{17.9}$$

where c_1 and c_2 are arbitrary constants.

Step 4. Substituting Eq. (17.5) into Eq. (17.4), using Eq. (17.8), and collecting all terms with the same order of (G'/G) together, Eq. (17.4) is converted into another polynomial in (G'/G). Then equating each coefficient of the resulting polynomial to zero, we obtain a set of algebraic equations for λ, μ, k, c, and a_i $(i = 0, 1, 2, ..., m)$.

Step 5. Solving the algebraic equations system in Step 4, subsequently substituting these constants λ, μ, k, c, and a_i $(i = 0, 1, 2, ..., m)$, and using Eq. (17.9), we can construct a variety of exact solutions of Eq. (17.1).

The following sections present two examples to demonstrate the usefulness of the (G'/G)-expansion technique to solve nonlinear fractional differential equations.

17.3 Application Problems

Here, we apply the present method to solve a nonlinear time-fractional Fisher's equation in Example 17.1 and the nonlinear time-fractional KdV-mKdV equation in Example 17.2.

Example 17.1   Let us consider the nonlinear time-fractional Fisher's equation (Veeresha et al. 2019)

$$\frac{\partial^\alpha u}{\partial t^\alpha} = \frac{\partial^2 u}{\partial x^2} + 6u(1-u), \quad t > 0, \quad x \in R, \quad 0 < \alpha \leq 1. \tag{17.10}$$

Solution

Equation (17.10) may be reduced to the ordinary differential equation in ξ by using the above-discussed transformation $u(x,t) = U(\xi), \xi = kx - \dfrac{ct^\alpha}{\Gamma(1+\alpha)}$ as

$$k^2 U'' + cU' + 6\,U - 6\,U^2 = 0, \tag{17.11}$$

where prime denotes the differentiation with respect to ξ. The highest order linear and nonlinear terms in Eq. (17.11) are U'' and U^2. So balancing the order of U'' and U^2 by using the concept Eq. (17.7), we get

$$\deg(U'') = \deg(U^2), \tag{17.12}$$
$$m + 2 = 2m \Rightarrow m = 2.$$

So, we assume the solution of the ordinary differential equation Eq. (17.11) as Eq. (17.5) at $m = 2$. This may be written as

$$U(\xi) = \sum_{i=0}^{2} a_i \left(\frac{G'}{G}\right)^i = a_0 + a_1 \left(\frac{G'}{G}\right)^1 + a_2 \left(\frac{G'}{G}\right)^2, a_2 \neq 0. \tag{17.13}$$

By using Eqs. (17.6) and (17.13) in Eq. (17.11), we have

$$U'(\xi) = -2a_2 \left(\frac{G'}{G}\right)^3 + (-2a_2\lambda - a_1)\left(\frac{G'}{G}\right)^2 + (-2a_2\mu - a_1\lambda)\left(\frac{G'}{G}\right) - a_1\mu, \tag{17.14}$$

$$U''(\xi) = 6a_2 \left(\frac{G'}{G}\right)^4 + (2a_1 + 10a_2\lambda)\left(\frac{G'}{G}\right)^3 + (8a_2\mu + 3a_1\lambda + 4a_2\lambda^2)\left(\frac{G'}{G}\right)^2 + \tag{17.15}$$
$$\left(6a_2\lambda\mu + 2a_1\mu + a_1\lambda^2\right)\left(\frac{G'}{G}\right) + a_1\lambda\mu + 2a_2\mu^2,$$

and

$$U^2(\xi) = a_2^2 \left(\frac{G'}{G}\right)^4 + 2a_1a_2 \left(\frac{G'}{G}\right)^3 + (2a_0a_2 + a_1^2)\left(\frac{G'}{G}\right)^2 + 2a_0a_1 \left(\frac{G'}{G}\right) + a_0^2. \tag{17.16}$$

By plugging Eqs. (17.13)–(17.16) into Eq. (17.11), collecting the coefficients of $\left(\dfrac{G'}{G}\right)^i, i = 0, ..., 4$ and setting these to zero, we obtain the following system

$$\left(\frac{G'}{G}\right)^0 : a_1 k^2 \lambda\mu + 2k^2\mu^2 a_2 - ca_1\mu - 6a_0^2 + 6a_0 = 0,$$

$$\left(\frac{G'}{G}\right)^1 : a_1 k^2\lambda^2 + 6a_2 k^2\lambda\mu + 2k^2 a_1\mu - c\lambda a_1 - 2ca_2\mu - 12a_0 a_1 + 6a_1 = 0,$$

$$\left(\frac{G'}{G}\right)^2 : 4a_2 k^2\lambda^2 + 3k^2\lambda a_1 + 8k^2 a_2\mu - 2ca_2\lambda - ca_1 - 12a_0 a_2 - 6a_1^2 + 6a_2 = 0, \tag{17.17}$$

$$\left(\frac{G'}{G}\right)^3 : 10k^2\lambda a_2 + 2k^2 a_1 - 2ca_2 - 12a_1 a_2 = 0,$$

$$\left(\frac{G'}{G}\right)^4 : 6k^2 a_2 - 6a_2^2 = 0.$$

When we solve a system of linear equations, the point of intersection of the two lines is the solution. However, in the case of a nonlinear system of equations, the graphs might be circles, parabolas, or hyperbolas, with several points of intersection and hence multiple solutions. By solving the above nonlinear system using Maple software, it gives different values of c, k, a_i, $i = 0, 1, 2$ in various cases. Few of them are provided here.

Case 1:

$$a_0 = -\frac{6\mu}{\lambda^2 - 4\mu}, \quad a_1 = -\frac{6\lambda}{\lambda^2 - 4\mu}, \quad a_2 = -\frac{6}{\lambda^2 - 4\mu}, \quad c = 0, \quad k = \frac{\sqrt{6}i}{\sqrt{\lambda^2 - 4\mu}}, \tag{17.18}$$

Case 2:

$$a_0 = -\frac{-\lambda^2 - 2\mu}{\lambda^2 - 4\mu}, \quad a_1 = \frac{6\lambda}{\lambda^2 - 4\mu}, \quad a_2 = \frac{6}{\lambda^2 - 4\mu}, \quad c = 0, \quad k = \frac{\sqrt{6}}{\sqrt{\lambda^2 - 4\mu}}, \tag{17.19}$$

Case 3:

$$a_0 = \frac{(\lambda^2 - 2\mu) \mp \lambda\sqrt{\lambda^2 - 4\mu}}{2(\lambda^2 - 4\mu)}, \quad a_1 = \frac{\lambda \mp \sqrt{\lambda^2 - 4\mu}}{\lambda^2 - 4\mu}, \quad a_2 = \frac{1}{\lambda^2 - 4\mu}, \quad c = 5\frac{\pm 1}{\sqrt{\lambda^2 - 4\mu}},$$
$$k = \frac{\pm 1}{\sqrt{\lambda^2 - 4\mu}}, \tag{17.20}$$

where λ and μ are arbitrary constants. The solution to the problem may be obtained by considering all of the cases one by one. However, only Case 3 is being considered for getting the solution in this case. Similarly, Cases 1 and 2 can also be used to find other solutions.

Substituting the values of constants of Eq. (17.20) into Eq. (17.13), we obtain

$$U(\xi) = \left(\frac{(\lambda^2 - 2\mu) \mp \lambda\sqrt{\lambda^2 - 4\mu}}{2(\lambda^2 - 4\mu)}\right) + \left(\frac{\lambda \mp \sqrt{\lambda^2 - 4\mu}}{\lambda^2 - 4\mu}\right)\left(\frac{G'}{G}\right)^1 + \left(\frac{1}{\lambda^2 - 4\mu}\right)\left(\frac{G'}{G}\right)^2, \tag{17.21}$$

where

$$\xi = \left(\frac{\pm 1}{\sqrt{\lambda^2 - 4\mu}}\right)x - \left(5\frac{\pm 1}{\sqrt{\lambda^2 - 4\mu}}\right)\left(\frac{t^a}{\Gamma(1 + \alpha)}\right). \tag{17.22}$$

Substituting the general solution of Eq. (17.6), which is given in Eq. (17.9) into the above Eq. (17.21), we have three types of traveling wave solution of the nonlinear time-fractional Fisher's equation as follows:

when $\lambda^2 - 4\mu > 0$,

$$U_{1,2}(\xi) = a_0 + a_1\left(-\frac{\lambda}{2} + \frac{\sqrt{\lambda^2 - 4\mu}}{2}\left(\frac{c_1 \sinh\left(\frac{\sqrt{\lambda^2 - 4\mu}}{2}\xi\right) + c_2 \cosh\left(\frac{\sqrt{\lambda^2 - 4\mu}}{2}\xi\right)}{c_1 \cosh\left(\frac{\sqrt{\lambda^2 - 4\mu}}{2}\xi\right) + c_2 \sinh\left(\frac{\sqrt{\lambda^2 - 4\mu}}{2}\xi\right)}\right)\right)^1 + a_2$$

$$\left(-\frac{\lambda}{2} + \frac{\sqrt{\lambda^2 - 4\mu}}{2}\left(\frac{c_1 \sinh\left(\frac{\sqrt{\lambda^2 - 4\mu}}{2}\xi\right) + c_2 \cosh\left(\frac{\sqrt{\lambda^2 - 4\mu}}{2}\xi\right)}{c_1 \cosh\left(\frac{\sqrt{\lambda^2 - 4\mu}}{2}\xi\right) + c_2 \sinh\left(\frac{\sqrt{\lambda^2 - 4\mu}}{2}\xi\right)}\right)\right)^2, \tag{17.23}$$

when $\lambda^2 - 4\mu < 0$,

$$U_{3,4}(\xi) = a_0 + a_1\left(-\frac{\lambda}{2} + \frac{\sqrt{4\mu - \lambda^2}}{2}\left(\frac{-c_1 \sin\left(\frac{\sqrt{4\mu - \lambda^2}}{2}\xi\right) + c_2 \cos\left(\frac{\sqrt{4\mu - \lambda^2}}{2}\xi\right)}{c_1 \cos\left(\frac{\sqrt{4\mu - \lambda^2}}{2}\xi\right) + c_2 \sin\left(\frac{\sqrt{4\mu - \lambda^2}}{2}\xi\right)}\right)\right)^1 + a_2$$

$$\left(-\frac{\lambda}{2} + \frac{\sqrt{4\mu - \lambda^2}}{2}\left(\frac{-c_1 \sin\left(\frac{\sqrt{4\mu - \lambda^2}}{2}\xi\right) + c_2 \cos\left(\frac{\sqrt{4\mu - \lambda^2}}{2}\xi\right)}{c_1 \cos\left(\frac{\sqrt{4\mu - \lambda^2}}{2}\xi\right) + c_2 \sin\left(\frac{\sqrt{4\mu - \lambda^2}}{2}\xi\right)}\right)\right)^2, \tag{17.24}$$

when $\lambda^2 - 4\mu = 0$,

$$U_{5,6}(\xi) = a_0 + a_1 \left(-\frac{\lambda}{2} + \frac{c_2}{c_1 + c_2\xi} \right)^1 + a_2 \left(-\frac{\lambda}{2} + \frac{c_2}{c_1 + c_2\xi} \right)^2, \tag{17.25}$$

where the values of a_0, a_1, a_2, and ξ are given in Eqs. (17.20) and (17.22). All the above Eqs. (17.23)–(17.25) are the exact solutions of the nonlinear time-fractional Fisher's equation. Putting the suitable values of the constants in Eqs. (17.23)–(17.25), we obtain the traveling wave solutions of the given model.

Example 17.2 Consider the following nonlinear time-fractional KdV-mKdV equation (Kaya et al. 2018)

$$\frac{\partial^\alpha u}{\partial t^\alpha} + 2u\frac{\partial u}{\partial x} + 3u^2\frac{\partial u}{\partial x} - \frac{\partial^3 u}{\partial x^3} = 0, \quad 0 < \alpha \leq 1. \tag{17.26}$$

Solution

Using the transformation $u(x,t) = U(\xi), \xi = kx - \dfrac{ct^\alpha}{\Gamma(1+\alpha)}$ in Eq. (17.26), it reduces to

$$cU' - 2kUU' - 3kU^2U' + k^3U''' = 0, \tag{17.27}$$

where prime denotes the differentiation with respect to ξ. The highest order linear and nonlinear terms in Eq. (17.27) are U''' and U^2U'. So balancing the order of U''' and U^2U' by using the concept Eq. (17.7), we get

$$\deg\left(U'''\right) = \deg(U^2U'), \tag{17.28}$$
$$m + 3 = 2m + m + 1 \Rightarrow m = 1.$$

So, we assume the solution of Eq. (17.27) as Eq (17.5) at $m = 1$. This may be written as

$$U(\xi) = \sum_{i=0}^{1} a_i \left(\frac{G'}{G} \right)^i = a_0 + a_1 \left(\frac{G'}{G} \right), a_1 \neq 0. \tag{17.29}$$

By using Eqs. (17.6) and (17.29) in Eq. (17.27), we have

$$U'(\xi) = -a_1\mu - a_1\lambda \left(\frac{G'}{G} \right) - a_1 \left(\frac{G'}{G} \right)^2, \tag{17.30}$$

$$U''(\xi) = 2a_1 \left(\frac{G'}{G} \right)^3 + 3a_1\lambda \left(\frac{G'}{G} \right)^2 + \left(2a_1\mu + a_1\lambda^2 \right) \left(\frac{G'}{G} \right) + a_1\lambda\mu, \tag{17.31}$$

$$U'''(\xi) = -6a_1 \left(\frac{G'}{G} \right)^4 - 12a_1\lambda \left(\frac{G'}{G} \right)^3 + \left(-8a_1\mu - 7a_1\lambda^2 \right) \left(\frac{G'}{G} \right)^2 + \left(-8a_1\lambda\mu - a_1\lambda^3 \right) \left(\frac{G'}{G} \right)$$
$$+ \left(-\lambda^2 a_1\mu - 2\mu^2 a_1 \right), \tag{17.32}$$

$$U(\xi)U'(\xi) = -a_1^2 \left(\frac{G'}{G} \right)^3 + \left(-a_1^2\lambda - a_0a_1 \right) \left(\frac{G'}{G} \right)^2 + \left(-a_1^2\mu - a_0a_1\mu \right) \left(\frac{G'}{G} \right)^1 + \left(-a_0a_1\mu \right), \tag{17.33}$$

and

$$U^2(\xi)U'(\xi) = -a_1^3 \left(\frac{G'}{G} \right)^4 + \left(-a_1^3\lambda - 2a_0a_1^2 \right) \left(\frac{G'}{G} \right)^3 + \left(-a_0^2a_1 - a_1^3\mu - 2a_0a_1^2\lambda \right) \left(\frac{G'}{G} \right)^2$$
$$+ \left(-a_0^2a_1\lambda - 2a_0a_1^2\mu \right) \left(\frac{G'}{G} \right) + \left(-a_0^2a_1\mu \right). \tag{17.34}$$

By substituting Eqs. (17.30)–(17.34) into Eq. (17.27), collecting the coefficients of $\left(\dfrac{G'}{G}\right)^i$, $i = 0, ..., 4$ and setting these to zero, we obtain the following system of the algebraic equation:

$$\left(\frac{G'}{G}\right)^0 : -a_1 k^3 \lambda^2 \mu - 2k^3 \mu^2 a_1 + 3k\mu a_0^2 a_1 + 2k\mu a_0 a_1 - c\mu a_1 = 0,$$

$$\left(\frac{G'}{G}\right)^1 : -a_1 k^3 \lambda^3 - 8a_1 k^3 \lambda \mu + 3k\lambda a_0^2 a_1 + 6k\mu a_0 a_1^2 + 2k\lambda a_0 a_1 + 2k\mu a_1^2 - c\lambda a_1 = 0,$$

$$\left(\frac{G'}{G}\right)^2 : -7a_1 k^3 \lambda^2 - 8k^3 \mu a_1 + 6k\lambda a_0 a_1^2 + 3k\mu a_1^3 + 2k\lambda a_1^2 + 3ka_0^2 a_1 + 2ka_0 a_1 - ca_1 = 0,$$ (17.35)

$$\left(\frac{G'}{G}\right)^3 : -12k^3 \lambda a_1 + 3k\lambda a_1^3 + 6ka_0 a_1^2 + 2ka_1^2 = 0,$$

$$\left(\frac{G'}{G}\right)^4 : -6k^3 a_1 + 3ka_1^3 = 0.$$

By solving the above nonlinear system using Maple software, it gives

$$a_0 = \frac{\pm k\lambda}{\sqrt{2}} - \frac{1}{3}, \quad a_1 = \pm \sqrt{2}k, \quad c = \frac{k^3}{2}(\lambda^2 - 4\mu) - \frac{k}{3}, \quad k = k,$$ (17.36)

where k is a free parameter and λ, μ are arbitrary constants. Using the values of constants from Eq. (17.36), Eq. (17.29) may be written as

$$U(\xi) = \left(\frac{\pm k\lambda}{\sqrt{2}} - \frac{1}{3}\right) + \left(\pm \sqrt{2}k\right)\left(\frac{G'}{G}\right)^1,$$ (17.37)

where

$$\xi = kx - \left(\frac{k^3}{2}(\lambda^2 - 4\mu) - \frac{k}{3}\right)\left(\frac{t^\alpha}{\Gamma(1 + \alpha)}\right).$$ (17.38)

By substituting the general solution of Eq. (17.6), which is provided in Eq. (17.9), into the above mentioned Eq. (17.37), we obtain three types of traveling wave solutions to the nonlinear time-fractional KdV-mKdV equation:

when $\lambda^2 - 4\mu > 0$,

$$U_{1,2}(\xi) = \left(\frac{\pm k\lambda}{\sqrt{2}} - \frac{1}{3}\right) + \left(\pm \sqrt{2}k\right)\left(-\frac{\lambda}{2} + \frac{\sqrt{\lambda^2 - 4\mu}}{2}\left(\frac{c_1 \sinh\left(\frac{\sqrt{\lambda^2 - 4\mu}}{2}\xi\right) + c_2 \cosh\left(\frac{\sqrt{\lambda^2 - 4\mu}}{2}\xi\right)}{c_1 \cosh\left(\frac{\sqrt{\lambda^2 - 4\mu}}{2}\xi\right) + c_2 \sinh\left(\frac{\sqrt{\lambda^2 - 4\mu}}{2}\xi\right)}\right)\right)$$ (17.39)

when $\lambda^2 - 4\mu < 0$,

$$U_{3,4}(\xi) = \left(\frac{\pm k\lambda}{\sqrt{2}} - \frac{1}{3}\right) + \left(\pm \sqrt{2}k\right)\left(-\frac{\lambda}{2} + \frac{\sqrt{4\mu - \lambda^2}}{2}\left(\frac{-c_1 \sin\left(\frac{\sqrt{4\mu - \lambda^2}}{2}\xi\right) + c_2 \cos\left(\frac{\sqrt{4\mu - \lambda^2}}{2}\xi\right)}{c_1 \cos\left(\frac{\sqrt{4\mu - \lambda^2}}{2}\xi\right) + c_2 \sin\left(\frac{\sqrt{4\mu - \lambda^2}}{2}\xi\right)}\right)\right)$$ (17.40)

when $\lambda^2 - 4\mu = 0$,

$$U_{5,6}(\xi) = \left(\frac{\pm k\lambda}{\sqrt{2}} - \frac{1}{3}\right) + \left(\pm \sqrt{2}k\right)\left(-\frac{\lambda}{2} + \frac{c_2}{c_1 + c_2\xi}\right).$$ (17.41)

Equations (17.39)–(17.41) are the exact solutions of the nonlinear time-fractional KdV-mKdV equation.

References

Behroozifar, M. and Ahmadpour, F. (2017). Comparative study on solving fractional differential equations via shifted Jacobi collocation method. *Iranian Mathematical Society* 43 (2): 535–560.

Bekir, A. and Guner, O. (2013). Exact solutions of nonlinear fractional differential equations by (G'/G)-expansion method. *Chinese Physics B* 22 (11): 110202.

Bekir, A. and Guner, O. (2014). The (G'/G)-expansion method using modified Riemann–Liouville derivative for some space-time fractional differential equations. *Ain Shams Engineering Journal* 5 (3): 959–965.

Bekir, A., Guner, O., Bhrawy, A.H., and Biswas, A. (2015). Solving nonlinear fractional differential equations using exp-function and G'/G- expansion methods. *Romanian Journal of Physics* 60 (3-4): 360–378.

Bin, Z. (2012). (G'/G)-Expansion method for solving fractional partial differential equations in the theory of mathematical physics. *Communications in Theoretical Physics* 58 (5): 623–630.

Diethelm, K., Ford, N.J., and Freed, A.D. (2002). A predictor- corrector approach for the numerical solution of fractional differential equations. *Nonlinear Dynamics* 29: 3–22.

Erturk, V. and Momani, S.M. (2008). Solving systems of fractional differential equations using differential transform method. *Journal of Computational and Applied Mathematics* 215 (1): 142–151.

Gepreel, K.A. and Omran, S. (2012). Exact solutions for nonlinear partial fractional differential equations. *Chinese Physics B* 21 (11): 110204.

Güner, O., Bekir, A., and Cevikel, A.C. (2015). A variety of exact solutions for the time fractional Cahn–Allen equation. *The European Physical Journal Plus* 130 (146): 1–13.

Jena, R.M., Chakraverty, S., and Jena, S.K. (2020a). Analysis of the dynamics of phytoplankton nutrient and whooping cough models with nonsingular kernel arising in the biological system. *Chaos, Solitons & Fractals* 141: 110373. (1-10).

Jena, R.M., Chakraverty, S., Rezazadeh, H., and Ganji, D.D. (2020b). On the solution of time-fractional dynamical model of Brusselator reaction-diffusion system arising in chemical reactions. *Mathematical Methods in the Applied Sciences* 43 (7): 3903–3913.

Kaya, D., Gulbahar, S., Yokus, A., and Gulbahar, M. (2018). Solutions of the fractional combined KdV–mKdV equation with collocation method using radial basis function and their geometrical obstructions. *Advances in Difference Equations* 77: 1–16.

Khan, H., Barak, S., Kumam, P., and Arif, M. (2019). Analytical solutions of fractional Klein–Gordon and gas dynamics equations, via the (G'/G)-expansion method. *Symmetry* 11: 566 (1-12).

Li, Z.B. and He, J.H. (2010). Fractional complex transformation for fractional differential equations. *Computer and Mathematics with Applications* 15: 970–973.

Shang, N. and Zheng, B. (2013). Exact solution for three fractional partial differential equations by the (G'/G) Method. *IAENG International Journal of Applied Mathematics* 43 (3): 1–6.

Veeresha, P., Prakasha, D.G., and Baskonus, H.M. (2019). Novel simulations to the tim-fractional Fisher's equation. *Mathematical Sciences* 13: 33–42.

18

(G'/G^2)-Expansion Method

18.1 Introduction

Exact solutions to nonlinear fractional differential equations (FDEs) play a vital role in nonlinear science. Several mathematical approaches (Diethelm et al. 2002; Erturk and Momani 2008; Behroozifar and Ahmadpour 2017) have been developed for solving FDEs. However, finding exact solutions to nonlinear FDEs remained challenging until Li and He (Li and He 2010) proposed a fractional complex transform to convert FDEs to ordinary differential equations (ODEs), allowing all analytical methods for solving ODEs to be used to partial differential equations (PDEs). Among these methods, the new analytical method, namely (G'/G^2)-expansion method (Arshed and Sadia 2018; Mohyud-Din and Bibi 2018; Ali et al. 2019; Hassaballa 2020), has been utilized to obtain the solutions of time and space FDEs. The (G'/G^2)-method is a comparatively new technique for finding the traveling wave solutions to nonlinear single and coupled equations that arise in physics, fluid mechanics, wave propagation, population dynamics, and other fields. This method is more efficient and reliable as compared to the (G'/G)-expansion method. This approach yields hyperbolic, trigonometric, and rational functions as solutions. These solutions are well suited to the investigation of nonlinear physical processes. Similar to the (G'/G)-expansion method, it also allows us to find the exact solution of FDEs without initial or boundary conditions. Again, the exact solutions can be achieved simply by finally solving a set of linear or nonlinear algebraic equations.

18.2 Description of the (G'/G^2)-Expansion Method

This section will briefly describe the main steps of this approach for solving fractional PDEs.

Step 1. In order to understand the (G'/G^2)-expansion method (Arshed and Sadia 2018; Mohyud-Din and Bibi 2018; Ali et al. 2019; Hassaballa 2020), we consider the following nonlinear fractional PDE in two independent variables x and t of the type

$$Q\left(u, u_x, u_{xx}, ..., D_t^\alpha u, ...\right) = 0, \quad 0 < \alpha \leq 1, \tag{18.1}$$

where $u = u(x, t)$ is an unknown function, and Q is a polynomial of u and its partial fractional derivatives, in which the highest order derivatives and the nonlinear terms are involved.

Step 2. The traveling wave variable is (Bekir et al. 2015)

$$u(x, t) = U(\xi), \tag{18.2}$$

$$\xi = kx - \frac{ct^\alpha}{\Gamma(1 + \alpha)}, \tag{18.3}$$

where c and k are nonzero constants to be determined later. By using the chain rule (Güner et al. 2015), we have

$$D_t^\alpha u = \Delta_t u_\zeta D_t^\alpha \zeta,$$

$$D_x^\alpha u = \Delta_x u_\zeta D_x^\alpha \zeta,$$

where Δ_t and Δ_x are the fractal indexes (Güner et al. 2015). Without loss of generality, we can consider $\Delta_t = \Delta_x = \kappa$, where κ is a constant. Using Eqs. (18.2) and (18.3), Eq. (18.1) can be rewritten in the following nonlinear ODE form

$$P\left(U, kU', k^2 U'', ..., -cU', ...\right) = 0, \tag{18.4}$$

where the prime denotes the derivative with respect to ξ.

Computational Fractional Dynamical Systems: Fractional Differential Equations and Applications, First Edition.
Snehashish Chakraverty, Rajarama Mohan Jena, and Subrat Kumar Jena.
© 2023 John Wiley & Sons, Inc. Published 2023 by John Wiley & Sons, Inc.

Step 3. We assume (G'/G^2)-expansion approach for the solution of Eq. (18.4) may be represented in the following form (Arshed and Sadia 2018; Ali et al. 2019; Hassaballa 2020)

$$U(\xi) = a_0 + \sum_{i=1}^{m} \left[a_i \left(\frac{G'}{G^2} \right)^i + b_i \left(\frac{G'}{G^2} \right)^{-i} \right],$$ (18.5)

where a_0, a_i, b_i ($i = 1, 2, ..., m$) are constants to be determined. It is worth noting that the constants a_m or b_m may be zero, but both cannot be zero simultaneously. The function $G = G(\xi)$ satisfies the following second-order linear differential equation (Arshed and Sadia 2018; Mohyud-Din and Bibi 2018; Ali et al. 2019; Hassaballa 2020)

$$\left(\frac{G'}{G^2} \right)' = \mu + \lambda \left(\frac{G'}{G^2} \right)^2,$$ (18.6)

where $\lambda \neq 0$ and $\mu \neq 1$ are integers. Then, the positive integer m may be calculated by considering the homogenous balance between the highest order derivatives and the nonlinear terms arising in Eq. (18.4). More precisely, we define the degree of $U(\xi)$ as $\deg[U(\xi)] = m$, which leads to the degrees of the other expressions as follows (Gepreel and Omran 2012):

$$\deg\left[\frac{d^p U}{d\xi^p} \right] = m + p,$$
$$\deg\left[U^s \left(\frac{d^p U}{d\xi^p} \right)^q \right] = ms + q\,(p + m).$$ (18.7)

Accordingly, we can obtain the value of m in Eq. (18.5).

By the generalized solutions of Eq. (18.6), we have (Arshed and Sadia 2018; Ali et al. 2019; Hassaballa 2020)

$$\left(\frac{G'}{G^2} \right) = \begin{cases} \sqrt{\frac{\mu}{\lambda}} \left(\frac{c_1 \cos\left(\sqrt{\lambda\mu}\xi\right) + c_2 \sin\left(\sqrt{\lambda\mu}\xi\right)}{c_2 \cos\left(\sqrt{\lambda\mu}\xi\right) - c_1 \sin\left(\sqrt{\lambda\mu}\xi\right)} \right), & \lambda\mu > 0, \\ -\frac{\sqrt{|\lambda\mu|}}{\lambda} \left(\frac{c_1 \sinh\left(2\sqrt{|\lambda\mu|}\xi\right) + c_1 \cosh\left(2\sqrt{|\lambda\mu|}\xi\right) + c_2}{c_1 \sinh\left(2\sqrt{|\lambda\mu|}\xi\right) + c_1 \cosh\left(2\sqrt{|\lambda\mu|}\xi\right) - c_2} \right), & \lambda\mu < 0, \\ -\frac{c_1}{\lambda(c_1\xi + c_2)}, & \lambda \neq 0, \mu = 0, \end{cases}$$ (18.8)

where c_1 and c_2 are nonzero constants.

Step 4. Substituting Eq. (18.5) into Eq. (18.4), using Eq. (18.6), and collecting the coefficients of like powers of $\left(\frac{G'}{G^2} \right)^i$, ($i = 0, \pm 1, \pm 2, \pm 3, ...$), Eq. (18.4) is converted into another polynomial in (G'/G^2). Then equating each coefficient of the resulting polynomial to zero, we obtain a set of algebraic equations for $\lambda, \mu, k, c, a_0, a_i,$ and b_i ($i = 1, 2, ..., m$).

Step 5. Solving the algebraic equations system in Step 4, after substituting these constants $\lambda, \mu, k, c, a_0, a_i,$ and b_i ($i = 1, 2, ..., m$), and using Eq. (18.8), we can construct a variety of exact solutions of Eq. (18.1).

Two examples are shown in the following sections to demonstrate the use of the (G'/G^2)-expansion method to solve nonlinear FDEs.

18.3 Numerical Examples

We apply the present method to solve a nonlinear time-fractional Fisher's equation in Example 18.1 and the nonlinear time-fractional classical Burgers equation in Example 18.2.

Example 18.1 Let us consider the nonlinear time-fractional Fisher's equation (Veeresha et al. 2019)

$$\frac{\partial^\alpha u}{\partial t^\alpha} = \frac{\partial^2 u}{\partial x^2} + 6u(1-u), \quad t > 0, \quad x \in R, \quad 0 < \alpha \leq 1.$$ (18.9)

Solution

Equation (18.9) can be reduced to the ODE in ξ by using the above-discussed transformation $u(x, t) = U(\xi), \xi = kx - \dfrac{ct^\alpha}{\Gamma(1 + \alpha)}$ as

$$k^2 U'' + cU' + 6 U - 6 U^2 = 0, \tag{18.10}$$

where prime denotes the differentiation with respect to ξ. The highest order linear and nonlinear terms in Eq. (18.10) are U'' and U^2. So balancing the order of U'' and U^2 by using the concept in Eq. (18.7), we get

$$\deg(U'') = \deg(U^2), \tag{18.11}$$
$$m + 2 = 2m \Rightarrow m = 2.$$

So, we assume the solution of Eq. (18.10) as Eq. (18.5) at $m = 2$. This may be written as

$$U(\xi) = a_0 + a_1 \left(\frac{G'}{G^2}\right) + b_1 \left(\frac{G'}{G^2}\right)^{-1} + a_2 \left(\frac{G'}{G^2}\right)^2 + b_2 \left(\frac{G'}{G^2}\right)^{-2}, \tag{18.12}$$

where $a_0, a_1, b_1, a_2,$ and b_2 are constant to be calculated.

By using Eqs. (18.6) and (18.12) in Eq. (18.10), we have

$$U'(\xi) = a_1 \mu - b_1 \lambda + 2a_2 \mu \left(\frac{G'}{G^2}\right) - 2b_2 \lambda \left(\frac{G'}{G^2}\right)^{-1} + a_1 \lambda \left(\frac{G'}{G^2}\right)^2 - b_1 \mu \left(\frac{G'}{G^2}\right)^{-2}$$
$$+ 2a_2 \lambda \left(\frac{G'}{G^2}\right)^3 - 2b_2 \mu \left(\frac{G'}{G^2}\right)^{-3}, \tag{18.13}$$

$$U''(\xi) = 2a_2 \mu^2 + 2b_2 \lambda^2 + 2a_1 \lambda \mu \left(\frac{G'}{G^2}\right) + 2b_1 \lambda \mu \left(\frac{G'}{G^2}\right)^{-1} + 8a_2 \lambda \mu \left(\frac{G'}{G^2}\right)^2 + 8b_2 \mu \lambda \left(\frac{G'}{G^2}\right)^{-2}$$
$$+ 2a_1 \lambda^2 \left(\frac{G'}{G^2}\right)^3 + 2b_1 \mu^2 \left(\frac{G'}{G^2}\right)^{-3} + 6a_2 \lambda^2 \left(\frac{G'}{G^2}\right)^4 + 6b_2 \mu^2 \left(\frac{G'}{G^2}\right)^{-4}, \tag{18.14}$$

and

$$U^2(\xi) = \left(a_0 + a_1 \left(\frac{G'}{G^2}\right) + b_1 \left(\frac{G'}{G^2}\right)^{-1} + a_2 \left(\frac{G'}{G^2}\right)^2 + b_2 \left(\frac{G'}{G^2}\right)^{-2}\right)^2, \tag{18.15}$$

Substituting Eqs. (18.12)–(18.15) into Eq. (18.10), collecting the coefficients of $\left(\dfrac{G'}{G^2}\right)^i, i = 0, \pm 1, \pm 2, \ldots$ and setting these to zero, we obtain the following system

$$\left(\frac{G'}{G^2}\right)^0 : 2k^2 \lambda^2 b_2 + 2k^2 \mu^2 a_2 - c\lambda b_1 + c\mu a_1 - 6a_0^2 - 12a_1 b_1 - 12a_2 b_2 + 6a_0 = 0,$$

$$\left(\frac{G'}{G^2}\right)^1 : 2k^2 \lambda \mu a_1 + 2c\mu a_2 - 12a_0 a_1 - 12a_2 b_1 + 6a_1 = 0,$$

$$\left(\frac{G'}{G^2}\right)^{-1} : 2k^2 \lambda \mu b_1 - 2c\lambda b_2 - 12a_0 b_1 - 12a_1 b_2 + 6b_1 = 0,$$

$$\left(\frac{G'}{G^2}\right)^2 : 8k^2 \lambda \mu a_2 + c\lambda a_1 - 12a_0 a_2 - 6a_1^2 + 6a_2 = 0,$$

$$\left(\frac{G'}{G^2}\right)^{-2} : 8k^2 \lambda \mu b_2 - c\mu b_1 - 12a_0 b_2 - 6b_1^2 + 6b_2 = 0, \tag{18.16}$$

$$\left(\frac{G'}{G^2}\right)^3 : 2k^2 \lambda^2 a_1 + 2c\lambda a_2 - 12a_1 a_2 = 0,$$

$$\left(\frac{G'}{G^2}\right)^{-3} : 2k^2 \mu^2 b_1 - 2c\mu b_2 - 12b_1 b_2 = 0,$$

$$\left(\frac{G'}{G^2}\right)^4 : 6k^2 \lambda^2 a_2 - 6a_2^2 = 0,$$

$$\left(\frac{G'}{G^2}\right)^{-4} : 6k^2 \mu^2 b_2 - 6b_2^2 = 0.$$

When we solve a system of linear equations, the point of intersection of the two lines is the solution. However, in the case of a nonlinear system of equations, the graphs might be circles, parabolas, or hyperbolas, with several points of intersection, and hence we may have multiple solutions. By solving the above nonlinear system, it gives different values of c, k, a_0, a_i, b_i, $i = 1, 2$ in various cases. Few of them are provided here.

Case 1:

$$a_0 = \frac{3}{2}, \quad a_1 = 0, \quad b_1 = 0, \quad a_2 = \frac{3\lambda}{2\mu}, \quad b_2 = 0, \quad k = \pm\sqrt{\frac{3}{2\lambda\mu}}, \quad c = 0, \tag{18.17}$$

Case 2:

$$a_0 = \frac{3}{2}, \quad a_1 = 0, \quad b_1 = 0, \quad a_2 = 0, \quad b_2 = \frac{3\mu}{2\lambda}, \quad k = \pm\sqrt{\frac{3}{2\lambda\mu}}, c = 0, \tag{18.18}$$

Case 3:

$$a_0 = \frac{-1}{2}, \quad a_1 = 0, \quad b_1 = 0, \quad a_2 = -\frac{3\lambda}{2\mu}, \quad b_2 = 0, \quad k = \pm\sqrt{\frac{3}{2\lambda\mu}}i, \quad c = 0, \tag{18.19}$$

Case 4:

$$a_0 = \frac{-1}{2}, a_1 = 0, \quad b_1 = 0, \quad a_2 = 0, \quad b_2 = -\frac{3\mu}{2\lambda}, \quad k = \pm\sqrt{\frac{3}{2\lambda\mu}}i, \quad c = 0, \tag{18.20}$$

Case 5:

$$a_0 = \frac{3}{4}, \quad a_1 = 0, \quad b_1 = 0, \quad a_2 = \frac{3\lambda}{8\mu}, \quad b_2 = \frac{3\mu}{8\lambda}, \quad k = \pm\sqrt{\frac{3}{8\lambda\mu}}, c = 0, \tag{18.21}$$

Case 6:

$$a_0 = \frac{1}{4}, \quad a_1 = 0, \quad b_1 = 0, \quad a_2 = -\frac{3\lambda}{8\mu}, \quad b_2 = -\frac{3\mu}{8\lambda}, \quad k = \pm\sqrt{\frac{3}{8\lambda\mu}}i, \quad c = 0, \tag{18.22}$$

Case 7:

$$a_0 = \frac{3}{4}, \quad a_1 = \pm\sqrt{\frac{1}{4\lambda\mu}}i, \quad b_1 = 0, \quad a_2 = \frac{\lambda}{4\mu}, \quad b_2 = 0, \quad k = \pm\sqrt{\frac{1}{4\lambda\mu}}, \quad c = 5\left(\pm\sqrt{\frac{1}{4\lambda\mu}}i\right), \tag{18.23}$$

where λ and μ are arbitrary constants. The solution to the problem may be obtained by considering all of the cases one by one. However, only case 1 is being considered for getting the exact solution. Similarly, one can take other cases to find the solutions.

From Case 1: Substituting the values of constants given from Eq. (18.17) into Eq. (18.12), we obtain

$$U(\xi) = \frac{3}{2} + \left(\frac{3\lambda}{2\mu}\right)\left(\frac{G'}{G^2}\right)^2. \tag{18.24}$$

Substituting the general solution of Eq. (18.6), which is given in Eq. (18.8) into the above Eq. (18.24), we have the following solution of the nonlinear time-fractional Fisher's equationwhen $\lambda\mu > 0$,

$$U_{1,2}(\xi) = \frac{3}{2} + \left(\frac{3\lambda}{2\mu}\right)\left(\sqrt{\frac{\mu}{\lambda}}\left(\frac{c_1\cos\left(\sqrt{\lambda\mu}\xi\right) + c_2\sin\left(\sqrt{\lambda\mu}\xi\right)}{c_2\cos\left(\sqrt{\lambda\mu}\xi\right) - c_1\sin\left(\sqrt{\lambda\mu}\xi\right)}\right)\right)^2, \tag{18.25}$$

when $\lambda\mu < 0$,

$$U_{3,4}(\xi) = \frac{3}{2} + \left(\frac{3\lambda}{2\mu}\right)\left(-\frac{\sqrt{|\lambda\mu|}}{\lambda}\left(\frac{c_1\sinh\left(2\sqrt{|\lambda\mu|}\xi\right) + c_1\cosh\left(2\sqrt{|\lambda\mu|}\xi\right) + c_2}{c_1\sinh\left(2\sqrt{|\lambda\mu|}\xi\right) + c_1\cosh\left(2\sqrt{|\lambda\mu|}\xi\right) - c_2}\right)\right)^2, \tag{18.26}$$

when $\lambda \neq 0, \mu = 0$,

$$U_{5,6}(\xi) = \frac{3}{2} + \left(\frac{3\lambda}{2\mu}\right)\left(-\frac{c_1}{\lambda\left(c_1\xi + c_2\right)}\right)^2, \tag{18.27}$$

where

$$\xi = \left(\pm \sqrt{\frac{3}{2\lambda\mu}} \right) x. \tag{18.28}$$

All the above Eqs. (18.25)–(18.27) are the exact solutions of the nonlinear time-fractional Fisher's equation for case 1. Putting the suitable values of the constants in Eqs. (18.25)–(18.27), we obtain the traveling wave solutions of the given model.

Example 18.2 Consider the following nonlinear time-fractional classical Burgers equation (Yildirim and Pinar 2010)

$$\frac{\partial^\alpha u}{\partial t^\alpha} = \frac{\partial^2 u}{\partial x^2} + u \frac{\partial u}{\partial x}, \quad 0 < \alpha \le 1. \tag{18.29}$$

Solution

Using the transformation $u(x, t) = U(\xi), \xi = kx - \dfrac{ct^\alpha}{\Gamma(1 + \alpha)}$ in Eq. (18.29), it reduces to

$$cU' + k^2 U'' + kUU' = 0, \tag{18.30}$$

where prime denotes the differentiation with respect to ξ. The highest order linear and nonlinear terms in Eq. (18.30) are U'' and UU'. So balancing the order of U'' and UU' by using the concept mentioned in Eq. (18.7), we get

$$\deg(U'') = \deg(UU'),$$
$$m + 2 = m + m + 1 \Rightarrow m = 1. \tag{18.31}$$

So, we assume the solution of Eq. (18.30) as Eq. (18.5) at $m = 1$. This may be written as

$$U(\xi) = a_0 + a_1 \left(\frac{G'}{G^2} \right) + b_1 \left(\frac{G'}{G^2} \right)^{-1}, \tag{18.32}$$

where a_0, a_1, and b_1 are constant to be calculated. By using Eqs. (18.6) and (18.32) in Eq. (18.30), we have

$$U'(\xi) = a_1\mu - b_1\lambda + a_1\lambda \left(\frac{G'}{G^2} \right)^2 - b_1\mu \left(\frac{G'}{G^2} \right)^{-2}, \tag{18.33}$$

and

$$U''(\xi) = 2a_1\lambda\mu \left(\frac{G'}{G^2} \right) + 2b_1\lambda\mu \left(\frac{G'}{G^2} \right)^{-1} + 2a_1\lambda^2 \left(\frac{G'}{G^2} \right)^3 + 2b_1\mu^2 \left(\frac{G'}{G^2} \right)^{-3}. \tag{18.34}$$

By substituting Eqs. (18.32)–(18.34) into Eq. (18.30), collecting the coefficients of $\left(\dfrac{G'}{G^2} \right)^i, i = 0, \pm 1, \pm 2, \dots$ and setting these to zero, we will have the following system of the algebraic equation:

$$\left(\frac{G'}{G^2} \right)^0 : -k\lambda a_0 b_1 + k\mu a_0 a_1 - c\lambda b_1 + c\mu a_1 = 0,$$

$$\left(\frac{G'}{G^2} \right)^1 : 2k^2\lambda\mu a_1 + k\mu a_1^2 = 0,$$

$$\left(\frac{G'}{G^2} \right)^{-1} : 2k^2\lambda\mu b_1 - k\lambda b_1^2 = 0,$$

$$\left(\frac{G'}{G^2} \right)^2 : k\lambda a_0 a_1 + c\lambda a_1 = 0, \tag{18.35}$$

$$\left(\frac{G'}{G^2} \right)^{-2} : -k\mu a_0 b_1 - c\mu b_1 = 0,$$

$$\left(\frac{G'}{G^2} \right)^3 : 2k^2\lambda^2 a_1 + k\lambda a_1^2 = 0,$$

$$\left(\frac{G'}{G^2} \right)^{-3} : 2k^2\mu^2 b_1 - k\mu b_1^2 = 0,$$

By solving the above nonlinear system, various values of parameters c, k, a_0, a_1, b_1 in different cases may be obtained as follows:

Case 1:

$$a_0 = \frac{-c}{k}, \quad a_1 = 0, \quad b_1 = 2k\mu, \quad k = k, \quad c = c, \tag{18.36}$$

Case 2:

$$a_0 = \frac{-c}{k}, \quad a_1 = -2k\lambda, \quad b_1 = 0, \quad k = k, \quad c = c, \tag{18.37}$$

Case 3:

$$a_0 = \frac{-c}{k}, \quad a_1 = -2k\lambda, \quad b_1 = 2k\mu, \quad k = k, \quad c = c, \tag{18.38}$$

where k and c are free parameters, and λ and μ are arbitrary constants. The solution to the problem may be obtained by going through all of the possible cases one by one. However, only case 1 is taken into account for determining the exact solution here. Similarly, other cases can be used to have the solutions to the problem.

From Case 1: Substituting the values of constants from Eq. (18.36) into Eq. (18.32), we obtain

$$U(\xi) = \frac{-c}{k} + 2k\mu \left(\frac{G'}{G^2}\right)^{-1}. \tag{18.39}$$

Substituting the general solution of Eq. (18.6), which is given in Eq. (18.8) into the above Eq. (18.39), we have the following solution of the time-fractional classical Burgers equation

when $\lambda\mu > 0$,

$$U_{1,2}(\xi) = \frac{-c}{k} + 2k\mu \left(\sqrt{\frac{\mu}{\lambda}} \left(\frac{c_1 \cos\left(\sqrt{\lambda\mu}\xi\right) + c_2 \sin\left(\sqrt{\lambda\mu}\xi\right)}{c_2 \cos\left(\sqrt{\lambda\mu}\xi\right) - c_1 \sin\left(\sqrt{\lambda\mu}\xi\right)}\right)\right)^{-1}, \tag{18.40}$$

when $\lambda\mu < 0$,

$$U_{3,4}(\xi) = \frac{-c}{k} + 2k\mu \left(-\frac{\sqrt{|\lambda\mu|}}{\lambda} \left(\frac{c_1 \sinh\left(2\sqrt{|\lambda\mu|}\xi\right) + c_1 \cosh\left(2\sqrt{|\lambda\mu|}\xi\right) + c_2}{c_1 \sinh\left(2\sqrt{|\lambda\mu|}\xi\right) + c_1 \cosh\left(2\sqrt{|\lambda\mu|}\xi\right) - c_2}\right)\right)^{-1}, \tag{18.41}$$

when $\lambda \neq 0$, $\mu = 0$,

$$U_{5,6}(\xi) = \frac{-c}{k} + 2k\mu \left(-\frac{c_1}{\lambda\left(c_1\xi + c_2\right)}\right)^{-1}, \tag{18.42}$$

where

$$\xi = k\,x - \frac{ct^\alpha}{\Gamma(1 + \alpha)}. \tag{18.43}$$

All the above Eqs. (18.40)–(18.42) are the exact solutions of the time-fractional classical Burgers equation for case 1. Putting the suitable values of the constants in Eqs. (18.40)–(18.42), we may obtain the traveling wave solutions of the given model.

References

Ali, M.N., Osman, M.S., and Husnine, S.M. (2019). On the analytical solutions of conformable time-fractional extended Zakharov–Kuznetsov equation through (G'/G^2)-expansion method and the modified Kudryashov method. *SeMA* 76: 15–25.

Aljahdaly, N.H. (2019). Some applications of the modified (G'/G^2)-expansion method in mathematical physics. *Results in Physics* 13: 102272 (1–7).

Arshed, S. and Sadia, M. (2018). (G'/G^2)-Expansion method: new traveling wave solutions for some nonlinear fractional partial differential equations. *Optical and Quantum Electronics* 50: 123(1–20).

Behroozifar, M. and Ahmadpour, F. (2017). Comparative study on solving fractional differential equations via shifted Jacobi collocation method. *Bulletin of Iranian Mathematical Society* 43 (2): 535–560.

Bekir, A., Guner, O., Bhrawy, A.H., and Biswas, A. (2015). Solving nonlinear fractional differential equations using exp-function and G′/G- expansion methods. *Romanian Journal of Physics* 60 (3-4): 360–378.

Diethelm, K., Ford, N.J., and Freed, A.D. (2002). A predictor-corrector approach for the numerical solution of fractional differential equations. *Nonlinear Dynamics* 29: 3–22.

Erturk, V. and Momani, S.M. (2008). Solving systems of fractional differential equations using differential transform method. *Journal of Computational and Applied Mathematics* 215 (1): 142–151.

Gepreel, K.A. and Omran, S. (2012). Exact solutions for nonlinear partial fractional differential equations. *Chinese Physics B* 21 (11): 110204.

Güner, O., Bekir, A., and Cevikel, A.C. (2015). A variety of exact solutions for the time fractional Cahn–Allen equation. *The European Physical Journal Plus* 130 (146): 1–13.

Hassaballa, A.A. (2020). The (G'/G^2) – expansion method for solving fractional burgers-fisher and burgers equations. *Applied and Computational Mathematics* 9 (3): 56–63.

Li, Z.B. and He, J.H. (2010). Fractional complex transformation for fractional differential equations. *Computer and Mathematics with Applications* 15: 970–973.

Mohyud-Din, S.T. and Bibi, S. (2018). Exact solutions for nonlinear fractional differential equations using (G'/G^2)-expansion method. *Alexandria Engineering Journal* 57: 1003–1008.

Veeresha, P., Prakasha, D.G., and Baskonus, H.M. (2019). Novel simulations to the tim-fractional Fisher's equation. *Mathematical Sciences* 13: 33–42.

Yildirim, A. and, Pinar, Z. (2010). Application of the exp-function method for solving nonlinear reaction-diffusion equations arising in mathematical biology. *Computers and Mathematics with Applications* 60: 1873–1880.

19

(G'/G, 1/G)-Expansion Method

19.1 Introduction

In recent decades, exact traveling wave solutions to nonlinear fractional differential equations have become more significant in studying complex physical and mechanical phenomena. The $(G'/G,1/G)$-expansion method (Zayed et al. 2012; Zayed and Alurrfi 2014; Yasar and Giresunlu 2016; Al-Shawba et al. 2018; Sirisubtawee et al. 2019; Duran 2021) has been utilized to obtain the solutions of time and space fractional differential equations. Recently, Jumarie (2006) proposed the modified Riemann–Liouville derivative. Using fractional complex transformation, one may convert the fractional differential equation into integer-order differential equations (Li and He 2010). The original (G'/G)-expansion approach assumes that a polynomial may describe the exact solutions of nonlinear PDEs in one variable (G'/G) satisfying the second-order ordinary differential equation $G''(\xi) + \lambda G'(\xi) + \mu = 0$, where λ and μ are constants. On the other hand, the two variables $(G'/G,1/G)$-expansion approach is an extension of the original (G'/G)-expansion method. The two variables $(G'/G,1/G)$-expansion approach is based on the assumption that a polynomial may describe exact traveling wave solutions to nonlinear PDEs in the two variables (G'/G) and $(1/G)$, in which $G = G(\xi)$ satisfies second-order linear ODE, namely, $G''(\xi) + \lambda G(\xi) = \mu$, where λ and μ are constants. The degree of this polynomial can be found by evaluating the homogeneous balance between the highest order derivatives and nonlinear terms in the given nonlinear PDEs, and the coefficients may be obtained by solving a set of algebraic equations generated by the procedure. Recently, (Ling-xiao et al. 2010) used the two variables $(G'/G,1/G)$-expansion approach to find the exact solutions of the Zakharov equations. Zayed and Abdelaziz (2012) used it to find the exact solutions of the nonlinear KdV-mKdV equation. This method is more efficient and reliable than the (G'/G) and (G'/G^2)-expansion methods. It also allows us to find the exact solutions of fractional differential equations without initial or boundary conditions.

19.2 Algorithm of the ($G'/G,1/G$)-Expansion Method

This section briefly describes the main steps of this approach for solving fractional partial differential equations (PDEs).

In order to understand the $(G'/G,1/G)$-expansion method (Zayed et al. 2012; Zayed and Alurrfi 2014; Yasar and Giresunlu 2016; Al-Shawba et al. 2018; Sirisubtawee et al. 2019; Duran 2021), we consider the following nonlinear fractional PDE in two independent variables x and t of the type

$$Q\left(u, u_x, u_{xx}, ..., D_t^\alpha u, ...\right) = 0, \quad 0 < \alpha \leq 1, \tag{19.1}$$

where $u = u(x, t)$ is an unknown function, and Q is a polynomial of u and its partial fractional derivatives, in which the highest order derivatives and the nonlinear terms are involved.

Step 1. First, the traveling wave variable is considered as (Li and He 2010)

$$u(x, t) = U(\xi), \tag{19.2}$$

$$\xi = kx - \frac{ct^\alpha}{\Gamma(1 + \alpha)}, \tag{19.3}$$

where c and k are nonzero constants to be determined later. By using the chain rule (Güner et al. 2015), we have

Computational Fractional Dynamical Systems: Fractional Differential Equations and Applications, First Edition.
Snehashish Chakraverty, Rajarama Mohan Jena, and Subrat Kumar Jena.
© 2023 John Wiley & Sons, Inc. Published 2023 by John Wiley & Sons, Inc.

$$D_t^\alpha u = \Delta_t u_\zeta D_t^\alpha \zeta,$$
$$D_x^\alpha u = \Delta_x u_\zeta D_x^\alpha \zeta,$$

where Δ_t and Δ_x are the fractal indexes (Güner et al. 2015). Without loss of generality, we can consider $\Delta_t = \Delta_x = \kappa$, where κ is a constant. Using Eqs. (19.2) and (19.3), Eq. (19.1) can be rewritten in the following nonlinear ordinary differential equation (ODE) form

$$P\left(U, kU', k^2 U'', ..., -cU', ...\right) = 0, \tag{19.4}$$

where P is a polynomial of $U(\xi)$ and its various derivatives. The prime ($'$) denotes the derivative with respect to ξ. The following ideas must be understood before going through the next steps of the $(G'/G, 1/G)$-expansion approach.

Let us consider the following linear ODE:

$$G''(\xi) + \lambda G(\xi) = \mu, \tag{19.5}$$

where the prime notation ($'$) denotes the derivative with respect to ξ and λ, μ are constants. Next, we set (Ling-xiao et al. 2010; Zayed and Abdelaziz 2012)

$$\phi(\xi) = \frac{G'(\xi)}{G(\xi)}, \quad \text{and} \quad \psi(\xi) = \frac{1}{G(\xi)}. \tag{19.6}$$

Equations (19.5) and (19.6) can be turned into a system of two nonlinear ODEs as follows:

$$\phi' = -\phi^2 + \mu\psi - \lambda, \quad \text{and} \quad \psi' = -\phi\psi. \tag{19.7}$$

The general solutions of Eq. (19.5) may be classified into the following three cases,

Case 1: If $\lambda < 0$, then the general solution of Eq. (19.5) is written as (Zayed et al. 2012; Zayed and Alurrfi 2014; Yasar and Giresunlu 2016; Al-Shawba et al. 2018; Sirisubtawee et al. 2019; Duran 2021)

$$G(\xi) = c_1 \sinh\left(\xi\sqrt{-\lambda}\right) + c_2 \cosh\left(\xi\sqrt{-\lambda}\right) + \frac{\mu}{\lambda}, \tag{19.8}$$

and we have

$$\psi^2 = \frac{-\lambda}{\lambda^2 \sigma_1 + \mu^2}\left(\phi^2 - 2\mu\psi + \lambda\right), \tag{19.9}$$

where c_1 and c_2 are arbitrary constants and $\sigma_1 = c_1^2 - c_2^2$.

Case 2: If $\lambda > 0$, then the general solution of Eq. (19.5) can be written as (Zayed et al. 2012; Zayed and Alurrfi 2014; Yasar and Giresunlu 2016; Al-Shawba et al. 2018; Sirisubtawee et al. 2019; Duran 2021)

$$G(\xi) = c_1 \sin\left(\xi\sqrt{\lambda}\right) + c_2 \cos\left(\xi\sqrt{\lambda}\right) + \frac{\mu}{\lambda}, \tag{19.10}$$

and we have the following relation

$$\psi^2 = \frac{\lambda}{\lambda^2 \sigma_2 - \mu^2}\left(\phi^2 - 2\mu\psi + \lambda\right), \tag{19.11}$$

where c_1 and c_2 are two arbitrary constants and $\sigma_2 = c_1^2 + c_2^2$.

Case 3: If $\lambda = 0$, then the general solution of Eq. (19.5) can be provided as (Zayed et al. 2012; Zayed and Alurrfi 2014; Yasar and Giresunlu 2016; Al-Shawba et al. 2018; Sirisubtawee et al. 2019; Duran 2021)

$$G(\xi) = \frac{\mu}{2}\xi^2 + c_1 \xi + c_2, \tag{19.12}$$

and the corresponding relation is

$$\psi^2 = \frac{1}{c_1^2 - 2\mu\, c_2}\left(\phi^2 - 2\mu\psi\right), \tag{19.13}$$

where c_1 and c_2 are two arbitrary constants.

Step 2. Let us assume that the solution to Eq. (19.4) can be expressed by a polynomial of two variables φ and ψ, as follows:

$$U(\xi) = a_0 + \sum_{i=1}^{m} \left[a_i \, \phi^i(\xi) + b_i \, \phi^{i-1}(\xi) \, \psi(\xi) \right], \tag{19.14}$$

where $a_0, a_i, b_i \, (i = 1, 2, ..., m)$ are constants to be determined later with $a_m^2 + b_m^2 \neq 0$ and the functions $\phi = \phi(\xi)$ and $\psi = \psi(\xi)$ satisfy the equation Eqs. (19.6) and (19.7).

Step 3. The positive integer m in Eq. (19.14) may be calculated by considering the homogenous balance between the highest order derivatives and the nonlinear terms in Eq. (19.4). More precisely, we define the degree of $U(\xi)$ as $\deg[U(\xi)] = m$, which leads to the degrees of the other expressions as follows (Sirisubtawee et al. 2019):

$$\deg\left[\frac{d^p U}{d\xi^p}\right] = m + p,$$
$$\deg\left[U^s \left(\frac{d^p U}{d\xi^p}\right)^q\right] = ms + q(p + m). \tag{19.15}$$

Accordingly, we can obtain the value of m in Eq. (19.14). Suppose that the balance number m of any nonlinear equation is not a positive integer (for example, a fraction or a negative integer). In that case, the specific transformations for $U(\xi)$ in Eq. (19.4) are used to get a new equation in terms of the new function $W(\xi)$ with a positive integer balance number (see details in (Zayed and Alurrfi 2016)).

Step 4. Substituting Eq. (19.14) into Eq. (19.4) and using Eqs. (19.7) and (19.9), the function P can be transferred into a polynomial in ϕ and ψ, in which the degree of ψ is not larger than one. Collecting the coefficients of like powers of ϕ^i ($i = 0, 1, ...$), $\psi^j(j = 0, 1)$, $\phi^i \psi$ ($i = 1, 2...$), and equating each coefficient to zero, we obtain a set of algebraic equations for $\phi, \psi, \lambda, \mu, k, c, a_0, a_i$, and b_i ($i = 1, 2, ..., m$). Solving the algebraic equations system, we obtain the values of the parameters k, c, a_0, a_i, and b_i ($i = 1, 2, ..., m$) for case 1. In this step, the hyperbolic functions are used to represent the resulting traveling wave solutions with the transformation given in Eqs. (19.2) and (19.3).

Step 5. Similar to Step 4, we can find the exact solutions of Eq. (19.1) by substituting Eq. (19.14) into Eq. (19.4) with the help of Eqs. (19.7) and (19.11) for $\lambda > 0$. The resulting solutions are written in the form of trigonometric functions.

Step 6. Again, substituting Eq. (19.14) into Eq. (19.4) with the aid of Eqs. (19.7) and (19.13) for $\lambda = 0$, we get the exact solutions of Eq. (19.1). The obtained exact solutions are expressed as rational functions.

Note: The two-variable $(G'/G, 1/G)$-expansion method reduces to the (G'/G)- expansion method by substituting $\mu = 0$ in Eq. (19.5) and $b_i = 0$ in Eq. (19.14). So, the (G'/G)- expansion method is a particular case of the $(G'/G, 1/G)$-expansion method.

One example problem is given in the following section to demonstrate the use of the $(G'/G, 1/G)$-expansion method to solve nonlinear fractional differential equations.

19.3 Illustrative Examples

We apply the present method to solve a nonlinear time-fractional classical Burgers equation in Example 19.1.

Example 19.1 Consider the following nonlinear time-fractional classical Burgers equation (Yildirim and Pinar 2010)

$$\frac{\partial^\alpha u}{\partial t^\alpha} = \frac{\partial^2 u}{\partial x^2} + u\frac{\partial u}{\partial x}, \quad 0 < \alpha \leq 1. \tag{19.16}$$

Solution

Using the transformation $u(x, t) = U(\xi), \xi = kx - \frac{ct^\alpha}{\Gamma(1 + \alpha)}$ in Eq. (19.16), it reduces to

$$cU' + k^2 U'' + kUU' = 0, \tag{19.17}$$

where prime denotes the differentiation with respect to ξ. Integrating Eq. (19.17) with respect to ξ, it gives

$$cU + k^2 U' + \frac{k}{2}U^2 + A = 0, \tag{19.18}$$

where A is an integration constant, and for simplicity, we have considered $A = 0$. So, Eq. (19.18) is now rewritten as

$$cU + k^2 U' + \frac{k}{2} U^2 = 0. \tag{19.19}$$

The highest order linear and nonlinear terms in Eq. (19.19) are U' and U^2. So balancing the order of U' and U^2 by using the concept mentioned in Eq. (19.15), we get

$$\deg(U') = \deg(U^2),$$
$$m + 1 = 2m \Rightarrow m = 1. \tag{19.20}$$

So, we assume the solution of Eq. (19.19) as Eq. (19.14) at $m = 1$. This can be written as

$$U(\xi) = a_0 + a_1 \phi(\xi) + b_1 \psi(\xi), \tag{19.21}$$

where a_0, a_1, and b_1 are constants to be calculated. As we discussed earlier in this method, three solution cases are obtained, which are now addressed as follows:

Case I. When $\lambda < 0$ (hyperbolic function solutions)
By substituting Eq. (19.21) into Eq. (19.19) and utilizing Eqs. (19.7) and (19.9), we have

$$U'(\xi) = -a_1 \phi^2 + a_1 \mu \psi - a_1 \lambda - b_1 \phi \psi, \tag{19.22}$$

and

$$U^2(\xi) = a_0^2 + a_1^2 \phi^2 + 2a_0 a_1 \phi + 2a_1 b_1 \phi \psi + 2a_0 b_1 \psi + \frac{2b_1^2 \mu \lambda \psi}{\lambda^2 \sigma_1 + \mu^2} - \frac{b_1^2 \lambda \phi^2}{\lambda^2 \sigma_1 + \mu^2} - \frac{b_1^2 \lambda^2}{\lambda^2 \sigma_1 + \mu^2}. \tag{19.23}$$

Substituting Eqs. (19.21)–(19.23) into Eq. (19.19), collecting the coefficients of $\phi^i, \psi^i, \phi^i \psi$ $(i = 1, 2, ...)$ and setting these to zero subject to the condition that $\lambda^2 \sigma_1 + \mu^2 \neq 0$, we obtain the following system

$$\phi^2: \quad -k^2 a_1 + \frac{k a_1^2}{2} - \frac{k b_1^2 \lambda}{2(\lambda^2 \sigma_1 + \mu^2)} = 0,$$

$$\phi \psi: \quad -k^2 b_1 + k a_1 b_1 = 0,$$

$$\phi: \quad k a_0 a_1 + c a_1 = 0,$$

$$\psi: \quad k^2 a_1 \mu + c b_1 + k a_0 b_1 + \frac{k b_1^2 \lambda \mu}{\lambda^2 \sigma_1 + \mu^2} = 0, \tag{19.24}$$

$$\text{Constant}: \quad -k^2 a_1 \lambda + c a_0 + \frac{k a_0^2}{2} - \frac{k b_1^2 \lambda^2}{2(\lambda^2 \sigma_1 + \mu^2)} = 0.$$

When we solve a system of linear equations, the point of intersection of the two lines is the solution. However, in the case of a nonlinear system of equations, the graphs might be circles, parabolas, or hyperbolas, with several points of intersection, and hence we may have multiple solutions. By solving the above nonlinear system, it gives different values of c, k, a_0, a_1, b_1 in various cases. Some of them are provided here.

Result 1:

$$a_0 = \frac{-2c}{k}, \quad a_1 = 0, \quad b_1 = 0, \quad c = c, \quad k = k, \tag{19.25}$$

where c and k are free parameters. Since $a_1^2 + b_1^2 = 0$, so Eq. (19.25) is not taken into account according to step 2.

Result 2:

$$a_0 = \frac{\pm b_1 \lambda}{\sqrt{\lambda^2 \sigma_1 + \mu^2}}, \quad a_1 = \frac{\mp b_1 \lambda}{\sqrt{\lambda} i \sqrt{\lambda^2 \sigma_1 + \mu^2}}, \quad b_1 = b_1, \quad c = \frac{\mp b_1^2 \lambda^{3/2} i}{\lambda^2 \sigma_1 + \mu^2}, \quad k = \frac{\pm \sqrt{\lambda} b_1 i}{\sqrt{\lambda^2 \sigma_1 + \mu^2}}, \tag{19.26}$$

where b_1 is a free parameter, λ, μ are arbitrary constants, and $\sigma_1 = c_1^2 - c_2^2$, where c_1 and c_2 are arbitrary constants. By substituting Eq. (19.26) into Eq. (19.21) and using Eqs. (19.6) and (19.8), we get the exact solution of Eq. (19.16) as follows:

$$U(\xi) = \frac{\pm b_1 \lambda}{\sqrt{\lambda^2 \sigma_1 + \mu^2}} + \frac{\mp b_1 \lambda \left(c_1 \cosh\left(\xi \sqrt{-\lambda}\right) + c_2 \sinh\left(\xi \sqrt{-\lambda}\right)\right) + b_1 \sqrt{\lambda^2 \sigma_1 + \mu^2}}{\left(\sqrt{\lambda^2 \sigma_1 + \mu^2}\right)\left(c_1 \sinh\left(\xi \sqrt{-\lambda}\right) + c_2 \cosh\left(\xi \sqrt{-\lambda}\right) + \frac{\mu}{\lambda}\right)}, \tag{19.27}$$

where

$$\xi = \left(\frac{\pm \sqrt{\lambda} b_1 i}{\sqrt{\lambda^2 \sigma_1 + \mu^2}} \right) x - \left(\frac{\mp b_1^2 \lambda^{\frac{3}{2}} i}{\lambda^2 \sigma_1 + \mu^2} \right) \frac{t^\alpha}{\Gamma(1 + \alpha)}. \tag{19.28}$$

In particular, by setting $c_1 = 0$, $c_2 > 0$, and $\mu = 0$ in Eq. (19.27), we have the solitary solution

$$U(\xi) = \mp b_1 i \pm b_1 i \tanh \left(\xi \sqrt{-\lambda} \right) + b_1 \mathrm{sech} \left(\xi \sqrt{-\lambda} \right), \tag{19.29}$$

While, if $c_2 = 0$, $c_1 > 0$, and $\mu = 0$, we have the solitary solution

$$U(\xi) = \mp b_1 \mp b_1 \coth \left(\xi \sqrt{-\lambda} \right) + b_1 \mathrm{csch} \left(\xi \sqrt{-\lambda} \right). \tag{19.30}$$

Case II. When $\lambda > 0$ (trigonometric function solutions)
By inserting Eq. (19.21) into Eq. (19.19) and along with the use of Eqs. (19.7) and (19.11), we have

$$U'(\xi) = -a_1 \phi^2 + a_1 \mu \psi - a_1 \lambda - b_1 \phi \psi, \tag{19.31}$$

and

$$U^2(\xi) = a_0^2 + a_1^2 \phi^2 + 2a_0 a_1 \phi + 2a_1 b_1 \phi \psi + 2a_0 b_1 \psi - \frac{2b_1^2 \mu \, \lambda \psi}{\lambda^2 \sigma_2 - \mu^2} + \frac{b_1^2 \lambda \phi^2}{\lambda^2 \sigma_2 - \mu^2} + \frac{b_1^2 \lambda^2}{\lambda^2 \sigma_2 - \mu^2}. \tag{19.32}$$

Substituting Eq. (19.21), and Eqs. (19.31)–(19.32) into Eq. (19.19), collecting the coefficients of ϕ^i, ψ^i, $\phi^i \psi$ ($i = 1, 2, ...$) and setting these to zero subject to the condition $\lambda^2 \sigma_2 - \mu^2 \neq 0$, we obtain the following system

$$\phi^2 : \quad -k^2 a_1 + \frac{k a_1^2}{2} + \frac{k b_1^2 \lambda}{2 (\lambda^2 \sigma_2 - \mu^2)} = 0,$$

$$\phi \psi : \quad -k^2 b_1 + k a_1 b_1 = 0,$$

$$\phi : \quad k a_0 a_1 + c a_1 = 0, \tag{19.33}$$

$$\psi : \quad k^2 a_1 \mu + c b_1 + k a_0 b_1 - \frac{k b_1^2 \lambda \mu}{\lambda^2 \sigma_2 - \mu^2} = 0,$$

$$\mathrm{Constant} : \quad -k^2 a_1 \lambda + c a_0 + \frac{k a_0^2}{2} + \frac{k b_1^2 \lambda^2}{2 (\lambda^2 \sigma_2 - \mu^2)} = 0.$$

By solving the above nonlinear system yields different values of c, k, a_0, a_1, b_1 in various cases. Some of them are again listed below:
Result 1:

$$a_0 = \frac{-2c}{k}, \quad a_1 = 0, \quad b_1 = 0, \quad c = c, \quad k = k, \tag{19.34}$$

where c and k are free parameters. Since $a_1^2 + b_1^2 = 0$, so Eq. (19.25) is not considered.
Result 2:

$$a_0 = \frac{\pm b_1 \lambda i}{\sqrt{\lambda^2 \sigma_2 - \mu^2}}, \quad a_1 = \frac{\pm b_1 \lambda^{1/2}}{\sqrt{\lambda^2 \sigma_2 - \mu^2}}, \quad b_1 = b_1, \quad c = \frac{\mp b_1^2 \lambda^{3/2} i}{\lambda^2 \sigma_2 - \mu^2}, \quad k = \frac{\pm b_1 \sqrt{\lambda}}{\sqrt{\lambda^2 \sigma_2 - \mu^2}}, \tag{19.35}$$

where b_1 is a free parameter, λ, μ are arbitrary constants, and $\sigma_2 = c_1^2 + c_2^2$, where c_1 and c_2 are arbitrary constants. Plugging Eq. (19.35) into Eq. (19.21) and using Eqs. (19.6) and (19.10), we have the exact solution of Eq. (19.16) as follows:

$$U(\xi) = \frac{\pm b_1 \lambda i}{\sqrt{\lambda^2 \sigma_2 - \mu^2}} + \frac{\pm b_1 \lambda \left(c_1 \cos \left(\xi \sqrt{\lambda} \right) - c_2 \sin \left(\xi \sqrt{\lambda} \right) \right) + b_1 \sqrt{\lambda^2 \sigma_2 - \mu^2}}{\left(\sqrt{\lambda^2 \sigma_2 - \mu^2} \right) \left(c_1 \sin \left(\xi \sqrt{\lambda} \right) + c_2 \cos \left(\xi \sqrt{\lambda} \right) + \frac{\mu}{\lambda} \right)}, \tag{19.36}$$

where

$$\xi = \left(\frac{\pm b_1 \sqrt{\lambda}}{\sqrt{\lambda^2 \sigma_2 - \mu^2}} \right) x - \left(\frac{\mp b_1^2 \lambda^{\frac{3}{2}} i}{\lambda^2 \sigma_2 - \mu^2} \right) \frac{t^\alpha}{\Gamma(1 + \alpha)}. \tag{19.37}$$

In particular, by setting $c_1 = 0$, $c_2 > 0$, and $\mu = 0$ in Eq. (19.36), we have the solitary solution

$$U(\xi) = \pm b_1 i \mp b_1 \tanh\left(\xi\sqrt{\lambda}\right) + b_1 \mathrm{sech}\left(\xi\sqrt{\lambda}\right), \tag{19.38}$$

While, if $c_2 = 0$, $c_1 > 0$, and $\mu = 0$, we have the solitary solution

$$U(\xi) = \pm b_1 i \pm b_1 \coth\left(\xi\sqrt{\lambda}\right) + b_1 \mathrm{csch}\left(\xi\sqrt{\lambda}\right). \tag{19.39}$$

Case III. When $\lambda = 0$ (rational function solutions)
By substituting Eq. (19.21) into Eq. (19.19) along with the use of Eqs. (19.7) and (19.13), we have

$$U'(\xi) = -a_1\phi^2 + a_1\mu\psi - b_1\phi\psi, \tag{19.40}$$

and

$$U^2(\xi) = a_0^2 + a_1^2\phi^2 + 2a_0a_1\phi + 2a_1b_1\phi\psi + 2a_0b_1\psi - \frac{2b_1^2\mu\,\psi}{c_1^2 - 2\mu\,c_2^2} + \frac{b_1^2\,\phi^2}{c_1^2 - 2\mu\,c_2^2}. \tag{19.41}$$

Substituting Eq. (19.21) and Eqs. (19.40)–(19.41) into Eq. (19.19), collecting the coefficients of φ^i, ψ^i, $\varphi^i\psi$ ($i = 1, 2, ...$) and setting these to zero provide that $c_1^2 - 2\mu\,c_2^2 \neq 0$, we obtain the following system

$$\begin{aligned}
\phi^2: &\quad -k^2a_1 + \frac{k\,a_1^2}{2} + \frac{k\,b_1^2}{2\left(c_1^2 - 2\mu\,c_2^2\right)} = 0, \\
\phi\psi: &\quad -k^2b_1 + k\,a_1b_1 = 0, \\
\phi: &\quad k\,a_0a_1 + c\,a_1 = 0, \\
\psi: &\quad k^2a_1\mu + cb_1 + k\,a_0b_1 - \frac{k\,b_1^2\mu}{c_1^2 - 2\mu\,c_2^2} = 0, \\
\text{Constant}: &\quad ca_0 + \frac{k\,a_0^2}{2} = 0.
\end{aligned} \tag{19.42}$$

By solving the above nonlinear system yields the following results:
Result 1:

$$a_0 = \frac{-2c}{k}, \quad a_1 = 0, \quad b_1 = 0, \quad c = c, \quad k = k, \tag{19.43}$$

where c and k are free parameters. Since $a_1^2 + b_1^2 = 0$, so Eq. (19.43) may not be considered.
Result 2:

$$a_0 = 0, \quad a_1 = k, \quad b_1 = \pm\sqrt{c_1^2 - 2\mu\,c_2^2}, \quad c = 0, \quad k = k, \tag{19.44}$$

where k is a free parameter, and c_1, μ, and c_2 are arbitrary constants. Plugging Eq. (19.44) into Eq. (19.21) and using Eqs. (19.6) and (19.12), we obtain the exact solution of Eq. (19.16) as follows:

$$U(\xi) = \left(\frac{k\,(\mu\,\xi + c_1) \pm \sqrt{c_1^2 - 2\mu\,c_2^2}}{\frac{\mu}{2}\xi^2 + c_1\xi + c_2}\right), \tag{19.45}$$

where

$$\xi = kx. \tag{19.46}$$

In particular, by setting $c_1 = 0$, $c_2 > 0$, and $\mu = 0$ in Eq. (19.45), we have the solitary solution

$$U(\xi) = 0, \tag{19.47}$$

while, if $c_2 = 0$, $c_1 > 0$, and $\mu = 0$, we have the solitary solution

$$U(\xi) = \frac{k \pm 1}{\xi}. \tag{19.48}$$

All the above Eqs. (19.27), (19.36), and (19.45) are the exact solutions of the nonlinear time-fractional classical Burgers equation. Putting the suitable values of the constants in these equations, we may obtain the traveling wave solutions of the given model.

References

Al-Shawba, A.A., Abdullah, F.A., Gepreel, K.A., and Azmi, A. (2018). Solitary and periodic wave solutions of higher-dimensional conformable time-fractional differential equations using the $(G'/G,1/G)$-expansion method. *Advances in Difference Equations* 2018: 362 (1–15).

Duran, S. (2021). Extractions of travelling wave solutions of $(2 + 1)$-dimensional Boiti–Leon–Pempinelli system via $(G'/G,1/G)$-expansion method. *Optical and Quantum Electronics* 53: 299 (1–13).

Güner, O., Bekir, A., and Cevikel, A.C. (2015). A variety of exact solutions for the time fractional Cahn–Allen equation. *The European Physical Journal Plus* 130 (146): 1–13.

Jumarie, G. (2006). Modified Riemann–Liouville derivative and fractional Taylor series of nondifferentiable functions further results. *Computers & Mathematics with Applications* 51 (9-10): 1367–1376.

Li, Z.B. and He, J.H. (2010). Fractional complex transformation for fractional differential equations. *Computer and Mathematics with Applications* 15: 970–973.

Ling-xiao, L., Er-qiang, L., and Ming-liang, W. (2010). The $(G'/G,1/G)$-expansion method and its application to travelling wave solutions of the Zakharov equations. *Applied Mathematics-A Journal of Chinese Universities* 25 (4): 454–462.

Sirisubtawee, S., Koonprasert, S., and Sungnul, S. (2019). Some applications of the $(G'/G,1/G)$-expansion method for finding exact traveling wave solutions of nonlinear fractional evolution equations. *Symmetry* 11: 952 (1–29).

Yasar, E. and Giresunlu, I.B. (2016). The $(G'/G,1/G)$-expansion method for solving nonlinear space-time fractional differential equations. *Pramana – Journal of Physics* 87: 17 (1–7).

Yildirim, A. and, Pinar, Z. (2010). Application of the exp-function method for solving nonlinear reaction-diffusion equations arising in mathematical biology. *Computers and Mathematics with Applications* 60: 1873–1880.

Zayed, E.M.E. and Abdelaziz, M.A.M. (2012). The two-variable $(G'/G,1/G)$-expansion method for solving the nonlinear KdV-mKdV equation. *Mathematical Problems in Engineering* 2012: 1–14.

Zayed, E.M.E. and Alurrfi, K.A.E. (2014). The $(G'/G,1/G)$-expansion method and its application to find the exact solutions of nonlinear PDEs for nanobiosciences. *Mathematical Problems in Engineering* 2014: 1–10.

Zayed, E. and Alurrfi, K. (2016). The $(G'/G,1/G)$-expansion method and its applications to two nonlinear Schrödinger equations describing the propagation of femtosecond pulses in nonlinear optical fibers. *Optik – International Journal for Light and Electron Optics* 127: 1581–1589.

Zayed, E.M.E., Ibrahim, S.A.H., and Abdelaziz, M.A.M. (2012). Traveling wave solutions of the nonlinear $(3 + 1)$-dimensional Kadomtsev–Petviashvili equation using the two variables $(G'/G,1/G)$-expansion method. *Journal of Applied Mathematics* 2012: 1–8.

20

The Modified Simple Equation Method

20.1 Introduction

In the last few years, the modified simple equation (MSE) method, which is an analytical method, has become very popular. This approach is robust because it uses a general solution form defined by the finite series sum containing the unknown function. The characteristic of the technique enables the derivation of new and more general solitary wave solutions by substituting particular values for arbitrary coefficients in the exact solutions. The methods, such as the modified extended Tanh function method, the sine-cosine method, the generalized Kudryashov method, the improved F-expansion method, etc., work on the principle of certain particular preset functions or a solution to the auxiliary equation. These approaches require lengthy computation to solve the system of algebraic equations. However, in the MSE method, any preset function is not predefined, or a solution of any predetermined equation is not considered. As a result, this strategy may yield some novel solutions. The MSE method is recently being used to obtain exact solutions to a variety of fractional partial differential equations, including the space–time fractional modified regularized long-wave equation, the space–time fractional modified Korteweg-de Vries equation, the space–time fractional coupled Burgers' equations (Kaplan et al. 2015), the nonlinear time-fractional Sharma-Tasso-Oliver equation (Zayed et al. 2016), the generalized fractional reaction Duffing equation, and the fractional nonlinear Cahn-Allen equation (Ali et al. 2018). The exact solutions of the time-fractional biological population model equation and nonlinear fractional Klein–Gordon equation have been obtained by Kaplan and Bekir (2016). Bakicierler et al. (2021) used the MSE method to find some new traveling wave solutions of the nonlinear conformable time-fractional approximate long water wave equation and coupled time-fractional Boussinesq–Burger equation. Other related works may be found in (Taghizadeh et al. 2012; Khan et al. 2013; Taghizadeh et al. 2013; Khan and Akbar 2014; Triki et al. 2015).

20.2 Procedure of the Modified Simple Equation Method

The main steps of this approach for solving fractional partial differential equations are briefly summarized in this section.

In order to understand the MSE method (Kaplan et al. 2015; Kaplan and Bekir 2016; Zayed et al. 2016; Ali et al. 2018; Bakicierler et al. 2021), we consider the following nonlinear fractional partial differential equation in two independent variables x and t of the type

$$Q(u, u_x, u_{xx}, ..., D_t^\alpha u, ...) = 0, \quad 0 < \alpha \leq 1, \tag{20.1}$$

where $u = u(x, t)$ is an unknown function, and Q is a polynomial of u and its partial fractional derivatives, in which the highest order derivatives and the nonlinear terms are involved.

Step 1. We use the nonlinear fractional complex transformation (Li and He 2010)

$$u(x, t) = U(\xi), \quad \xi = kx + \frac{ct^\alpha}{\Gamma(1 + \alpha)}, \tag{20.2}$$

where c (angular wave number) and k (wave frequency) are nonzero constants to be determined later. By using the chain rule (Güner et al. 2015), we have

$$D_t^\alpha u = \Delta_t u_\zeta D_t^\alpha \zeta, \tag{20.3}$$

where Δ_t is the fractal index (Güner et al. 2015). Without loss of generality, we may consider $\Delta_t = \kappa$, where κ is a constant. Using Eq. (20.2), Eq. (20.1) reduces to the following nonlinear ODE form

$$P\left(U, kU', k^2 U'', ..., cU', ...\right) = 0, \tag{20.4}$$

where the $' = \dfrac{d}{d\xi}$, $'' = \dfrac{d^2}{d\xi^2}$, and so on.

Step 2. Equation (20.4) is integrated as long as all terms contain derivatives. This procedure ends when one of the terms includes no derivatives. The associated integration constants are taken to be zero.

Step 3. We suppose that the MSE method is based on the assumption that the solution of Eq. (20.4) may be expressed as (Taghizadeh et al. 2012; Khan et al. 2013; Taghizadeh et al. 2013; Khan and Akbar 2014; Kaplan et al. 2015; Triki et al. 2015; Kaplan and Bekir 2016; Zayed et al. 2016; Ali et al. 2018; Bakicierler et al. 2021)

$$U(\xi) = \sum_{i=0}^{N} A_i \left(\frac{\psi'(\xi)}{\psi(\xi)}\right)^i, \quad A_N \neq 0, \tag{20.5}$$

where A_i $(i = 0, 1, 2, ..., N)$ are constants to be determined and $\psi = \psi(\xi)$ is an unknown function to be determined later such that $\psi'(\xi) \neq 0$. In the tanh function method, the (G'/G)-expansion method, and the exp function method, etc., the solutions are represented in terms of some predefined functions. However, the MSE method does not represent the solution in terms of any predefined functions or $\psi(\xi)$ does not represent a solution of any predefined equation. As a result, this strategy may lead to identifying some new solutions.

Step 4. The positive integer N in Eq. (20.5) may be calculated by considering the homogenous balance between the highest order derivatives and the nonlinear terms arising in Eq. (20.4). More precisely, we define the degree of $U(\xi)$ as $\deg[U(\xi)] = N$, which leads to the degrees of the other expressions as follows (Gepreel and Omran 2012):

$$\deg\left[\frac{d^p U}{d\xi^p}\right] = N + p,$$

$$\deg\left[U^s \left(\frac{d^p U}{d\xi^p}\right)^q\right] = Ns + q(p + N). \tag{20.6}$$

Therefore, we can obtain the value of N in Eq. (20.5).

Step 5. Next, we substitute Eq. (20.5) into Eq. (20.4), calculate all necessary derivatives $U', U'', ...$ of the unknown function U, and account for the function $\psi(\xi)$. As a result of the substitutions, we get a polynomial of ψ^{-i} $(i = 0, 1, 2, ...)$. In this polynomial, all the terms of the same power of ψ^{-i} $(i = 0, 1, 2, ...)$ are gathered. We then equate with zero all the coefficients of this polynomial. This operation yields a system of algebraic equations that can be solved to find A_i and $\psi(\xi)$. Consequently, we can get the exact explicit solutions of Eq. (20.1).

The following section offers two examples that show the applicability of the MSE approach for solving nonlinear fractional differential equations.

20.3 Application Problems

Here, we apply the present method to solve a nonlinear time-fractional KdV equation in Example 20.1 and the mKdV equation in Example 20.2.

Example 20.1 We consider the following nonlinear time-fractional KdV equation (Wazwaz 2009)

$$\frac{\partial^\alpha u}{\partial t^\alpha} + 6u\frac{\partial u}{\partial x} + \frac{\partial^3 u}{\partial x^3} = 0, \quad t > 0, \quad x \in R, \quad 0 < \alpha \le 1. \tag{20.7}$$

Solution

Equation (20.7) may be reduced to the ordinary differential equation in ξ by using the transformation
$u(x,t) = U(\xi), \quad \xi = kx - \dfrac{ct^\alpha}{\Gamma(1+\alpha)}$ as

$$-cU' + 6kUU' + k^3 U''' = 0, \tag{20.8}$$

where prime denotes the differentiation with respect to ξ. Integrating Eq. (20.8) with respect to ξ and substituting the integration constant to zero, we have

$$-cU + 3kU^2 + k^3 U'' = 0. \tag{20.9}$$

The highest order derivative and nonlinear terms in Eq. (20.9) are U'' and U^2. So balancing the order of U'' and U^2 by using the concept Eq. (20.6), we get

$$\begin{aligned}
&\deg(U'') = \deg(U^2), \\
&N + 2 = 2N \Rightarrow N = 2.
\end{aligned} \tag{20.10}$$

So, for $N = 2$, the solution of Eq. (20.9) may be written as

$$U(\xi) = \sum_{i=0}^{2} A_i \left(\frac{\psi'(\xi)}{\psi(\xi)}\right)^i = A_0 + A_1 \left(\frac{\psi'}{\psi}\right)^1 + A_2 \left(\frac{\psi'}{\psi}\right)^2, \tag{20.11}$$

where A_0, A_1, and A_2 are constants such that $A_2 \neq 0$ and $\psi(\xi)$ is the unknown function to be determined. From Eq. (20.11), we have

$$\begin{aligned}
U''(\xi) = \left(6A_2 \psi'^4\right) \psi^{-4} + \left(2A_1 \psi'^3 - 10A_2 \psi'^2 \psi''\right) \psi^{-3} \\
+ \left(2A_2 \psi''^2 + 2A_2 \psi' \psi''' - 3A_1 \psi' \psi''\right) \psi^{-2} + \left(A_1 \psi'''\right) \psi^{-1}.
\end{aligned} \tag{20.12}$$

By plugging Eqs. (20.11) and (20.12) into Eq. (20.9), and then equating the coefficients of ψ^{-i}, $i = 0, \dots, 4$ to zero, we obtain the following system:

$$\psi^0 : 3kA_0^2 - cA_0 = 0, \tag{20.13}$$

$$\psi^{-1} : k^3 A_1 \psi''' + 6kA_0 A_1 \psi' - cA_1 \psi' = 0, \tag{20.14}$$

$$\psi^{-2} : -3k^3 A_1 \psi' \psi'' + 2k^3 A_2 \psi' \psi''' + 2k^3 A_2 \psi''^2 + 6kA_0 A_2 \psi'^2 + 3kA_1^2 \psi'^2 - cA_2 \psi'^2 = 0, \tag{20.15}$$

$$\psi^{-3} : 2k^3 A_1 \psi'^3 - 10k^3 A_2 \psi'^2 \psi'' + 6kA_1 A_2 \psi'^3 = 0, \tag{20.16}$$

$$\psi^{-4} : 6k^3 A_2 \psi'^4 + 3kA_2^2 \psi'^4 = 0. \tag{20.17}$$

Solving Eqs. (20.13), (20.14), and (20.17), we deduce four cases of solutions as

Set 1: $A_0 = 0$, $A_1 = 0$, $A_2 = 0$, **Set 2:** $A_0 = 0$, $A_1 = 0$, $A_2 = -2k^2$,
Set 3: $A_0 = \dfrac{c}{3k}$, $A_1 = 0$, $A_2 = 0$, **Set 4:** $A_0 = \dfrac{c}{3k}$, $A_1 = 0$, $A_2 = -2k^2$.
According to Eq. (20.5), Sets 1 and 3 are not acceptable as $A_2 = 0$. Now, we have two cases

Case 1: $A_0 = 0$, $A_1 = 0$, $A_2 = -2k^2$.
In this case, from Eq. (20.16), we obtain the following three sets of solutions

$$\psi(\xi) = c_1 \xi + c_2, \tag{20.18}$$

$$\psi(\xi) = c_1 \left(\frac{-1 - \sqrt{3}i}{2}\right) x + c_2, \tag{20.19}$$

$$\psi(\xi) = c_1 \left(\frac{-1 + \sqrt{3}i}{2}\right) x + c_2, \tag{20.20}$$

where c_1 and c_2 are arbitrary constants of integration.

By considering Eq. (20.18), the solution of Eq. (20.7) may be written as

$$u(x,t) = A_0 + A_1 \left(\frac{\psi'}{\psi}\right)^1 + A_2 \left(\frac{\psi'}{\psi}\right)^2 = -2k^2 \left(\frac{c_1}{c_1\xi + c_2}\right)^2, \tag{20.21}$$

where $\xi = kx - \dfrac{ct^\alpha}{\Gamma(1+\alpha)}$.

Similarly, we can consider Eqs. (20.19) and (20.20) to obtain other solutions of the given model.

Case 2: $A_0 = \dfrac{c}{3k}, \quad A_1 = 0, \quad A_2 = -2k^2.$

In this case, we get the same sets of solutions as Case 1, viz. Eqs. (20.18)–(20.20). So, using Eq. (20.18), we have

$$u(x,t) = A_0 + A_1 \left(\frac{\psi'}{\psi}\right)^1 + A_2 \left(\frac{\psi'}{\psi}\right)^2 = \frac{c}{3k} - 2k^2 \left(\frac{c_1}{c_1\xi + c_2}\right)^2, \xi = kx - \frac{ct^\alpha}{\Gamma(1+\alpha)}. \tag{20.22}$$

Other forms of solutions of the model can be determined by using Eqs. (20.19) and (20.20).

Putting the suitable values of the constants in Eqs. (20.21) and (20.22), we obtain the traveling wave solutions of the time-fractional KdV equation.

Example 20.2 Let us consider the following nonlinear time-fractional mKdV equation (Wazwaz 2009)

$$\frac{\partial^\alpha u}{\partial t^\alpha} - 6u^2 \frac{\partial u}{\partial x} + \frac{\partial^3 u}{\partial x^3} = 0, \quad 0 < \alpha \le 1, t > 0. \tag{20.23}$$

Solution

The traveling wave transformation Eq. (20.2) reduces Eq. (20.23) to the following ODE,

$$-cU' - 6kU^2U' + k^3U''' = 0. \tag{20.24}$$

Integrating Eq. (20.24) with respect to ξ and substituting integration constant to zero, we have

$$-cU - 2kU^3 + k^3U'' = 0. \tag{20.25}$$

The highest order derivative and nonlinear terms in Eq. (20.25) are U'' and U^3. So balancing the order of U'' and U^3, we get

$$\deg(U'') = \deg(U^3),$$
$$N + 2 = 3N \Rightarrow N = 1. \tag{20.26}$$

So, we assume the solution of Eq. (20.25) at $N = 1$. This may be written as

$$U(\xi) = \sum_{i=0}^1 A_i \left(\frac{\psi'(\xi)}{\psi(\xi)}\right)^i = A_0 + A_1 \left(\frac{\psi'}{\psi}\right), \tag{20.27}$$

where A_0 and A_1 are constants such that $A_1 \ne 0$. Now, from Eq. (20.27), we get

$$U''(\xi) = \left(2A_1\psi'^3\right)\psi^{-3} - (3A_1\psi'\psi'')\psi^{-2} + \left(A_1\psi'''\right)\psi^{-1}. \tag{20.28}$$

By substituting Eqs. (20.27) and (20.28) into Eq. (20.25), and equating the coefficients of $\psi^{-i}, i = 0, ..., 3$ to zero, we get the following system of equations

$$\psi^0 : -2kA_0^3 - cA_0 = 0, \tag{20.29}$$
$$\psi^{-1} : k^3A_1\psi''' - 6kA_0^2A_1\psi' - cA_1\psi' = 0, \tag{20.30}$$
$$\psi^{-2} : -3k^3A_1\psi'\psi'' - 6kA_0A_1^2\psi'^2 = 0, \tag{20.31}$$
$$\psi^{-3} : 2k^3A_1\psi'^3 - 2kA_1^3\psi'^3 = 0, \tag{20.32}$$

From Eqs. (20.29), and (20.32), we have

Set 1: $A_0 = 0$, $A_1 = 0$, **Set 2:** $A_0 = 0$, $A_1 = k$, **Set 3:** $A_0 = 0$, $A_1 = -k$,

Set 4: $A_0 = \sqrt{\dfrac{-c}{2k}}$, $A_1 = 0$, **Set 5:** $A_0 = \sqrt{\dfrac{-c}{2k}}$, $A_1 = k$, **Set 6:** $A_0 = \sqrt{\dfrac{-c}{2k}}$, $A_1 = -k$.

Solution sets 1 and 4 contradict the assumption that $A_2 \neq 0$. Thus, these two sets are rejected. Now, we have four cases.

Case 1: If $A_0 = 0$, $A_1 = k$, then from Eq. (20.30), we get

$$\psi(\xi) = c_1 + c_2 \exp\left(\frac{\sqrt{c}}{k^{\frac{3}{2}}}\xi\right) + c_3 \exp\left(-\frac{\sqrt{c}}{k^{\frac{3}{2}}}\xi\right), \tag{20.33}$$

where c_1, c_2, and c_3 are arbitrary constants.

So, the solution of Eq. (20.23) can be written as

$$u(x,t) = A_0 + A_1\left(\frac{\psi'}{\psi}\right) = \sqrt{\frac{c}{k}}\left(\frac{c_2 \exp\left(\frac{\sqrt{c}}{k^{\frac{3}{2}}}\xi\right) - c_3 \exp\left(-\frac{\sqrt{c}}{k^{\frac{3}{2}}}\xi\right)}{c_1 + c_2 \exp\left(\frac{\sqrt{c}}{k^{\frac{3}{2}}}\xi\right) + c_3 \exp\left(-\frac{\sqrt{c}}{k^{\frac{3}{2}}}\xi\right)}\right). \tag{20.34}$$

If we put $c_1 = 0$, $c_2 = c_3$, into Eq. (20.34), we obtain the kink solitary 1-wave solution

$$u(x,t) = \sqrt{\frac{c}{k}}\tanh\left(\frac{\sqrt{c}}{k^{\frac{3}{2}}}\xi\right), \tag{20.35}$$

while, if we set $c_1 = 0$, $c_2 = -c_3$, then we get the anti-kink solitary 1-wave solution

$$u(x,t) = \sqrt{\frac{c}{k}}\cot\left(\frac{\sqrt{c}}{k^{\frac{3}{2}}}\xi\right), \quad \xi = kx - \frac{ct^\alpha}{\Gamma(1+\alpha)} \tag{20.36}$$

Case 2: $A_0 = \sqrt{\dfrac{-c}{2k}}, A_1 = k$.

In this case, from Eq. (20.31), we obtain

$$\psi(\xi) = c_1 + c_2 \exp\left(-\sqrt{-\frac{2c}{k^{\frac{5}{2}}}}\xi\right), \tag{20.37}$$

So, the solution of the given model Eq. (20.23) can be written as

$$u(x,t) = A_0 + A_1\left(\frac{\psi'}{\psi}\right) = \sqrt{\frac{-c}{2k}} + \sqrt{-\frac{2c}{k}}\left(\frac{c_2 \cosh\left(\sqrt{-\frac{2c}{k^{\frac{5}{2}}}}\xi\right) - c_2 \sinh\left(\sqrt{-\frac{2c}{k^{\frac{5}{2}}}}\xi\right)}{c_1 + c_2 \cosh\left(\sqrt{-\frac{2c}{k^{\frac{5}{2}}}}\xi\right) - c_2 \sinh\left(\sqrt{-\frac{2c}{k^{\frac{5}{2}}}}\xi\right)}\right), \tag{20.38}$$

where $\xi = kx - \dfrac{ct^\alpha}{\Gamma(1+\alpha)}$.

Likewise, we may derive various solutions using solution sets 3 and 6.

References

Ali, A.M., Ali, N.M.H., and Wazwaz, A.M. (2018). Closed form traveling wave solutions of non-linear fractional evolution equations through the modified simple equation method. *Thermal Science* 22: 341–352.

Bakicierler, G., Alfaqeih, S., and Mısırlı, E. (2021). Application of the modified simple equation method for solving two nonlinear time-fractional long water wave equations. *Revista Mexicana de Física* 67: 1–7.

Gepreel, K.A. and Omran, S. (2012). Exact solutions for nonlinear partial fractional differential equations. *Chinese Physics B* 21 (11): 110204.

Güner, O., Bekir, A., and Cevikel, A.C. (2015). A variety of exact solutions for the time fractional Cahn–Allen equation. *The European Physical Journal Plus* 130 (146): 1–13.

Kaplan, M. and Bekir, A. (2016). The modified simple equation method for solving some fractional-order nonlinear equations. *Pramana* 87 (1): 1–5.

Kaplan, M., Bekir, A., Akbulut, A., and Aksoy, E. (2015). The modified simple equation method for nonlinear fractional differential equations. *Romanian Journal of Physics* 60 (9-10): 1374–1383.

Khan, K. and Akbar, M.A. (2014). Exact solutions of the (2+ 1)-dimensional cubic Klein–Gordon equation and the (3+ 1)-dimensional Zakharov–Kuznetsov equation using the modified simple equation method. *Journal of the Association of Arab Universities for Basic and Applied Sciences* 15: 74–81.

Khan, K., Akbar, M.A., and Ali, N.H. (2013). The modified simple equation method for exact and solitary wave solutions of nonlinear evolution equation: the GZK-BBM equation and right-handed noncommutative Burgers equations. *International Scholarly Research Notices* 2013: 1–5.

Li, Z.B. and He, J.H. (2010). Fractional complex transformation for fractional differential equations. *Computer and Mathematics with Applications* 15: 970–973.

Taghizadeh, N., Mirzazadeh, M., Paghaleh, A.S., and Vahidi, J. (2012). Exact solutions of nonlinear evolution equations by using the modified simple equation method. *Ain Shams Engineering Journal* 3 (3): 321–325.

Taghizadeh, N., Mirzazadeh, M., Rahimian, M., and Akbari, M. (2013). Application of the simplest equation method to some time-fractional partial differential equations. *Ain Shams Engineering Journal* 4 (4): 897–902.

Triki, H., Mirzazadeh, M., Bhrawy, A.H. et al. (2015). Solitons and other solutions to long-wave short-wave interaction equation. *Romanian Journal of Physics* 60 (1–2): 72–86.

Wazwaz, A.M. (2009). *Partial Differential Equations and Solitary Waves Theory*. Springer, Berlin, Heidelberg: Nonlinear Physical Science.

Zayed, E.M., Amer, Y.A., and Al-Nowehy, A.G. (2016). The modified simple equation method and the multiple exp-function method for solving nonlinear fractional Sharma-Tasso-Olver equation. *Acta Mathematicae Applicatae Sinica, English Series* 32 (4): 793–812.

21

Sine-Cosine Method

21.1 Introduction

Nonlinear fractional differential equations have been studied using several approaches in recent decades. These methods may require a significant size of computational work. Several strategies have been developed to find single soliton solutions of the fractional models. It may be noted that there is no single approach that can be applied to all nonlinear fractional models. The sine-cosine method (Wazwaz 2004b, 2004c, 2005, 2017) proposed by Wazwaz (Wazwaz 2004a) is a powerful strategy that has recently been used in various researches. Using the sine-cosine approach, Sabi'u et al. (2019) found the exact solution for the (3 + 1) conformable space–time fractional modified Korteweg–de-Vries equations. Alquran (2012) has obtained the periodic and bell-shaped solitons solutions to the Benjamin-Bona-Mahony, the Gardner equations, and the Cassama-Holm equation. Yusufoglu and Bekir (2006) have successfully derived many new families of exact traveling wave solutions of the (2 + 1)-dimensional Konopelchenko–Dubrovsky equations and the coupled nonlinear Klein–Gordon and Nizhnik–Novikov–Veselov equations. Bekir (2008) successfully established exact traveling wave solutions of the symmetric regularized long-wave (SRLW) and Klein–Gordon–Zakharov (KGZ) equations with the help of this method. The sine-cosine method has been used to solve a broad range of nonlinear problems. Various nonlinear dispersive and dissipative equations have also been successfully solved using this method.

21.2 Details of the Sine-Cosine Method

The main steps of this method for solving fractional partial differential equations are briefly described in this section.

In order to understand the sine-cosine method (Wazwaz 2004a, 2004b, 2004c, 2005, 2017), we consider the following non-linear fractional partial differential equation in two independent variables x and t of the type

$$Q(u, u_x, u_{xx}, u_{xxx}..., D_t^\alpha u, ...) = 0, \quad 0 < \alpha \le 1, \tag{21.1}$$

where $u = u(x, t)$ is an unknown function, and Q is a polynomial of u and its partial fractional derivatives, in which the highest order derivatives and the nonlinear terms are involved.

Step 1. First, the traveling wave variable is considered as (Li and He 2010)

$$u(x, t) = u(\xi), \tag{21.2}$$

$$\xi = kx - \frac{ct^\alpha}{\Gamma(1 + \alpha)}, \tag{21.3}$$

where c and k are nonzero constants to be determined later.

Step 2. Next, we use the following results:

$$\frac{\partial^\alpha u}{\partial t^\alpha} = -c\frac{du}{d\xi}, \quad \frac{\partial u}{\partial x} = k\frac{du}{d\xi}, \quad \frac{\partial^2 u}{\partial x^2} = k^2\frac{d^2 u}{d\xi^2}, \quad \frac{\partial^3 u}{\partial x^3} = k^3\frac{d^3 u}{d\xi^3}, \tag{21.4}$$

Computational Fractional Dynamical Systems: Fractional Differential Equations and Applications, First Edition.
Snehashish Chakraverty, Rajarama Mohan Jena, and Subrat Kumar Jena.
© 2023 John Wiley & Sons, Inc. Published 2023 by John Wiley & Sons, Inc.

and so on for other derivatives. Using Eq. (21.4), Eq. (21.1) can be rewritten in the following nonlinear ordinary differential equation (ODE) form

$$P\left(u, ku', k^2 u'', k^3 u''' ..., -cu', ...\right) = 0, \tag{21.5}$$

where P is a polynomial of $u(\xi)$ and its various derivatives. The prime $(')$ denotes the derivative with respect to ξ.

Step 3. The ordinary differential equation Eq. (21.5) is integrated as many times as needed, and the integration constant is set to zero.

Step 4. The solution of Eq. (21.5) may be expressed in the form (Wazwaz 2004b, 2004c, 2005, 2017)

$$u(x,t) = \begin{cases} \lambda \sin^\beta(\mu\xi), & |\xi| \le \dfrac{\pi}{\mu}, \\ 0, & \text{otherwise}, \end{cases} \tag{21.6}$$

or in the form

$$u(x,t) = \begin{cases} \lambda \cos^\beta(\mu\xi), & |\xi| \le \dfrac{\pi}{2\mu}, \\ 0, & \text{otherwise}, \end{cases} \tag{21.7}$$

where λ, μ, and β are parameters that are to be determined.

Step 5. As a consequence, the derivatives of Eq. (21.6) becomes

$$
\begin{aligned}
u(\xi) &= \lambda \sin^\beta(\mu\xi), \\
u^n(\xi) &= \lambda^n \sin^{n\beta}(\mu\xi), \\
(u^n)_\xi &= n\mu\beta\lambda^n \cos(\mu\xi) \sin^{n\beta-1}(\mu\xi), \\
(u^n)_{\xi\xi} &= -n^2\mu^2\beta^2\lambda^n \sin^{n\beta}(\mu\xi) + n\mu^2\lambda^n\beta(n\beta-1)\sin^{n\beta-2}(\mu\xi).
\end{aligned} \tag{21.8}
$$

The derivatives of Eq. (21.7) may be written as

$$
\begin{aligned}
u(\xi) &= \lambda \cos^\beta(\mu\xi), \\
u^n(\xi) &= \lambda^n \cos^{n\beta}(\mu\xi), \\
(u^n)_\xi &= -n\mu\beta\lambda^n \sin(\mu\xi) \cos^{n\beta-1}(\mu\xi), \\
(u^n)_{\xi\xi} &= -n^2\mu^2\beta^2\lambda^n \cos^{n\beta}(\mu\xi) + n\mu^2\lambda^n\beta(n\beta-1)\cos^{n\beta-2}(\mu\xi),
\end{aligned} \tag{21.9}
$$

and so on for the other derivatives.

Step 6. We substitute Eq. (21.8) or Eq. (21.9) into Eq. (21.5) and balance the terms of the sine functions when Eq. (21.8) is used, or balance the terms of the cosine functions when Eq. (21.9) is used.

Step 7. Collecting all terms with the same power in $\cos^k(\mu\xi)$ or $\sin^k(\mu\xi)$ and equating each coefficient to zero, we obtain a set of algebraic equations in λ, μ, β, k, and c. Solving the algebraic equations system, we obtain the values of the parameters λ, μ, and β.

The key benefits of this method are that it may be directly applied to any differential problems. Another benefit of the approach is its ability to reduce the size of computational work.

21.3 Numerical Examples

We apply the present method to solve the nonlinear time-fractional Korteweg de Vries (KdV) equation in Example 21.1 and the modified Korteweg de Vries (mKdV) equation in Example 21.2.

Example 21.1 Consider the following nonlinear time-fractional KdV equation (Wazwaz 2009)

$$\frac{\partial^\alpha u}{\partial t^\alpha} + 6u\frac{\partial u}{\partial x} + \frac{\partial^3 u}{\partial x^3} = 0, \quad 0 < \alpha \le 1, t > 0. \tag{21.10}$$

Solution

Using the transformation $u(x,t) = u(\xi), \xi = kx - \dfrac{ct^\alpha}{\Gamma(1+\alpha)}$ in Eq. (21.10), it reduces to

$$-cu' + 6ku\,u' + k^3 u''' = 0, \tag{21.11}$$

where prime denotes the differentiation with respect to ξ. Integrating Eq. (21.11) with respect to ξ and substituting the integration constant to zero, it gives

$$-cu + 3ku^2 + k^3 u'' = 0. \tag{21.12}$$

Using Eq. (21.8), we have

$$u(\xi) = \lambda \sin^\beta(\mu\xi), \tag{21.13}$$

$$u^2(\xi) = \lambda^2 \sin^{2\beta}(\mu\xi), \tag{21.14}$$

and

$$u''(\xi) = -\mu^2\beta^2\lambda \sin^\beta(\mu\xi) + \mu^2\lambda\beta(\beta-1)\sin^{\beta-2}(\mu\xi). \tag{21.15}$$

Plugging Eqs. (21.13)–(21.15) into Eq. (21.12), we have

$$k^3\lambda\mu^2\beta(\beta-1)\sin^{\beta-2}(\mu\xi) + \left(-k^3\lambda\mu^2\beta^2 - c\lambda\right)\sin^\beta(\mu\xi) + 3k\lambda^2\sin^{2\beta}(\mu\xi) = 0. \tag{21.16}$$

We use the balance between the exponents of the sine functions. This indicates that Eq. (21.16) is satisfied only if the following algebraic equations are valid.

$$\begin{aligned} &\beta - 1 \neq 0, \\ &\beta - 2 - 2\beta = 0, \\ &-3k\lambda^2 - k^3\lambda\mu^2\beta(\beta-1) = 0, \\ &-c\lambda - k^3\lambda\mu^2\beta^2 = 0. \end{aligned} \tag{21.17}$$

Solving Eq. (21.17), it gives

$$\begin{aligned} \beta &= -2, \\ \mu &= \frac{1}{2\sqrt{k^3}}\sqrt{-c}, \\ \lambda &= \frac{c}{2k}. \end{aligned} \tag{21.18}$$

Similarly, using cosine function Eq. (21.9), we have

$$u(\xi) = \lambda \cos^\beta(\mu\xi), \tag{21.19}$$

$$u^2(\xi) = \lambda^2 \cos^{2\beta}(\mu\xi), \tag{21.20}$$

and

$$u''(\xi) = -\mu^2\beta^2\lambda \cos^\beta(\mu\xi) + \mu^2\lambda\beta(\beta-1)\cos^{\beta-2}(\mu\xi). \tag{21.21}$$

Plugging Eqs. (21.19)–(21.21) into Eq. (21.12), we have

$$k^3\lambda\mu^2\beta(\beta-1)\cos^{\beta-2}(\mu\xi) + \left(-k^3\lambda\mu^2\beta^2 - c\lambda\right)\cos^\beta(\mu\xi) + 3k\lambda^2\cos^{2\beta}(\mu\xi) = 0. \tag{21.22}$$

Equating the exponents and coefficients of each pair of cosine functions, we obtain the same system of algebraic equations Eq. (21.17) and solutions Eq. (21.18).

So, the solution of Eq. (21.10) in terms of sine and cosine functions, respectively, may be written as

$$u(x,t) = \begin{cases} \dfrac{c}{2k}\csc^2\left(\dfrac{1}{2\sqrt{k^3}}\sqrt{-c}\left(kx - c\dfrac{t^\alpha}{\Gamma(1+\alpha)}\right)\right), & \left|kx - c\dfrac{t^\alpha}{\Gamma(1+\alpha)}\right| \le \dfrac{\pi}{\mu}, \\ 0, & \text{otherwise,} \end{cases} \tag{21.23}$$

and

$$u(x,t) = \begin{cases} \dfrac{c}{2k} \sec^2 \left(\dfrac{1}{2\sqrt{k^3}} \sqrt{-c} \left(kx - c\dfrac{t^\alpha}{\Gamma(1+\alpha)} \right) \right), & \left| kx - c\dfrac{t^\alpha}{\Gamma(1+\alpha)} \right| \leq \dfrac{\pi}{2\mu}, \\ 0, & \text{otherwise.} \end{cases} \tag{21.24}$$

Example 21.2 Consider the following nonlinear time-fractional mKdV equation (Wazwaz 2009)

$$\frac{\partial^\alpha u}{\partial t^\alpha} - 6u^2 \frac{\partial u}{\partial x} + \frac{\partial^3 u}{\partial x^3} = 0, \quad 0 < \alpha \leq 1, t > 0. \tag{21.25}$$

Solution

Using the transformation Eqs. (21.2) and (21.3) in Eq. (21.25), it reduces to

$$-cu' - 6ku^2 u' + k^3 u''' = 0, \tag{21.26}$$

where prime denotes the differentiation with respect to ξ. Integrating Eq. (21.26) with respect to ξ and substituting the integration constant to zero, it gives

$$-cu - 2ku^3 + k^3 u'' = 0. \tag{21.27}$$

Using Eq. (21.8), we have

$$u^3(\xi) = \lambda^3 \sin^{3\beta}(\mu\xi), \tag{21.28}$$

and

$$u''(\xi) = -\mu^2 \beta^2 \lambda \sin^\beta(\mu\xi) + \mu^2 \lambda \beta(\beta - 1) \sin^{\beta-2}(\mu\xi). \tag{21.29}$$

Substituting Eqs. (21.13), (21.28) and (21.29) into Eq. (21.26), we have

$$k^3 \lambda \mu^2 \beta(\beta - 1) \sin^{\beta-2}(\mu\xi) + \left(-k^3 \lambda \mu^2 \beta^2 - c\lambda \right) \sin^\beta(\mu\xi) - 2k\lambda^3 \sin^{3\beta}(\mu\xi) = 0. \tag{21.30}$$

We may obtain the following system of algebraic equations by equating the exponents and coefficients of each pair of sine functions

$$\begin{aligned} &\beta - 1 \neq 0, \\ &\beta - 2 - 3\beta = 0, \\ &-2k\lambda^3 - k^3 \lambda \mu^2 \beta(\beta - 1) = 0, \\ &-c\lambda - k^3 \lambda \mu^2 \beta^2 = 0. \end{aligned} \tag{21.31}$$

Solving Eq. (21.31), it gives

$$\begin{aligned} \beta &= -1, \\ \mu &= \sqrt{\frac{-c}{k^3}}, \\ \lambda &= \pm \sqrt{\frac{c}{k}}. \end{aligned} \tag{21.32}$$

Similarly, using cosine function Eq. (21.9), we have

$$u(\xi) = \lambda \cos^\beta(\mu\xi), \tag{21.33}$$

$$u^3(\xi) = \lambda^3 \cos^{3\beta}(\mu\xi), \tag{21.34}$$

and

$$u''(\xi) = -\mu^2 \beta^2 \lambda \cos^\beta(\mu\xi) + \mu^2 \lambda \beta(\beta - 1) \cos^{\beta-2}(\mu\xi). \tag{21.35}$$

Plugging Eqs. (21.19)–(21.21) into Eq. (21.12), we have

$$k^3\lambda\mu^2\beta(\beta-1)\cos^{\beta-2}(\mu\xi) + \left(-k^3\lambda\mu^2\beta^2 - c\lambda\right)\cos^{\beta}(\mu\xi) - 2k\lambda^3\cos^{3\beta}(\mu\xi) = 0. \tag{21.36}$$

We derive the same system of algebraic equations Eq. (21.31) and solution Eq. (21.32) by equating the exponents and coefficients of each pair of cosine functions.

So, in terms of sine and cosine functions, Eq. (21.25) may be represented as follows:

$$u(x,t) = \begin{cases} \pm\sqrt{\dfrac{c}{k}}\csc\left(\sqrt{\dfrac{-c}{k^3}}\left(kx - c\dfrac{t^\alpha}{\Gamma(1+\alpha)}\right)\right), & \left|kx - c\dfrac{t^\alpha}{\Gamma(1+\alpha)}\right| \leq \dfrac{\pi}{\mu}, \\ 0, & \text{otherwise}, \end{cases} \tag{21.37}$$

and

$$u(x,t) = \begin{cases} \pm\sqrt{\dfrac{c}{k}}\sec\left(\sqrt{\dfrac{-c}{k^3}}\left(kx - c\dfrac{t^\alpha}{\Gamma(1+\alpha)}\right)\right), & \left|kx - c\dfrac{t^\alpha}{\Gamma(1+\alpha)}\right| \leq \dfrac{\pi}{2\mu}, \\ 0, & \text{otherwise}. \end{cases} \tag{21.38}$$

All the above Eqs. (21.23), (21.24), (21.37), and (21.38) are the exact solutions of the nonlinear time-fractional KdV and mKdV equations, respectively. Putting the suitable values of the constants in these equations, we may obtain the traveling wave solutions of the given models.

References

Alquran, M.T. (2012). Solitons and periodic solutions to nonlinear partial differential equations by the sine-cosine method. *Applied Mathematics & Information Sciences* 6 (1): 85–88.

Bekir, A. (2008). New solitons and periodic wave solutions for some nonlinear physical models by using the sine–cosine method. *Physica Scripta* 77: 045008 (1-4).

Li, Z.B. and He, J.H. (2010). Fractional complex transformation for fractional differential equations. *Computer and Mathematics with Applications* 15: 970–973.

Sabi'u, J., Jibril, A., and Gadu, A.M. (2019). New exact solution for the (3 + 1) conformable space–time fractional modified Korteweg–de-Vries equations via sine-cosine method. *Journal of Taibah University for Science* 13 (1): 91–95.

Wazwaz, A.M. (2004a). A sine-cosine method for handling nonlinear wave equations. *Mathematical and Computer Modelling* 40: 499–508.

Wazwaz, A.M. (2004b). New compactons, solitons and periodic solutions for nonlinear variants of the KdV and the KP equations. *Chaos, Solitons and Fractals* 22: 249–260.

Wazwaz, A.M. (2004c). The sine–cosine method for obtaining solutions with compact and noncompact structures. *Applied Mathematics and Computation* 159: 559–576.

Wazwaz, A.M. (2005). Compactons, solitons and periodic solutions for variants of the KdV and the KP equations. *Applied Mathematics and Computation* 161: 561–575.

Wazwaz, A.M. (2009). *Partial Differential Equations and Solitary Waves Theory*. Springer, Berlin, Heidelberg: Nonlinear Physical Science.

Wazwaz, A.M. (2017). Exact soliton and kink solutions for New (3+1)-dimensional nonlinear modified equations of wave propagation. *Open Engineering* 7 (1): 169–174.

Yusufoglu, E. and Bekir, A. (2006). Solitons and periodic solutions of coupled nonlinear evolution equations by using the sine–cosine method. *International Journal of Computer Mathematics* 83 (12): 915–924.

22

Tanh Method

22.1 Introduction

Nonlinear phenomena have an essential role in various scientific fields. Exact solutions of these nonlinear phenomena are obtained using several approaches, including the tanh method, the inverse scattering method, Hirota's bilinear methodology, and the truncated Painleve expansion. Among them, the tanh method (Malfliet 1992; Malfliet 1996a; Malfliet 1996b; Wazwaz 2004; Wazwaz 2005; Wazwaz 2008) is a robust method for finding the exact traveling wave solutions. Huibin and Kelin (1990) proposed a power series in tanh as a solution and directly substituted this expansion into a higher order KdV equation. The coefficients of the power series were obtained from the resulting algebraic equations. The tanh method is one of the most straightforward technique for getting exact solutions to nonlinear diffusion equations (Khater et al. 2002). Various forms of the tanh method have been developed, and then a power series in tanh was utilized as an ansatz to get analytical solutions of certain nonlinear evolution equations (Malfliet 1996a).

In order to reduce the complexity of the tanh method, Malfliet (1992) modified the tanh approach by introducing tanh as a new variable. After that, a straightforward analysis was performed to ensure that the approach might be applicable to a wide range of equations (Khater et al. 2002). Later, in (Malfliet 1992; Malfliet 1996a; Malfliet 1996b), this approach was improved by adding the boundary conditions into the series expansion. Fan and Hon (2002) developed a generalized tanh approach for getting multiple traveling wave solutions, in which the Riccati equation solution is used to replace the hyperbolic tan function in the tanh method. Pandir and Yildirim (2018) applied the generalized tanh method to obtain the exact traveling wave solution of the space-time fractional foam drainage equation, the nonlinear time-space fractional Korteweg–de Vries equation, and time-fractional reaction-diffusion equation.

22.2 Description of the Tanh Method

This section summarises the major steps of this approach for solving fractional partial differential equations.

In order to understand the Tanh method (Malfliet 1992, 1996a, 1996b; Wazwaz 2004, 2005, 2008), we consider the following nonlinear fractional partial differential equation in two independent variables x and t of the type

$$Q\left(u, u_x, u_{xx}, u_{xxx}..., D_t^\alpha u,. ...\right) = 0, \quad 0 < \alpha \le 1, \tag{22.1}$$

where $u = u(x, t)$ is an unknown function, and Q is a polynomial of u and its partial fractional derivatives, in which the highest order derivatives and the nonlinear terms are involved.

Step 1. First, the traveling wave variable is considered as (Li and He 2010)

$$u(x, t) = u(\xi), \tag{22.2}$$

$$\xi = kx - \frac{ct^\alpha}{\Gamma(1 + \alpha)}, \tag{22.3}$$

where c and k are nonzero constants which are to be determined later.

Step 2. From Eqs. (22.2) and (22.3), we obtain the following results:

$$\frac{\partial^\alpha u}{\partial t^\alpha} = -c\frac{du}{d\xi}, \quad \frac{\partial u}{\partial x} = k\frac{du}{d\xi}, \quad \frac{\partial^2 u}{\partial x^2} = k^2\frac{d^2 u}{d\xi^2}, \quad \frac{\partial^3 u}{\partial x^3} = k^3\frac{d^3 u}{d\xi^3}, \tag{22.4}$$

Computational Fractional Dynamical Systems: Fractional Differential Equations and Applications, First Edition.
Snehashish Chakraverty, Rajarama Mohan Jena, and Subrat Kumar Jena.
© 2023 John Wiley & Sons, Inc. Published 2023 by John Wiley & Sons, Inc.

and so on for other derivatives. Using Eq. (22.4), Eq. (22.1) can be rewritten in the following nonlinear ordinary differential equation (ODE) form

$$P\left(u, ku', k^2u'', k^3u'''..., -cu', ...\right) = 0,\tag{22.5}$$

where P is a polynomial of $u(\xi)$ and its various derivatives. The prime $(')$ denotes the derivative with respect to ξ.

Step 3. Equation (22.5) is integrated as long as all terms contain derivatives. This procedure comes to an end when one of the terms includes no derivatives. The associated integration constants are taken to be zero.

Step 4. Now, we introduce a new independent variable (Pandir and Yildirim 2018)

$$Y = \tanh\left(\mu\xi\right).\tag{22.6}$$

The corresponding derivatives are then derived as follows:

$$\frac{du}{d\xi} = \frac{dY}{d\xi}\frac{du}{dY} = \mu\sec h^2(\mu\xi)\frac{du}{dY} = \mu\left(1 - \tanh^2(\mu\xi)\right)\frac{du}{dY} = \mu\left(1 - Y^2\right)\frac{du}{dY},\tag{22.7}$$

$$\begin{aligned}\frac{d^2u}{d\xi^2} &= \frac{d}{d\xi}\frac{du}{d\xi} = \frac{d}{d\xi}\left(\mu\left(1 - Y^2\right)\frac{du}{dY}\right) = \frac{d}{dY}\left(\mu\left(1 - Y^2\right)\frac{du}{dY}\right)\frac{dY}{d\xi}\\ &= \mu^2\left(1 - Y^2\right)^2\frac{d^2u}{dY^2} - 2\mu^2 Y\left(1 - Y^2\right)\frac{du}{dY},\end{aligned}\tag{22.8}$$

$$\frac{d^3}{d\xi^3} = 2\mu^3\left(1 - Y^2\right)\left(3Y^2 - 1\right)\frac{du}{dY} - 6\mu^3 Y\left(1 - Y^2\right)^2\frac{d^2u}{dY^2} + \mu^3\left(1 - Y^2\right)^3\frac{d^3u}{dY^3},\tag{22.9}$$

and so on, where μ is a parameter that is to be determined.

Step 5. We assume the tanh method for the solution of Eq. (22.5) in the following finite series expansion form (Malfliet 1992, 1996a, 1996b; Wazwaz 2004, 2005, 2008)

$$u(\xi) = S(Y) = \sum_{i=0}^{M} a_i Y^i.\tag{22.10}$$

The value of M may be calculated by balancing the highest order derivatives with the nonlinear terms arising in Eq. (22.5). More precisely, we define the degree of $u(\xi)$ as $\deg[u(\xi)] = M$, which leads to the degrees of the other expressions as follows (Gepreel and Omran 2012)

$$\begin{aligned}\deg\left[\frac{d^p u}{d\xi^p}\right] &= M + p,\\ \deg\left[u^s\left(\frac{d^p u}{d\xi^p}\right)^q\right] &= Ms + q\left(p + M\right).\end{aligned}\tag{22.11}$$

Accordingly, we can obtain the value of M in Eq. (22.10).

Step 6. Substituting Eqs. (22.7)–(22.9) into Eq. (22.5), using Eq. (22.10), and collecting the coefficients of like powers of Y^i, $(i = 0, 1, 2, ...)$, Eq. (22.5) is converted into another polynomial in Y. Then equating each coefficient of the resulting polynomial to zero, we obtain a set of algebraic equations for μ, k, c, and $a_i(i = 0, 1, ..., M)$. Solving the algebraic equations system and substituting the values of these constants μ, k, c, and $a_i(i = 0, 1, ..., M)$ into Eq. (22.10), we can have a variety of exact solutions of Eq. (22.1).

Remark: It is not always possible to obtain the value of M as a positive integer. In some cases, the values of M can be obtained as negative numbers too. So, in order to avoid the singularity for $Y \to 0$, $(-1 \leq Y \leq 1)$, the series expansion Eq. (22.10) may be modified as follows:

$$u(\xi) = S(Y) = \left(\sum_{i=0}^{M'} a_i Y^i\right)^{-1}, \quad M' = -M(> 0).\tag{22.12}$$

22.3 Numerical Examples

We apply the present method to solve the nonlinear time-fractional KdV equation in Example 22.1 and the mKdV equation in Example 22.2.

Example 22.1   Consider the following nonlinear time-fractional KdV equation (Wazwaz 2009)

$$\frac{\partial^\alpha u}{\partial t^\alpha} + 6u\frac{\partial u}{\partial x} + \frac{\partial^3 u}{\partial x^3} = 0, \quad 0 < \alpha \leq 1, t > 0 \tag{22.13}$$

Solution

Using the transformation $u(x, t) = u(\xi)$, $\xi = kx - \dfrac{ct^\alpha}{\Gamma(1+\alpha)}$ in Eq. (22.13), it reduces to

$$-cu' + 6ku\,u' + k^3 u''' = 0, \tag{22.14}$$

where prime denotes the differentiation with respect to ξ. Integrating Eq. (22.14) with respect to ξ and substituting the integration constant to zero, it gives

$$-cu + 3ku^2 + k^3 u'' = 0. \tag{22.15}$$

The highest order linear and nonlinear terms in Eq. (22.15) are u'' and u^2. So balancing the order of u'' and u^2 by using the concept given in Eq. (22.11), we get

$$\begin{aligned} \deg(u'') &= \deg(u^2), \\ M + 2 &= 2M \Rightarrow M = 2. \end{aligned} \tag{22.16}$$

Now, we assume the solution of Eq. (22.15) as Eq. (22.10) at $M = 2$. This may be written as

$$u(\xi) = S(Y) = a_0 + a_1\,Y + a_2\,Y^2, \tag{22.17}$$

where a_0, a_1, and a_2 are constants to be calculated. Substituting Eqs. (22.7) and (22.8) into Eq. (22.15), we have

$$-cu + 3ku^2 + k^3\left(\mu^2\left(1 - Y^2\right)^2 \frac{d^2 u}{dY^2} - 2\mu^2 Y\left(1 - Y^2\right)\frac{du}{dY}\right) = 0. \tag{22.18}$$

From Eq. (22.17), we have

$$\begin{aligned} u'(\xi) &= S'(Y) = a_1 + 2a_2 Y, \\ u''(\xi) &= S''(Y) = 2a_2. \end{aligned} \tag{22.19}$$

Substituting Eqs. (22.17) and (22.19) into Eq. (22.18) yields a polynomial in Y with the parameters μ, k, c, a_0, a_1, and a_2 as follows:

$$-c\left(a_0 + a_1\,Y + a_2\,Y^2\right) + 3k\left(a_0 + a_1\,Y + a_2\,Y^2\right)^2 + k^3\left\{\begin{array}{l}\mu^2(1-Y^2)^2(2a_2) - 2\mu^2 Y \\ (1-Y^2)(a_1\,Y + 2a_2\,Y)\end{array}\right\} = 0. \tag{22.20}$$

Now, collecting the coefficients of Y^i, $i = 0, 1, 2, \ldots$ and setting these to zero, we obtain the following system:

$$\begin{aligned} Y^4 &: \ 6k^3\mu^2 a_2 + 3ka_2^2, \\ Y^3 &: 2k^3\mu^2 a_1 + 6ka_1 a_2, \\ Y^2 &: -8k^3\mu^2 a_2 + 6ka_0 a_2 + 3ka_1^2 - ca_2, \\ Y^1 &: -2k^3\mu^2 a_2 c + 6ka_0 a_1 c - c^2 a_1, \\ Y^0 &: 2k^3\mu^2 a_2 + 3ka_0^2 - ca_0. \end{aligned} \tag{22.21}$$

It may be noted that the point of intersection of the two lines is the solution when solving a system of linear equations. In the case of a nonlinear system of equations, however, the diagrams might be circles, parabolas, or hyperbolas with many points of intersection, resulting in multiple solutions. By solving the above nonlinear system, it gives different values of c, k, μ, a_0, a_1, a_2 in various cases. Some of them are provided here.

Case 1:

$$a_0 = \frac{2}{3}\mu^2 k^2, \quad a_1 = 0, \quad a_2 = -2\mu^2 k^2, \quad c = -4\mu^2 k^3, \quad \mu = \mu, \quad k = k, \tag{22.22}$$

where μ and k are free parameters. Substituting the values of constants given in Eq. (22.22) into Eq. (22.17), we obtain

$$u(x, t) = \frac{2}{3}\mu^2 k^2 - 2\mu^2 k^2 \tanh^2\left(\mu\left(kx + 4\mu^2 k^3 \frac{t^\alpha}{\Gamma(1+\alpha)}\right)\right). \tag{22.23}$$

Case 2:

$$a_0 = 2\mu^2 k^2, \quad a_1 = 0, \quad a_2 = -2\mu^2 k^2, \quad c = 4\mu^2 k^3, \quad \mu = \mu, \quad k = k, \tag{22.24}$$

where μ and k are free parameters. Using the values of constants from Eq. (22.24), Eq. (22.17) is written as

$$u(x, t) = 2\mu^2 k^2 - 2\mu^2 k^2 \tanh^2\left(\mu\left(kx - 4\mu^2 k^3 \frac{t^\alpha}{\Gamma(1+\alpha)}\right)\right). \tag{22.25}$$

All the above Eqs. (22.23) and (22.25) are the exact solutions of the nonlinear time-fractional KdV equation for case 1 and case 2. Putting the suitable values of μ and k in Eqs. (22.23) and (22.25), we obtain the solutions of the given model.

Example 22.2 Let us consider the following nonlinear time-fractional mKdV equation (Wazwaz 2009)

$$\frac{\partial^\alpha u}{\partial t^\alpha} - 6u^2 \frac{\partial u}{\partial x} + \frac{\partial^3 u}{\partial x^3} = 0, \quad 0 < \alpha \le 1, t > 0. \tag{22.26}$$

Solution

Using the transformation Eqs. (22.2) and (22.3) in Eq. (22.26), it reduces to

$$-cu' - 6ku^2 u' + k^3 u''' = 0. \tag{22.27}$$

Integrating Eq. (22.27) with respect to ξ and substituting the integration constant to zero, it gives

$$-cu - 2ku^3 + k^3 u'' = 0. \tag{22.28}$$

The highest order linear and nonlinear terms in Eq. (22.28) are u'' and u^3. So balancing the order of u'' and u^3 by using the concept in Eq. (22.11), we get

$$\deg(u'') = \deg(u^3),$$
$$M + 2 = 3M \Rightarrow M = 1. \tag{22.29}$$

By assuming the solution of Eq. (22.28) as Eq. (22.10) at $M = 1$, we have

$$u(\xi) = S(Y) = a_0 + a_1 Y, \tag{22.30}$$

where a_0 and a_1 are constants. Plugging Eqs. (22.7) and (22.8) into Eq. (22.28), we have

$$-cu - 2ku^3 + k^3\left(\mu^2(1-Y^2)^2 \frac{d^2u}{dY^2} - 2\mu^2 Y(1-Y^2)\frac{du}{dY}\right) = 0. \tag{22.31}$$

From Eq. (22.30), we have

$$u'(\xi) = S'(Y) = a_1,$$
$$u''(\xi) = S''(Y) = 0. \tag{22.32}$$

Substituting Eqs. (22.30) and (22.32) into Eq. (22.31), we get

$$-c(a_0 + a_1 Y) - 2k(a_0 + a_1 Y)^3 + k^3\{-2\mu^2 Ya_1(1-Y^2)\} = 0. \tag{22.33}$$

Now, collecting the coefficients of equal power of Y and setting these to zero, we obtain the following system

$$
\begin{aligned}
Y^3 &: \ 2k^3\mu^2 a_1 - 2ka_1^3, \\
Y^2 &: \ -6ka_0 a_1^2, \\
Y^1 &: \ -2k^3\mu^2 a_1 c - 6ka_0^2 a_1 c - c^2 a_1, \\
Y^0 &: \ -2ka_0^3 - ca_0.
\end{aligned}
\tag{22.34}
$$

Solving the above nonlinear system gives various cases, as mentioned in the previous example, and a few of those are included here.

Case 1:

$$
a_0 = 0, \quad a_1 = \mu k, \quad c = -2\mu^2 k^3, \quad \mu = \mu, \quad k = k,
\tag{22.35}
$$

where μ and k are free parameters. By substituting the constant values from Eq. (22.35) into Eq. (22.30), we get

$$
u(x, t) = \mu k \ \tanh\left(\mu \left(kx + 2\mu^2 k^3 \frac{t^\alpha}{\Gamma(1+\alpha)} \right) \right).
\tag{22.36}
$$

Case 2:

$$
a_0 = 0, \quad a_1 = -\mu k, \quad c = -2\mu^2 k^3, \quad \mu = \mu, \quad k = k,
\tag{22.37}
$$

where μ and k are free parameters. Using the values of constants from Eq. (22.37), Eq. (22.30) is written as

$$
u(x, t) = -\mu k \ \tanh\left(\mu \left(kx + 2\mu^2 k^3 \frac{t^\alpha}{\Gamma(1+\alpha)} \right) \right).
\tag{22.38}
$$

Eqs. (22.23), (22.25), (22.36), and (22.38) are the exact solutions to the nonlinear time-fractional KdV and mKdV equations. We may find the solutions of the given models by substituting appropriate constant values into these equations.

References

Fan, E. and Hon, Y.C. (2002). Generalized tanh method extended to special types of nonlinear equations. *Zeitschrift für Naturforschung* 57a: 692–700.

Gepreel, K.A. and Omran, S. (2012). Exact solutions for nonlinear partial fractional differential equations. *Chinese Physics B* 21 (11): 110204.

Huibin, L. and Kelin, W. (1990). Exact solutions for two nonlinear equations: I. *Journal of Physics A: Mathematical and General* 23: 3923–3928.

Khater, A.H., Malfliet, W., Callebaut, D.K., and Kamel, E.S. (2002). The tanh method, a simple transformation and exact analytical solutions for nonlinear reaction–diffusion equations. *Chaos, Solitons & Fractals* 14: 513–522.

Li, Z.B. and He, J.H. (2010). Fractional complex transformation for fractional differential equations. *Computer and Mathematics with Applications* 15: 970–973.

Malfliet, W. (1992). Solitary wave solutions of nonlinear wave equations. *Am. J. Phys.* 60 (7): 650–654.

Malfliet, W. (1996a). The tanh method: I. Exact solutions of nonlinear evolution and wave equations. *Physica Scripta* 54: 563–568.

Malfliet, W. (1996b). The tanh method: II. Perturbation technique for conservative systems. *Physica Scripta* 54: 569–575.

Pandir, Y. and Yildirim, A. (2018). Analytical approach for the fractional differential equations by using the extended tanh method. *Waves in Random and Complex Media* 28 (3): 399–410.

Wazwaz, A.M. (2004). The tanh method for traveling wave solutions of nonlinear equations. *Applied Mathematics and Computation* 154: 713–723.

Wazwaz, A.M. (2005). The tanh method: solitons and periodic solutions for the Dodd–Bullough–Mikhailov and the Tzitzeica–Dodd–Bullough equations. *Chaos, Solitons and Fractals* 25: 55–63.

Wazwaz, A.M. (2008). The tanh method for travelling wave solutions to the Zhiber–Shabat equation and other related equations. *Communications in Nonlinear Science and Numerical Simulation* 13: 584–592.

Wazwaz, A.M. (2009). *Partial Differential Equations and Solitary Waves Theory*, Nonlinear Physical Science. Berlin, Heidelberg: Springer.

23

Fractional Subequation Method

23.1 Introduction

Based on the results obtained by (Zhang et al. 2010), Zhang and Zhang (2011) recently developed a novel algebraic approach called the fractional subequation method for determining the traveling wave solutions to nonlinear fractional partial differential equations (FPDEs). The homogeneous balancing principle (Wang 1999) and Jumarie's modified Riemann-Liouville derivative of fractional order (Jumarie 2006) is used in this technique. Zhang et al. Zhang and Zhang 2011) used this technique to derive traveling wave solutions for the nonlinear time fractional biological population model and the (4 + 1)-dimensional space–time fractional Fokas equation. The traveling wave solutions of the space–time fractional mBBM equation and the ZKBBM equation (Alzaidy 2013a) were obtained using this approach. The space–time fractional Potential Kadomtsev–Petviashvili (PKP) equation and the space–time fractional symmetric regularized long wave (SRLW) equation have both been solved using the subequation approach (Alzaidy 2013b). This approach was utilized by (Yépez-Martínez et al. 2014) to establish analytical solutions for the space–time fractional coupled Hirota-Satsuma Korteweg de Vries (KdV) and modified Korteweg de Vries (mKdV) equations. The fractional subequation method employing Jumarie's modified Riemann-Liouville derivative was used by (Mohyud-Dina et al. 2017) to produce analytical solutions of the space–time fractional Calogero-Degasperis (CD) and potential Kadomtsev-Petviashvili (PKP) equations. Yépez-Martínez and Gómez-Aguilar (2019) examined the space–time fractional Hirota–Satsuma-coupled KdV and mKdV equations using the fractional subequation approach described in Atangana's conformable derivative sense. Bekir et al. (2014) used this approach to obtain exact traveling wave solutions to the space–time fractional fifth-order Sawada-Kotera and (2 + 1)-dimensional dispersive long-wave equations.

23.2 Implementation of the Fractional Subequation Method

Now, we will discuss the fractional subequation method (Zhang and Zhang 2011; Alzaidy 2013a, 2013b; Mohyud-Dina et al. 2017; Yépez-Martínez and Gómez-Aguilar 2019) for obtaining the solution of the nonlinear fractional differential equation. In order to understand this method, we consider the following nonlinear fractional partial differential equation in two independent variables x and t of the type

$$Q(u, u_x, u_t, D_x^\alpha u, D_t^\alpha u, ...) = 0, \quad 0 < \alpha \leq 1, \tag{23.1}$$

where $D_x^\alpha u$ and $D_t^\alpha u$ are Jumarie's modified Reimann–Liouville derivatives of u, $u = u(x, t)$ is an unknown function, and Q is a polynomial of u and its partial derivatives, in which the highest order derivatives and the nonlinear terms are involved.

The main steps of this method are given as follows:

Step 1. First, we consider the traveling wave transformation as (Zhang and Zhang 2011; Bekir et al. 2014; Yépez-Martínez et al. 2014; Mohyud-Dina et al. 2017; Yépez-Martínez and Gómez-Aguilar 2019)

$$u(x, t) = u(\xi), \quad \xi = kx + ct, \tag{23.2}$$

where c and k are nonzero constants to be determined later.

Computational Fractional Dynamical Systems: Fractional Differential Equations and Applications, First Edition.
Snehashish Chakraverty, Rajarama Mohan Jena, and Subrat Kumar Jena.
© 2023 John Wiley & Sons, Inc. Published 2023 by John Wiley & Sons, Inc.

Step 2. Using the chain rule concept (Guner et al. 2015) and Eq. (23.2), we obtain the following results:

$$\frac{\partial^\alpha u}{\partial t^\alpha} = \frac{\partial^\alpha u}{\partial \xi^\alpha}\frac{\partial^\alpha \xi}{\partial t^\alpha} = c^\alpha \frac{d^\alpha u}{d\xi^\alpha} = c^\alpha D_\xi^\alpha u,$$

$$\frac{\partial^\alpha u}{\partial x^\alpha} = \frac{\partial^\alpha u}{\partial \xi^\alpha}\frac{\partial^\alpha \xi}{\partial x^\alpha} = k^\alpha \frac{d^\alpha u}{d\xi^\alpha} = k^\alpha D_\xi^\alpha u,$$

(23.3)

and so on for other derivatives. Using Eqs. (23.2) and (23.3), Eq. (23.1) is converted into the following nonlinear fractional ordinary differential equation (ODE) for $u = u(\xi)$

$$P\left(u, ku', cu', k^\alpha D_\xi^\alpha u, c^\alpha D_\xi^\alpha u, \dots\right) = 0,$$

(23.4)

where P is a polynomial of $u(\xi)$ and its various derivatives. The prime (') denotes the derivative with respect to ξ.

Step 3. According to the fractional subequation method, we suppose that the solution of Eq. (23.4) can be expressed in the following finite series expansion form (Zhang and Zhang 2011; Alzaidy 2013a; Yépez-Martínez et al. 2014; Mohyud-Dina et al. 2017; Yépez-Martínez and Gómez-Aguilar 2019)

$$u(\xi) = \sum_{i=0}^{M} a_i(\varphi(\xi))^i,$$

(23.5)

where $a_i (i = 0, 1, 2, \dots, M)$ are the constants to be determined later and $\varphi(\xi)$ satisfies the following fractional Riccati equation (Yépez-Martínez et al. 2014; Mohyud-Dina et al. 2017; Yépez-Martínez and Gómez-Aguilar 2019)

$$D_\xi^\alpha \varphi(\xi) = \sigma + \varphi^2(\xi),$$

(23.6)

where σ is a constant.

Step 4. The value of M in Eq. (23.5) may be determined by balancing the highest order derivatives and the nonlinear terms in Eq. (23.1) or (23.4). Let us assume that the degree of $u(\xi)$ as $\deg[u(\xi)] = M$, which leads to the degrees of the other expressions as follows (Bekir et al. 2014):

$$\deg\left[\frac{d^p u}{d\xi^p}\right] = M + p,$$

$$\deg\left[u^s\left(\frac{d^p u}{d\xi^p}\right)^q\right] = Ms + q(p + M).$$

(23.7)

So, using the concept of Eq. (23.7), one can find the value of M in Eq. (23.5).

Step 5. By employing the generalized Exp-function approach via Mittag–Leffler functions, Zhang et al. (2010) first derived the following solutions of the fractional Riccati equation Eq. (23.6)

$$\varphi(\xi) = \begin{cases} -\sqrt{-\sigma}\tanh_\alpha(\sqrt{-\sigma}\xi), & \sigma < 0, \\ -\sqrt{-\sigma}\coth_\alpha(\sqrt{-\sigma}\xi), & \sigma < 0, \\ \sqrt{\sigma}\tan_\alpha(\sqrt{\sigma}\xi), & \sigma > 0, \\ -\sqrt{\sigma}\cot_\alpha(\sqrt{\sigma}\xi), & \sigma > 0, \\ -\dfrac{\Gamma(1+\alpha)}{\xi^\alpha + \omega}, & \omega = \text{constant}, \ \sigma = 0, \end{cases}$$

(23.8)

where the generalized hyperbolic and trigonometric functions are defined as

$$\sinh_\alpha(x) = \frac{E_\alpha(x^\alpha) - E_\alpha(-x^\alpha)}{2}, \quad \cosh_\alpha(x) = \frac{E_\alpha(x^\alpha) + E_\alpha(-x^\alpha)}{2}, \quad \tanh_\alpha(x) = \frac{\sinh_\alpha(x)}{\cosh_\alpha(x)},$$

$$\coth_\alpha(x) = \frac{\cosh_\alpha(x)}{\sinh_\alpha(x)}, \quad \sin_\alpha(x) = \frac{E_\alpha(ix^\alpha) - E_\alpha(-ix^\alpha)}{2i}, \quad \cos_\alpha(x) = \frac{E_\alpha(ix^\alpha) + E_\alpha(-ix^\alpha)}{2},$$

$$\tan_\alpha(x) = \frac{\sin_\alpha(x)}{\cos_\alpha(x)}, \quad \cot_\alpha(x) = \frac{\cos_\alpha(x)}{\sin_\alpha(x)},$$

where $E_\alpha(z)$ denotes the Mittag–Leffler function, which is given as

$$E_\alpha(z) = \sum_{i=0}^{\infty} \frac{z^i}{\Gamma(1+i\alpha)}.$$

Step 6. Substituting Eq. (23.5) along with Eq. (23.6) into Eq. (23.4) and using the properties of Jumarie's modified Reimann–Liouville derivative (Jumarie 2006), we can get a polynomial in $\varphi(\xi)$. Setting all the coefficients of φ^i ($i = 0, 1, 2, ...,$) to zero yields a set of nonlinear algebraic equations in a_i($i = 0, 1, 2, ..., M$), k, c and σ.

Step 7. By solving the nonlinear algebraic equations obtained in step 6, we get the values of the constants. Substituting these constants and the solution of Eq. (23.6) into Eq. (23.5), we can obtain the explicit solutions of Eq. (23.1).

Remark: If $\alpha \to 1$, the Riccati equation becomes $\varphi'(\xi) = \sigma + \varphi^2(\xi)$. As a result, the present method can solve the integer-order differential equation. It is worth mentioning that the tanh-function method, which has been discussed in the previous chapter, is a special case of this method.

23.3 Numerical Examples

This section uses the fractional subequation method to solve two nonlinear fractional PDEs: space–time fractional nonlinear KdV equation in Example 23.1 and space–time fractional nonlinear mKdV equation in Example 23.2.

Example 23.1 We first consider the space–time fractional nonlinear KdV equation (Wazwaz 2009) in the form

$$\frac{\partial^\alpha u}{\partial t^\alpha} + 6u\frac{\partial^\alpha u}{\partial x^\alpha} + \frac{\partial^{3\alpha} u}{\partial x^{3\alpha}} = 0, \quad 0 < \alpha \le 1, t > 0. \tag{23.9}$$

Solution

By using the transformation Eqs. (23.2) and (23.3), Eq. (23.9) is reduced to the following fractional-order ODE

$$c^\alpha D_\xi^\alpha u + 6u\, k^\alpha D_\xi^\alpha u + k^{3\alpha} D_\xi^{3\alpha} u = 0. \tag{23.10}$$

Considering the homogenous balance between the highest order linear term $D_\xi^{3\alpha} u$ and nonlinear term $u\, D_\xi^\alpha u$ in Eq. (23.10) and using the concept of Eq. (23.7), we get

$$M + 3 = M + M + 1 \Rightarrow M = 2. \tag{23.11}$$

Now, we assume Eq. (23.5) as the solution of Eq. (23.10) at $M = 2$. This may be written as

$$u(\xi) = a_0 + a_1\varphi(\xi) + a_2\,\varphi^2(\xi), \tag{23.12}$$

where a_0, a_1, and a_2 are constants to be calculated and $\varphi(\xi)$ satisfies Eq. (23.6). Now, by using Eq. (23.12) along with Eq. (23.6) and using the properties of Jumarie's modified Reimann–Liouville derivative (Jumarie 2006), we have

$$D_\xi^\alpha u(\xi) = a_1\sigma + a_1\varphi^2(\xi) + 2a_2\varphi(\xi)\,\sigma + 2\,a_2\varphi^3(\xi), \tag{23.13}$$

and

$$D_\xi^{3\alpha} u(\xi) = 2a_1\sigma^2 + 8a_1\sigma\varphi^2(\xi) + 6a_1\varphi^4(\xi) + 16a_2\varphi(\xi)\,\sigma^2 + 40\,a_2\varphi^3(\xi)\sigma + 24\,a_2\varphi^5(\xi). \tag{23.14}$$

Substituting Eqs. (23.12)–(23.14) into Eq. (23.10) gives a polynomial in $\varphi^n(\xi)$, $n = 0, 1, 2, 3, 4, 5$. with the parameters $k, c, \sigma, a_0, a_1,$ and a_2 as follows:

$$\begin{aligned} &c^\alpha a_1\sigma + c^\alpha a_1\varphi^2(\xi) + 2c^\alpha a_2\varphi(\xi)\sigma + 2c^\alpha a_2\varphi^3(\xi) + 6\,k^\alpha\big(a_0 + a_1\varphi(\xi) + a_2\,\varphi^2(\xi)\big)\\ &\big(a_1\sigma + a_1\varphi^2(\xi) + 2a_2\varphi(\xi)\,\sigma + 2\,a_2\varphi^3(\xi)\big) + 2k^{3\alpha}a_1\sigma^2 + 8k^{3\alpha}a_1\sigma\varphi^2(\xi) +\\ &6k^{3\alpha}a_1\varphi^4(\xi) + 16k^{3\alpha}a_2\,\varphi(\xi)\,\sigma^2 + 40\,k^{3\alpha}a_2\varphi^3(\xi)\,\sigma + 24\,k^{3\alpha}a_2\varphi^5(\xi) = 0. \end{aligned} \tag{23.15}$$

Now, collecting the coefficients of $\varphi^n(\xi)$, $n = 0, 1, 2, 3, 4, 5$ and setting these to zero, we obtain the following nonlinear system of equations

$$\varphi^5(\xi) : 12k^\alpha a_2^2 + 24k^{3\alpha}a_2,$$

$$\varphi^4(\xi) : 18k^\alpha a_1 a_2 + 6k^{3\alpha}a_1,$$

$$\varphi^3(\xi) :\ 2c^\alpha a_2 + 12k^\alpha a_0 a_2 + 6k^\alpha a_1^2 + 12k^\alpha a_2^2\sigma + 40k^{3\alpha}a_2\sigma,$$

$$\varphi^2(\xi) : c^\alpha a_1 + 6k^\alpha a_0 a_1 + 18k^\alpha a_1 a_2\sigma + 8k^{3\alpha}a_1\sigma,$$

$$\varphi^1(\xi) : 2c^\alpha a_2\sigma + 12k^\alpha a_0 a_2\sigma + 6k^\alpha a_1^2\sigma + 16k^\alpha a_2\sigma^2,$$

$$\varphi^0(\xi) : c^\alpha a_1\sigma + 6k^\alpha a_0 a_1\sigma + 2k^{3\alpha}a_1\sigma^2. \tag{23.16}$$

Solving the above nonlinear system, we obtain the values of a_0, a_1, and a_2 as follows:

$$a_0 = -\frac{1}{6}\frac{8k^{3\alpha}\sigma + c^\alpha}{k^\alpha}, \quad a_1 = 0, \quad a_2 = -2k^{2\alpha}. \tag{23.17}$$

As a result, Eqs. (23.8), (23.12), and (23.17) provide three types of exact solutions to Eq. (23.9), namely, two generalized hyperbolic function solutions, two generalized trigonometric function solutions, and one rational solution as follows:

$$u_1(x, t) = -\frac{1}{6}\frac{8k^{3\alpha}\sigma + c^\alpha}{k^\alpha} + 2k^{2\alpha}\sigma\tanh_\alpha^2\left(\sqrt{-\sigma}\xi\right), \quad \sigma < 0, \tag{23.18}$$

$$u_2(x, t) = -\frac{1}{6}\frac{8k^{3\alpha}\sigma + c^\alpha}{k^\alpha} + 2k^{2\alpha}\sigma\coth_\alpha^2\left(\sqrt{-\sigma}\xi\right), \quad \sigma < 0, \tag{23.19}$$

$$u_3(x, t) = -\frac{1}{6}\frac{8k^{3\alpha}\sigma + c^\alpha}{k^\alpha} - 2k^{2\alpha}\sigma\tan_\alpha^2\left(\sqrt{\sigma}\xi\right), \quad \sigma > 0, \tag{23.20}$$

$$u_4(x, t) = -\frac{1}{6}\frac{8k^{3\alpha}\sigma + c^\alpha}{k^\alpha} - 2k^{2\alpha}\sigma\cot_\alpha^2\left(\sqrt{\sigma}\xi\right), \quad \sigma > 0, \tag{23.21}$$

$$u_5(x, t) = -\frac{1}{6}\frac{8k^{3\alpha}\sigma + c^\alpha}{k^\alpha} - 2k^{2\alpha}\frac{\Gamma^2(1+\alpha)}{(\xi^\alpha + \omega)^2}, \quad \omega = \text{constant}, \quad \sigma = 0, \tag{23.22}$$

where $\xi = kx + ct$ and k, c are arbitrary constants.

Example 23.2 We next consider the following nonlinear space–time fractional mKdV equation (Wazwaz 2009)

$$\frac{\partial^\alpha u}{\partial t^\alpha} - 6u^2\frac{\partial^\alpha u}{\partial x^\alpha} + \frac{\partial^{3\alpha}u}{\partial x^{3\alpha}} = 0, \quad 0 < \alpha \le 1, t > 0. \tag{23.23}$$

Solution

Using the transformation Eqs. (23.2) and (23.3) in Eq. (23.23), it reduces to

$$c^\alpha D_\xi^\alpha u - 6u^2 k^\alpha D_\xi^\alpha u + k^{3\alpha}D_\xi^{3\alpha}u = 0. \tag{23.24}$$

By balancing the highest order linear term $D_\xi^{3\alpha}u$ and nonlinear term $u^2 D_\xi^\alpha u$ in Eq. (23.24), we obtain

$$M + 3 = 2M + M + 1 \Rightarrow M = 1. \tag{23.25}$$

So, the solution of Eq. (23.24) at $M = 1$ may be written as

$$u(\xi) = a_0 + a_1\varphi(\xi), \tag{23.26}$$

where a_0 and a_1 are constants and and $\varphi(\xi)$ satisfies Eq. (23.6). By using the properties of modified Reimann–Liouville derivative (Jumarie 2006), Eqs. (23.6) and (23.26), we have

$$D_\xi^\alpha u(\xi) = a_1\sigma + a_1\varphi^2(\xi), \tag{23.27}$$

and

$$D_\xi^{3\alpha}u(\xi) = 2a_1\sigma^2 + 8a_1\sigma\varphi^2(\xi) + 6a_1\varphi^4(\xi). \tag{23.28}$$

Plugging Eqs. (23.26)–(23.28) into Eq. (23.24), it gives a polynomial in $\varphi^n(\xi)$, $n = 0, 1, 2, 3, 4$ as follows:

$$c^\alpha a_1 \sigma + c^\alpha a_1 \varphi^2(\xi) - 6 k^\alpha (a_0 + a_1 \varphi(\xi))^2 (a_1 \sigma + a_1 \varphi^2(\xi)) + 2a_1 \sigma^2 + 8a_1 \sigma \varphi^2(\xi) \\ + 6a_1 \varphi^4(\xi) = 0. \tag{23.29}$$

Equating the coefficients of equal power of $\varphi^n(\xi)$, $n = 0, 1, 2, 3, 4$ to zero, the following nonlinear system of equations is obtained

$$\begin{aligned}
\varphi^4(\xi) &: \ -6k^\alpha a_1^3 + 6k^{3\alpha} a_1, \\
\varphi^3(\xi) &: \ -12k^\alpha a_0 a_1^2, \\
\varphi^2(\xi) &: \ c^\alpha a_1 - 6k^\alpha a_0^2 a_1 - 6k^\alpha a_1^3 \sigma + 8k^{3\alpha} a_1 \sigma, \\
\varphi^1(\xi) &: \ -12k^\alpha a_0 a_1^2 \sigma, \\
\varphi^0(\xi) &: \ c^\alpha a_1 \sigma - 6k^\alpha a_0^2 a_1 \sigma + 2k^{3\alpha} a_1 \sigma^2.
\end{aligned} \tag{23.30}$$

Solving Eq. (23.30), we obtain the values of a_0 and a_1 in two different cases as follows:
Case 1:

$$a_0 = 0, \quad a_1 = k^\alpha, \quad \sigma = -\frac{1}{2} \frac{c^\alpha}{k^{3\alpha}}. \tag{23.31}$$

Substituting Eq. (23.31) into Eq. (23.26) and using Eqs (23.8), we achieve three types of exact solutions to Eq. (23.23) as given below

$$u_{1,1}(x, t) = -k^\alpha \left(\sqrt{\frac{1}{2} \frac{c^\alpha}{k^{3\alpha}}} \tanh_\alpha \left(\sqrt{\frac{1}{2} \frac{c^\alpha}{k^{3\alpha}}} (kx + ct) \right) \right), \quad c^\alpha > 0, \quad k^{3\alpha} > 0, \tag{23.32}$$

$$u_{1,2}(x, t) = -k^\alpha \left(\sqrt{\frac{1}{2} \frac{c^\alpha}{k^{3\alpha}}} \coth_\alpha \left(\sqrt{\frac{1}{2} \frac{c^\alpha}{k^{3\alpha}}} (kx + ct) \right) \right), \quad c^\alpha > 0, \quad k^{3\alpha} > 0, \tag{23.33}$$

$$u_{1,3}(x, t) = k^\alpha \left(\sqrt{-\frac{1}{2} \frac{c^\alpha}{k^{3\alpha}}} \tan_\alpha \left(\sqrt{-\frac{1}{2} \frac{c^\alpha}{k^{3\alpha}}} (kx + ct) \right) \right), \quad c^\alpha < 0, \quad k^{3\alpha} > 0, \tag{23.34}$$

$$u_{1,4}(x, t) = -k^\alpha \left(\sqrt{-\frac{1}{2} \frac{c^\alpha}{k^{3\alpha}}} \cot_\alpha \left(\sqrt{-\frac{1}{2} \frac{c^\alpha}{k^{3\alpha}}} (kx + ct) \right) \right), \quad c^\alpha < 0, \quad k^{3\alpha} > 0, \tag{23.35}$$

$$u_{1,5}(x, t) = -k^\alpha \left(\frac{\Gamma(1 + \alpha)}{(kx + ct)^\alpha + \omega} \right), \quad c^\alpha = 0, \quad \omega = \text{constant}. \tag{23.36}$$

Case 2:

$$a_0 = 0, \quad a_1 = -k^\alpha, \quad \sigma = -\frac{1}{2} \frac{c^\alpha}{k^{3\alpha}}. \tag{23.37}$$

By substituting Eq. (23.37) into Eq. (23.26) and utilizing Eqs (23.8), we get the exact solutions to Eq. (23.23) as follows:

$$u_{2,1}(x, t) = k^\alpha \left(\sqrt{\frac{1}{2} \frac{c^\alpha}{k^{3\alpha}}} \tanh_\alpha \left(\sqrt{\frac{1}{2} \frac{c^\alpha}{k^{3\alpha}}} (kx + ct) \right) \right), \quad c^\alpha > 0, \quad k^{3\alpha} > 0, \tag{23.38}$$

$$u_{2,2}(x, t) = k^\alpha \left(\sqrt{\frac{1}{2} \frac{c^\alpha}{k^{3\alpha}}} \coth_\alpha \left(\sqrt{\frac{1}{2} \frac{c^\alpha}{k^{3\alpha}}} (kx + ct) \right) \right), \quad c^\alpha > 0, \quad k^{3\alpha} > 0, \tag{23.39}$$

$$u_{2,3}(x, t) = -k^\alpha \left(\sqrt{-\frac{1}{2} \frac{c^\alpha}{k^{3\alpha}}} \tan_\alpha \left(\sqrt{-\frac{1}{2} \frac{c^\alpha}{k^{3\alpha}}} (kx + ct) \right) \right), \quad c^\alpha < 0, \quad k^{3\alpha} > 0, \tag{23.40}$$

$$u_{2,4}(x, t) = k^\alpha \left(\sqrt{-\frac{1}{2} \frac{c^\alpha}{k^{3\alpha}}} \cot_\alpha \left(\sqrt{-\frac{1}{2} \frac{c^\alpha}{k^{3\alpha}}} (kx + ct) \right) \right), \quad c^\alpha < 0, \quad k^{3\alpha} > 0, \tag{23.41}$$

$$u_{2,5}(x, t) = k^{\alpha} \left(\frac{\Gamma(1 + \alpha)}{(kx + ct)^{\alpha} + \omega} \right), \quad c^{\alpha} = 0, \quad \omega = \text{constant}, \tag{23.42}$$

where k and c are arbitrary constants.

References

Alzaidy, J.F. (2013a). Fractional sub-equation method and its applications to the space–time fractional differential equations in mathematical physics. *British Journal of Mathematics & Computer Science* 3 (2): 153–163.

Alzaidy, J.F. (2013b). The fractional sub-equation method and exact analytical solutions for some nonlinear fractional PDEs. *American Journal of Mathematical Analysis* 1 (1): 14–19.

Bekir, A., Aksoy, E., and Güner, Ö. (2014). A generalized fractional sub-equation method for nonlinear fractional differential equations. *AIP Conference Proceedings* 1611: 78–83.

Guner, O., Bekir, A., and Cevikel, A.C. (2015). A variety of exact solutions for the time fractional Cahn-Allen equation. *The European Physical Journal Plus* 130: 146 (1-13).

Jumarie, G. (2006). Modified Riemann-Liouville derivative and fractional Taylor series of nondifferentiable functions further results. *Computers & Mathematics with Applications* 51 (9–10): 1367–1376.

Mohyud-Dina, S.T., Nawaza, T., Azharb, E., and Akbar, M.A. (2017). Fractional sub-equation method to space–time fractional Calogero–Degasperis and potential Kadomtsev–Petviashvili equations. *Journal of Taibah University for Science* 11: 258–263.

Wang, M. (1999). Solitary wave solutions for variant Boussinesq equations. *Physics Letters A* 199 (3-4): 169–172.

Wazwaz, A.M. (2009). *Partial Differential Equations and Solitary Waves Theory*, Nonlinear Physical Science. Berlin, Heidelberg: Springer.

Yépez-Martínez, H. and Gómez-Aguilar, J.F. (2019). Fractional sub-equation method for Hirota–Satsuma-coupled KdV equation and coupled mKdV equation using the Atangana's conformable derivative. *Waves in Random and Complex Media* 29 (4): 678–693.

Yépez-Martínez, H., Reyes, J.M., and Sosa, I.O. (2014). Fractional sub-equation method and analytical solutions to the Hirota-Satsuma Coupled KdV equation and coupled mKdV equation. *British Journal of Mathematics & Computer Science* 4 (4): 572–589.

Zhang, S. and Zhang, H.Q. (2011). Fractional sub-equation method and its applications to nonlinear fractional PDEs. *Physics Letters A* 375 (7): 1069–1073.

Zhang, S., Zong, Q.A., Liu, D., and Gao, Q. (2010). A generalized exp-function method for fractional Riccati differential equations. *Communication Fraction Calculator* 1 (1): 48–52.

24

Exp-Function Method

24.1 Introduction

He and Wu (2006) were the first to suggest the exp-function method, which was effectively used to find solitary and periodic solutions to nonlinear partial differential equations. Further, researchers have also utilized this strategy in their studies to deal with different other equations like stochastic modified Korteweg de Vries (mKDV) equation (Dai and Zhang 2009a), one-dimensional fractional wave equation, and fractional reaction-diffusion problem (Yildirim and Pinar 2010; Bekir et al. 2015), space–time fractional Fokas, and the nonlinear fractional Sharma-Tasso-Olver equations (Zheng 2013), fractional Fitzhugh-Nagumo and KdV equations (Bekir et al. 2017), and so on (He and Abdou 2007; Wu and He 2007; Zhang 2007; Zhu 2008). This approach may be used to solve difference-differential equations (Zhu 2007a, 2007b) and equations with variable coefficients (El-wakil et al. 2007; Zhang 2008). The method can also be used to create n-soliton solutions and rational solutions (Dai and Zhang 2009a; Dai and Zhang 2009b). It converts fractional partial differential equations into ordinary differential equations via fractional complex transform, simplifying the solution process. The exp-function approach is simple and effective in obtaining generalized solitary and periodic solutions to nonlinear evolution equations. The key advantage of this approach over others is that it produces more general solutions with certain free parameters. Simplified exact solutions are generally obtained by using the exp-function approach. However, it is not always straightforward to get simplifications while considering complex expressions. Thus, this is the primary shortcoming of this method.

24.2 Procedure of the Exp-Function Method

This technique will be presented here for solving fractional partial differential equations. In order to understand the exp-function method (He and Wu 2006; Dai and Zhang 2009a; Yildirim and Pinar 2010; Bekir et al. 2015), let us consider the following nonlinear fractional partial differential equation of the type

$$Q\left(u, D_t^\alpha u, D_t^\alpha D_t^\alpha u, \ldots\right) = 0, \quad 0 < \alpha \le 1, \tag{24.1}$$

where u is an unknown function, and Q is a polynomial of u and its partial fractional derivatives, in which the highest order derivatives and the nonlinear terms are involved.

The traveling wave variable is considered as (Bekir et al. 2015)

$$u(x, t) = U(\xi), \tag{24.2}$$

$$\xi = kx - \frac{ct^\alpha}{\Gamma(1 + \alpha)}, \tag{24.3}$$

where c and k are nonzero constants.

Using Eqs. (24.2) and (24.3), Eq. (24.1) can be rewritten in the following nonlinear ODE form:

$$P\left(U, U', U'', U''', \ldots\right) = 0, \tag{24.4}$$

where the prime denotes the derivative with respect to ξ.

Computational Fractional Dynamical Systems: Fractional Differential Equations and Applications, First Edition.
Snehashish Chakraverty, Rajarama Mohan Jena, and Subrat Kumar Jena.
© 2023 John Wiley & Sons, Inc. Published 2023 by John Wiley & Sons, Inc.

We suppose that the exp-function approach is based on the assumption that the wave solution of Eq. (24.4) may be represented in the following form (Chakraverty et al. 2019)

$$U(\xi) = \frac{\sum\limits_{n=-c}^{d} a_n \exp\,(n\xi)}{\sum\limits_{n=-p}^{q} b_n \exp\,(n\xi)}, \tag{24.5}$$

where c, d, p, and q are positive integers that are determined later, and a_n, b_n are unknown constants. Now, Eq. (24.5) can be rewritten in the following equivalent form:

$$U(\xi) = \frac{a_{-c} \exp\,(-c\xi) + \cdots + a_d \exp\,(d\xi)}{b_{-p} \exp\,(-p\xi) + \cdots + b_q \exp\,(q\xi)}. \tag{24.6}$$

This analogous formulation is essential for determining the analytic solution to the problems. The values of c and p are obtained by balancing the exp-functions of the lowest order linear term with the lowest- order nonlinear term in Eq. (24.4). Similarly, the values for d and q can be determined by balancing the exp-functions of the highest order linear term in Eq. (24.4) with the highest order nonlinear term (He and Wu 2006; He and Abdou 2007).

We can then derive an equation in terms of $\exp\,(n\xi)$ by putting Eq. (24.5) into Eq. (24.4) along with the previously determined values of c, d, p, and q. When all the coefficients of the different powers of $\exp\,(n\xi)$ are set to zero, a series of algebraic equations in terms of a_n, b_n, k, and c are obtained. The values of a_n, b_n, k, and c are determined by solving these algebraic equations. Furthermore, the solution of Eq. (24.1) may be derived by incorporating these values into Eq. (24.5).

In the following sections, we present two examples to demonstrate the usefulness of the exp-function technique to solving nonlinear fractional differential equations.

24.3 Numerical Examples

Here, we apply the present method to solve a time-fractional nonlinear advection equation in Example 24.1 and nonlinear time-fractional fisher's equation in Example 24.2.

Example 24.1   Consider the following nonlinear time-fractional advection equation (Wazwaz 2007)

$$\frac{\partial^\alpha u}{\partial t^\alpha} + u\frac{\partial u}{\partial x} = 0, \quad 0 < \alpha \leq 1. \tag{24.7}$$

Solution

Using the transformation $u(x, t) = U(\xi)$, $\xi = kx - \dfrac{ct^\alpha}{\Gamma(1 + \alpha)}$ in Eq. (24.7), it reduces to

$$-cU' + kUU' = 0, \tag{24.8}$$

where prime denotes the differentiation with respect to ξ. Let us assume that the ordinary differential equation Eq. (24.8) has a solution similar to Eq (24.6). As a result, we may suppose that the solution is as follows:

$$U(\xi) = \frac{a_{-c} \exp\,[-c\xi] + \cdots + a_d \exp\,[d\xi]}{b_{-p} \exp\,[-p\xi] + \cdots + b_q \exp\,[q\xi]}. \tag{24.9}$$

The highest order linear and nonlinear terms in Eq. (24.8) are U' and UU'.
Differentiating and simplifying Eq. (24.9), we get

$$U' = \frac{c_1 \exp\,[-(p+c)\xi] + \cdots + d_1 \exp\,[(d+q)\xi]}{c_2 \exp\,[-2p\xi] + \cdots + d_2 \exp\,[2q\xi]}, \tag{24.10}$$

and

$$UU' = \frac{c_3 \exp\,[-(p+2c)\xi] + \cdots + d_3 \exp\,[(2d+q)\xi]}{c_4 \exp\,[-3p\xi] + \cdots + d_4 \exp\,[3q\xi]}. \tag{24.11}$$

It is worth noting that after differentiation and simplification, we will obtain various terms containing different combinations of a_{-c}, a_d, b_{-p}, ... etc., which for simplicity we have considered as c_1, ...d_1, c_2, ...d_2, c_3, ...d_3 and c_4, ...d_4.

The lowest order exp-function term of Eq. (24.10) is

$$\frac{c_1 \exp\left[-(p+c)\xi\right] + \cdots}{c_2 \exp\left[-2p\xi\right] + \cdots}.$$
(24.12)

Multiplying and dividing by $\exp(-p\xi)$, we get

$$\frac{c_1 \exp\left[-(2p+c)\xi\right] + \cdots}{c_2 \exp\left[-3p\xi\right] + \cdots}.$$
(24.13)

Similarly, the highest order exp-function term of Eq. (24.10) is

$$\frac{\cdots + d_1 \exp\left[(d+q)\xi\right]}{\cdots + d_2 \exp\left[2q\xi\right]}.$$
(24.14)

Multiplying and dividing by $\exp(q\xi)$, we obtain

$$\frac{\cdots + d_1 \exp\left[(d+2q)\xi\right]}{\cdots + d_2 \exp\left[3q\xi\right]}.$$
(24.15)

Balancing the lowest order exp-functions in Eqs. (24.13) and (24.11), we get

$$-2p - c = -2c - p$$
$$\Rightarrow p = c.$$

Balancing the highest order exp-functions in Eqs. (24.15) and (24.11), we get

$$d + 2q = 2d + q$$
$$\Rightarrow q = d.$$

For simplicity, we choose the values $p = c = 1$ and $q = d = 1$ (He and Wu 2006; Dai and Zhang 2009a; Yildirim and Pinar 2010; Zheng 2013; Bekir et al. 2015; Bekir et al. 2017). The trial solution Eq. (24.5) therefore takes the following form:

$$U(\xi) = \frac{a_{-1}\exp(-\xi) + a_0 + a_1\exp(\xi)}{b_{-1}\exp(-\xi) + b_0 + b_1\exp(\xi)}.$$
(24.16)

Here, a_{-1}, a_0, a_1, b_{-1}, b_0, and b_1 are the coefficients to be determined next.
Plugging Eq. (24.16) in Eq. (24.8), we have

$$\frac{1}{A}[R_2\exp(2\xi) + R_1\exp(\xi) + R_0 + R_{-1}\exp(-\xi) + R_{-2}\exp(-2\xi)] = 0,$$
(24.17)

where

$$A = (b_{-1}\exp(-\xi) + b_0 + b_1\exp(\xi))^3,$$
$$R_2 = (a_0b_1 - a_1b_0)(cb_1 - ka_1),$$
$$R_1 = 2ka_1^2 b_{-1} + ((-2cb_{-1} - 2ka_{-1})b_1 - cb_0^2 + ka_0b_0)a_1 + b_1(2ca_{-1}b_1 + ca_0b_0 - ka_0^2),$$
$$R_0 = 3(a_{-1}b_1 - a_1b_{-1})(cb_0 - ka_0),$$
$$R_{-1} = -2ka_{-1}^2 b_1 + ((2cb_1 + 2ka_1)b_{-1} + cb_0^2 - ka_0b_0)a_{-1} - b_{-1}(2ca_1b_{-1} + ca_0b_0 - ka_0^2),$$
$$R_{-2} = (a_{-1}b_0 - a_0b_{-1})(cb_{-1} - ka_{-1}),$$
(24.18)

By equating the coefficients of $\exp(n\xi)$ in Eq. (24.18) to zero, we get the following system of equations:

$$R_2 = 0, \quad R_1 = 0, \quad R_0 = 0, \quad R_{-1} = 0, \quad \text{and} \quad R_{-2} = 0.$$
(24.19)

When we solve a system of linear equations, the solution of the system is the point of intersection of the two lines. But in the case of systems of nonlinear equations, the graphs may be circles, parabolas, or hyperbolas, and there may be several points of intersection, and so several solutions. By solving the above nonlinear system of algebraic Eq. (24.19), we may get many cases of the solution, some of them are given as follows:

Case 1

$$a_0 = \frac{b_0 a_{-1}}{b_{-1}}, \quad b_0 = b_0, \quad a_1 = \frac{b_1 a_{-1}}{b_{-1}}, \quad b_1 = b_1, \quad a_{-1} = a_{-1}, \quad b_{-1} = b_{-1}, \quad c = c, \quad k = k,$$
(24.20)

where b_0, b_1, a_{-1}, b_{-1}, c, and k are free parameters. Substituting the above values into Eq. (24.16), we will obtain the following exact solution:

$$u_{1,2}(x,t) = \frac{a_{-1}\exp\left[-\left(kx-\frac{ct^\alpha}{\Gamma(1+\alpha)}\right)\right] + \frac{b_0 a_{-1}}{b_{-1}} + \frac{b_1 a_{-1}}{b_{-1}}\exp\left(kx-\frac{ct^\alpha}{\Gamma(1+\alpha)}\right)}{b_{-1}\exp\left[-\left(kx-\frac{ct^\alpha}{\Gamma(1+\alpha)}\right)\right] + b_0 + b_1\exp\left(kx-\frac{ct^\alpha}{\Gamma(1+\alpha)}\right)}.$$ (24.21)

Case 2

$$a_0 = 0, \quad b_0 = 0, \quad a_1 = a_1, \quad b_1 = b_1, \quad a_{-1} = 0, \quad b_{-1} = 0, \quad c = c, \quad k = k,$$ (24.22)

where a_1, b_1, c, and k are free parameters. Substituting Eq. (24.22) into Eq. (24.16), we get the below exact solution

$$u_{3,4}(x,t) = \frac{a_1\exp\left(kx-\frac{ct^\alpha}{\Gamma(1+\alpha)}\right)}{b_1\exp\left(kx-\frac{ct^\alpha}{\Gamma(1+\alpha)}\right)} = \frac{a_1}{b_1}.$$ (24.23)

Case 3

$$a_0 = a_0, \quad b_0 = b_0, \quad a_1 = \frac{a_0 b_1}{b_0}, \quad b_1 = b_1, \quad a_{-1} = 0, \quad b_{-1} = 0, \quad c = c, \quad k = k,$$ (24.24)

where a_0, b_0, b_1, c, and k are free parameters. Plugging Eq. (24.24) in Eq. (24.16), we get

$$u_{5,6}(x,t) = \frac{a_0 + \frac{a_0 b_1}{b_0}\exp\left(kx-\frac{ct^\alpha}{\Gamma(1+\alpha)}\right)}{b_0 + b_1\exp\left(kx-\frac{ct^\alpha}{\Gamma(1+\alpha)}\right)}.$$ (24.25)

All the above are the exact solutions of the time-fractional nonlinear advection equation.

Example 24.2 Let us consider the nonlinear time-fractional Fisher's equation (Veeresha et al. 2019)

$$\frac{\partial^\alpha u}{\partial t^\alpha} = \frac{\partial^2 u}{\partial x^2} + 6u(1-u), \quad t>0, \quad x\in R, \quad 0<\alpha\le 1.$$ (24.26)

Solution

Equation (24.26) may be reduced to the ordinary differential equation in ξ by using the above-discussed transformation $u(x,t) = U(\xi)$, $\xi = kx - \frac{ct^\alpha}{\Gamma(1+\alpha)}$ as

$$k^2 U'' + cU' + 6U - 6U^2 = 0,$$ (24.27)

where prime denotes the differentiation with respect to ξ.

Next, we assume the solution of the ordinary differential equation Eq. (24.27) as Eq (24.6). This may be written as

$$U(\xi) = \frac{a_{-c}\exp\left[-c\xi\right] + \cdots + a_d\exp\left[d\xi\right]}{b_{-p}\exp\left[-p\xi\right] + \cdots + b_q\exp\left[q\xi\right]}.$$ (24.28)

The highest order linear and nonlinear terms in Eq. (24.27) are U'' and U^2. Differentiating and simplifying Eq. (24.28), we get

$$U'' = \frac{c_1\exp\left[-(3p+c)\xi\right] + \cdots + d_1\exp\left[(d+3q)\xi\right]}{c_2\exp\left[-4p\xi\right] + \cdots + d_2\exp\left[4q\xi\right]},$$ (24.29)

and

$$U^2 = \frac{c_3\exp\left[-2c\xi\right] + \cdots + d_3\exp\left[2d\xi\right]}{c_4\exp\left[-2p\xi\right] + \cdots + d_4\exp\left[2q\xi\right]}.$$ (24.30)

Again as mentioned in Example 24.1, after differentiation and simplification, we will get various terms containing different combinations of a_{-c}, a_d, b_{-p}, ... etc., which for simplicity, we assumed as c_1, ...d_1, c_2, ...d_2, c_3, ...d_3 and c_4, ...d_4.

The lowest order exp-function term of Eq. (24.30) is

$$\frac{c_3 \exp\left[-2c\xi\right] + \ldots}{c_4 \exp\left[-2p\xi\right] + \ldots}. \tag{24.31}$$

Multiplying and dividing by $\exp(-2p\xi)$, we get

$$\frac{c_3 \exp\left[-(2c + 2p)\xi\right] + \cdots}{c_4 \exp\left[-4p\xi\right] + \cdots}. \tag{24.32}$$

Similarly, the highest order exp-function term of Eq. (24.30) is

$$\frac{\cdots + d_3 \exp\left[2d\xi\right]}{\cdots + d_4 \exp\left[2q\xi\right]}. \tag{24.33}$$

Multiplying and dividing by $\exp(2q\xi)$, we obtain

$$\frac{\cdots + d_3 \exp\left[(2d + 2q)\xi\right]}{\cdots + d_4 \exp\left[4q\xi\right]}. \tag{24.34}$$

Next, we balance the lowest order exp-functions in Eqs. (24.29) and (24.32) to have a relation between p and c as below

$$-3p - c = -2c - 2p$$
$$\Rightarrow p = c.$$

Similarly, we balance the highest order exp-functions in Eqs. (24.29) and (24.34), and we get

$$d + 3q = 2d + 2q$$
$$\Rightarrow q = d.$$

Here, for simplicity, we consider $p = c = 1$ and $q = d = 1$ (He and Wu 2006; Dai and Zhang 2009a; Yildirim and Pinar 2010; Zheng 2013; Bekir et al. 2015; Bekir et al. 2017). Then, the trial solution Eq. (24.5) takes the following form:

$$U(\xi) = \frac{a_{-1} \exp\left(-\xi\right) + a_0 + a_1 \exp\left(\xi\right)}{b_{-1} \exp\left(-\xi\right) + b_0 + b_1 \exp\left(\xi\right)}. \tag{24.35}$$

Here, a_{-1}, a_0, a_1, b_{-1}, b_0, and b_1 are the coefficients as explained in Example 24.1, which are to be determined next. Plugging Eq. (24.35) in Eq. (24.27) and simplifying, we have

$$\frac{1}{A}\left[\begin{array}{l} R_3 \exp\left(3\xi\right) + R_2 \exp\left(2\xi\right) + R_1 \exp\left(\xi\right) + R_0 + R_{-1} \exp\left(-\xi\right) \\ + R_{-2} \exp\left(-2\xi\right) + R_{-3} \exp\left(-3\xi\right) \end{array}\right] = 0, \tag{24.36}$$

where

$$A = \left(b_{-1} \exp\left(-\xi\right) + b_0 + b_1 \exp\left(\xi\right)\right)^3,$$
$$R_3 = -6(a_1 - b_1)a_1 b_1,$$
$$R_2 = -a_0\left(-k^2 + c - 6\right)b_1^2 + a_1 b_1\left(-12a_0 + \left(-k^2 + c + 12\right)b_0\right) - 6a_1^2 b_0,$$
$$R_1 = -2a_{-1}\left(-2k^2 + c - 3\right)b_1^2 + \left(\begin{array}{l} \left(-12a_{-1} + \left(-4k^2 + 2c + 12\right)b_{-1}\right)a_1 - \\ \left(\left(k^2 + c - 12\right)b_0 + 6a_0\right)a_0 \end{array}\right)b_1 +,$$
$$\left(-6b_{-1}a_1 + b_0\left(\left(k^2 + c + 6\right)b_0 - 12a_0\right)\right)a_1$$
$$R_0 = 6a_0 b_0^2 - 6\left(2a_{-1}b_1 + \left(k^2 b_1 + 2a_1 - 2b_1\right)b_{-1}\right)a_0 + \left(\begin{array}{l} -12a_{-1}a_1 - 3\left(-k^2 + c - 4\right)a_{-1}b_1 \\ -6a_0^2 + 3\left(k^2 + c + 4\right)a_1 b_{-1} \end{array}\right)b_0, \tag{24.37}$$
$$R_{-1} = 2\left(2k^2 + c + 3\right)a_1 b_{-1}^2 + \left(\left(-12a_1 + \left(-4k^2 - 2c + 12\right)b_1\right)a_{-1} + a_0\left(\left(-k^2 + c + 3\right)b_0 - 6a_0\right)\right)b_{-1}$$
$$- \left(6a_{-1}b_1 + b_0\left(\left(-k^2 + c - 6\right)b_0 + 12a_0\right)\right)a_{-1},$$
$$R_{-2} = a_0\left(k^2 + c + 6\right)b_{-1}^2 - \left(12a_0 + \left(k^2 + c - 12\right)b_0\right)a_{-1}b_{-1} - 6a_{-1}^2 b_0,$$
$$R_{-3} = -6(a_{-1} - b_{-1})a_{-1}b_{-1}.$$

Equating the coefficients of $\exp(n\xi)$ in Eq. (24.37) to zero, we have

$$R_3 = 0, \quad R_2 = 0, \quad R_1 = 0, \quad R_0 = 0, \quad R_{-1} = 0, \quad R_{-2} = 0, \quad \text{and} \quad R_{-3} = 0. \tag{24.38}$$

By solving the above nonlinear system of algebraic Eq. (24.38), we may obtain many cases of the solution. However, some of them are as follows:

Case 1

$$a_0 = \frac{\left(b_0 \pm \sqrt{b_0^2 - 4b_1 b_{-1}}\right)}{2}, \, b_0 = b_0, \, a_1 = 0, \, b_1 = b_1, \, a_{-1} = b_{-1}, \, b_{-1} = b_{-1}, \, c = 6, \, k = 0, \tag{24.39}$$

where b_0, b_1, and b_{-1} are free parameters. Substituting the above values into Eq. (24.35), we will get the following exact solution

$$u_{1,2}(x, t) = \frac{b_{-1} \exp\left(\dfrac{6t^\alpha}{\Gamma(1+\alpha)}\right) + \left(\dfrac{\left(b_0 \pm \sqrt{b_0^2 - 4b_1 b_{-1}}\right)}{2}\right)}{b_{-1} \exp\left(\dfrac{6t^\alpha}{\Gamma(1+\alpha)}\right) + b_0 + b_1 \exp\left(-\dfrac{6t^\alpha}{\Gamma(1+\alpha)}\right)}. \tag{24.40}$$

Case 2

$$a_0 = \frac{\left(b_0 \pm \sqrt{b_0^2 - 4b_1 b_{-1}}\right)}{2}, \, b_0 = b_0, \, a_1 = b_1, \, b_1 = b_1, \, a_{-1} = 0, \, b_{-1} = b_{-1}, \, c = -6, \, k = 0, \tag{24.41}$$

where b_0, b_1, and b_{-1} are free parameters. Substituting Eq. (24.41) into Eq. (24.35), we get the below exact solution

$$u_{3,4}(x, t) = \frac{\dfrac{\left(b_0 \pm \sqrt{b_0^2 - 4b_1 b_{-1}}\right)}{2} + b_1 \exp\left(\dfrac{6t^\alpha}{\Gamma(1+\alpha)}\right)}{b_{-1} \exp\left(-\dfrac{6t^\alpha}{\Gamma(1+\alpha)}\right) + b_0 + b_1 \exp\left(\dfrac{6t^\alpha}{\Gamma(1+\alpha)}\right)}. \tag{24.42}$$

Case 3

$$a_0 = -4\left(\pm \sqrt{b_1 b_{-1}}\right), \, b_0 = 2\left(\pm \sqrt{b_1 b_{-1}}\right), \, a_1 = b_1, \, b_1 = b_1, \, a_{-1} = b_{-1}, \, b_{-1} = b_{-1}, \, c = 0, \, k = \pm \sqrt{6}, \tag{24.43}$$

where b_1, and b_{-1} are free parameters. Plugging Eq. (24.43) in Eq. (24.35), we get

$$u_{5,6}(x, t) = \frac{b_{-1} \exp\left[-\left(\pm \sqrt{6}x\right)\right] - 4\left(\pm \sqrt{b_1 b_{-1}}\right) + b_1 \exp\left(\pm \sqrt{6}x\right)}{b_{-1} \exp\left[-\left(\pm \sqrt{6}x\right)\right] + 2\left(\pm \sqrt{b_1 b_{-1}}\right) + b_1 \exp\left(\pm \sqrt{6}x\right)}. \tag{24.44}$$

Case 4

$$a_0 = 2\left(\pm \sqrt{b_1 b_{-1}}\right), \, b_0 = 2\left(\pm \sqrt{b_1 b_{-1}}\right), \, a_1 = 0, \, b_1 = b_1, \, a_{-1} = b_{-1}, \, b_{-1} = b_{-1}, \, c = 5, \, k = i, \tag{24.45}$$

where b_1, and b_{-1} are free parameters. Plugging Eq. (24.45) in Eq. (24.35), we have

$$u_{7,8}(x, t) = \frac{b_{-1} \exp\left[-\left(ix - \dfrac{5t^\alpha}{\Gamma(1+\alpha)}\right)\right] + 2\left(\pm \sqrt{b_1 b_{-1}}\right)}{b_{-1} \exp\left[-\left(ix - \dfrac{5t^\alpha}{\Gamma(1+\alpha)}\right)\right] + 2\left(\pm \sqrt{b_1 b_{-1}}\right) + b_1 \exp\left(ix - \dfrac{5t^\alpha}{\Gamma(1+\alpha)}\right)}. \tag{24.46}$$

Case 5

$$a_0 = 2\left(\pm \sqrt{b_1 b_{-1}}\right), \, b_0 = 2\left(\pm \sqrt{b_1 b_{-1}}\right), \, a_1 = b_1, \, b_1 = b_1, \, a_{-1} = 0, \, b_{-1} = b_{-1}, \, c = -5, \, k = i, \tag{24.47}$$

where b_1, and b_{-1} are free parameters. Using Eq. (24.47) in Eq. (24.35), we get

$$u_{9,10}(x, t) = \frac{2\left(\pm \sqrt{b_1 b_{-1}}\right) + b_1 \exp\left(ix + \dfrac{5t^\alpha}{\Gamma(1+\alpha)}\right)}{b_{-1} \exp\left[-\left(ix + \dfrac{5t^\alpha}{\Gamma(1+\alpha)}\right)\right] + 2\left(\pm \sqrt{b_1 b_{-1}}\right) + b_1 \exp\left(ix + \dfrac{5t^\alpha}{\Gamma(1+\alpha)}\right)}. \tag{24.48}$$

Case 6

$$a_0 = 6\left(\pm \sqrt{b_1 b_{-1}} \right), b_0 = 2\left(\pm \sqrt{b_1 b_{-1}} \right), a_1 = 0, b_1 = b_1, a_{-1} = 0, b_{-1} = b_{-1}, c = 0, k = 6i, \tag{24.49}$$

where b_1, and b_{-1} are free parameters. Plugging Eq. (24.49) in Eq. (24.35), we get

$$u_{11,12}(x, t) = \frac{6\left(\pm \sqrt{b_1 b_{-1}} \right)}{b_{-1} \exp\left[-(6ix) \right] + 2\left(\pm \sqrt{b_1 b_{-1}} \right) + b_1 \exp(6ix)}. \tag{24.50}$$

Case 7

$$a_0 = 0, b_0 = 2\left(\pm \sqrt{b_1 b_{-1}} \right), a_1 = 0, b_1 = b_1, a_{-1} = b_{-1}, b_{-1} = b_{-1}, c = 5, k = -1, \tag{24.51}$$

where b_1, and b_{-1} are free parameters. Plugging Eq. (24.51) in Eq. (24.35), we get

$$u_{13,14}(x, t) = \frac{b_{-1} \exp\left[-\left(-x - \frac{5t^\alpha}{\Gamma(1+\alpha)} \right) \right]}{b_{-1} \exp\left[-\left(-x - \frac{5t^\alpha}{\Gamma(1+\alpha)} \right) \right] + 2\left(\pm \sqrt{b_1 b_{-1}} \right) + b_1 \exp\left(-x - \frac{5t^\alpha}{\Gamma(1+\alpha)} \right)}. \tag{24.52}$$

Case 8

$$a_0 = 0, b_0 = 2\left(\pm \sqrt{b_1 b_{-1}} \right), a_1 = 0, b_1 = b_1, a_{-1} = b_{-1}, b_{-1} = b_{-1}, c = 5, k = 1, \tag{24.53}$$

where b_1, and b_{-1} are free parameters. Substituting Eq. (24.53) into Eq. (24.35), we obtain

$$u_{15,16}(x, t) = \frac{b_{-1} \exp\left[-\left(x - \frac{5t^\alpha}{\Gamma(1+\alpha)} \right) \right]}{b_{-1} \exp\left[-\left(x - \frac{5t^\alpha}{\Gamma(1+\alpha)} \right) \right] + 2\left(\pm \sqrt{b_1 b_{-1}} \right) + b_1 \exp\left(x - \frac{5t^\alpha}{\Gamma(1+\alpha)} \right)}. \tag{24.54}$$

Case 9

$$a_0 = 0, b_0 = 0, a_1 = 0, b_1 = b_1, a_{-1} = b_{-1}, b_{-1} = b_{-1}, c = 3, k = 0, \tag{24.55}$$

where b_1, and b_{-1} are free parameters. Plugging Eq. (24.55) in Eq. (24.35), we get

$$u_{17,18}(x, t) = \frac{b_{-1} \exp\left[-\left(-\frac{3t^\alpha}{\Gamma(1+\alpha)} \right) \right]}{b_{-1} \exp\left[-\left(-\frac{3t^\alpha}{\Gamma(1+\alpha)} \right) \right] + b_1 \exp\left(-\frac{3t^\alpha}{\Gamma(1+\alpha)} \right)}. \tag{24.56}$$

Case 10

$$a_0 = 0, b_0 = 2\left(\pm \sqrt{b_1 b_{-1}} \right), a_1 = b_1, b_1 = b_1, a_{-1} = 0, b_{-1} = b_{-1}, c = -5, k = -1, \tag{24.57}$$

where b_1, and b_{-1} are free parameters. Plugging Eq. (24.57) in Eq. (24.35), we get

$$u_{19,20}(x, t) = \frac{b_1 \exp\left(-x + \frac{5t^\alpha}{\Gamma(1+\alpha)} \right)}{b_{-1} \exp\left[-\left(-x + \frac{5t^\alpha}{\Gamma(1+\alpha)} \right) \right] + 2\left(\pm \sqrt{b_1 b_{-1}} \right) + b_1 \exp\left(-x + \frac{5t^\alpha}{\Gamma(1+\alpha)} \right)}. \tag{24.58}$$

Case 11

$$a_0 = 0, b_0 = 2\left(\pm \sqrt{b_1 b_{-1}} \right), a_1 = b_1, b_1 = b_1, a_{-1} = 0, b_{-1} = b_{-1}, c = -5, k = 1, \tag{24.59}$$

where b_1, and b_{-1} are free parameters. Plugging Eq. (24.59) in Eq. (24.35), we get

$$u_{21,22}(x, t) = \frac{b_1 \exp\left(x + \frac{5t^\alpha}{\Gamma(1 + \alpha)}\right)}{b_{-1} \exp\left[-\left(x + \frac{5t^\alpha}{\Gamma(1 + \alpha)}\right)\right] + 2\left(\pm \sqrt{b_1 b_{-1}}\right) + b_1 \exp\left(x + \frac{5t^\alpha}{\Gamma(1 + \alpha)}\right)}. \tag{24.60}$$

Case 12

$$a_0 = 0, \, b_0 = 0, \, a_1 = b_1, \, b_1 = b_1, \, a_{-1} = 0, \, b_{-1} = b_{-1}, \, c = -3, \, k = 0, \tag{24.61}$$

where b_1, and b_{-1} are free parameters. Plugging Eq. (24.61) in Eq. (24.35), we get

$$u_{23,24}(x, t) = \frac{b_1 \exp\left(\frac{3t^\alpha}{\Gamma(1 + \alpha)}\right)}{b_{-1} \exp\left[-\left(\frac{3t^\alpha}{\Gamma(1 + \alpha)}\right)\right] + b_1 \exp\left(\frac{3t^\alpha}{\Gamma(1 + \alpha)}\right)}. \tag{24.62}$$

All the above are the exact solutions of the nonlinear time-fractional Fisher's equation.

References

Bekir, A., Guner, O., Bhrawy, A.H., and Biswas, A. (2015). Solving nonlinear fractional differential equations using exp-function and G′/G-expansion methods. *Romanian Journal of Physics* 60 (3–4): 360–378.

Bekir, A., Guner, O., and Cevikel, A. (2017). The exp-function method for some time-fractional differential equations. *IEEE/CAA Journal of Automatica Sinica* 4 (2): 315–321.

Chakraverty, S., Mahato, N.R., Perumandla, K., and Rao, T.D. (2019). *Advanced Numerical and Semi-Analytical Methods for Differential Equations*. USA: John Wiley & Sons.

Dai, C.Q. and Zhang, J.F. (2009a). Application of He's exp-function method to the stochastic mKdV equation. *International Journal of Nonlinear Sciences and Numerical Simulation* 10 (5): 675–680.

Dai, C.Q. and Zhang, J.F. (2009b). Stochastic exact solutions and two-soliton solution of the Wick-type stochastic KdV equation. *Europhysics Letters* 86: 40006.

El-wakil, S.A., Madkour, M.A., and Abdou, M.A. (2007). Application of the exp-function method for nonlinear evolution equations with variable coefficients. *Physics Letters A* 369: 62–69.

He, J.H. and Abdou, M.A. (2007). New periodic solutions for nonlinear evolution equations using exp-function method. *Chaos, Solitons and Fractals* 34: 1421–1429.

He, J.H. and Wu, X.H. (2006). Exp-function method for nonlinear wave equations. *Chaos, Solitons and Fractals* 30 (3): 700–708.

Veeresha, P., Prakasha, D.G., and Baskonus, H.M. (2019). Novel simulations to the tim-fractional Fisher's equation. *Mathematical Sciences* 13: 33–42.

Wazwaz, A.M. (2007). A comparison between the variational iteration method and adomian decomposition method. *Journal of Computational and Applied Mathematics* 207: 129–136.

Wu, X.H. and He, J.H. (2007). Solitary solutions, periodic solutions and compacton-like solutions using the exp-function method. *Computers & Mathematcs with Applications* 54: 966–986.

Yildirim, A. and Pinar, Z. (2010). Application of the exp-function method for solving nonlinear reaction-diffusion equations arising in mathematical biology. *Computers and Mathematics with Applications* 60: 1873–1880.

Zhang, S. (2007). Exp-function method for solving Maccari's system. *Physics Letters A* 371: 65–71.

Zhang, S. (2008). Application of the exp-function method to Riccati equation and new exact solutions with three arbitrary functions of Broer-Kaup-Kupershmidt equations. *Physics Letters A* 372: 1873–1880.

Zheng, B. (2013). Exp-function method for solving fractional partial differential equations. *The Scientific World Journal* 2013: 465723 (1-8).

Zhu, S.D. (2007a). Exp-function method for the hybrid lattice system. *International Journal of Nonlinear Sciences and Numerical Simulation* 8: 461–464.

Zhu, S.D. (2007b). Exp-function method for the discrete mKdV lattice. *International Journal of Nonlinear Sciences and Numerical Simulation* 8: 465–468.

Zhu, S. (2008). Discrete (2+1)-dimensional Toda lattice equation via exp-function method. *Physics Letters A* 372: 654–657.

25

Exp($-\varphi(\xi)$)-Expansion Method

25.1 Introduction

The exp($-\varphi(\xi)$)-expansion method is used for finding solitary and periodic solutions to nonlinear partial differential equations. This method was utilized by Akbulut et al. (Akbulut et al. 2017) to obtain the solution of the Zakharov Kuznetsov–Benjamin Bona Mahony (ZK-BBM) equation and ill-posed Boussinesq equation. Hafez (2016) implemented this method to construct the new exact traveling wave solutions of the (3 + 1)-dimensional coupled Klein–Gordon–Zakharov equation arising in mathematical physics and engineering. With the help of the exp($-\varphi(\xi)$)-expansion method, Islam (2015) examined traveling wave solutions of the Benney–Luke problem. Hafez and Akbar (2015) obtained new explicit and exact traveling wave solutions of the (1 + 1)-dimensional nonlinear Klein–Gordon–Zakharov equation, which describes the interaction of the Langmuir wave and the ion-acoustic wave in high-frequency plasma. The exact traveling wave solutions of the Zhiber–Shabat equation have been studied by Hafez et al. (2014). Alam et al. (2015a) found the exact traveling wave solutions to the (3 + 1)-dimensional modified Korteweg de Vries (mKdV)–ZK and the (2 + 1)-dimensional Burgers equations using this approach. The solutions to the space–time fractional nonlinear Whitham–Broer–Kaup and generalized nonlinear Hirota–Satsuma coupled Korteweg de Vries (KdV) equations were obtained using this technique by Moussa and Alhakim (2020). Alam et al. (2015b) used this approach to solve the (2 + 1)-dimensional Boussinesq equation, which is an important equation in mathematical physics. For the first time, Abdelrahman et al. (2015) addressed the wave solution of a nonlinear dynamical system in a new double-chain model of DNA and a diffusive predator–prey system by utilizing the said method.

The following section describes the main steps of the exp($-\varphi(\xi)$)-expansion method for obtaining the solution of the nonlinear fractional partial differential equation.

25.2 Methodology of the Exp($-\varphi(\xi)$)-Expansion Method

Let us discuss the exp($-\varphi(\xi)$)-expansion method (Hafez et al. 2014; Abdelrahman et al. 2015; Hafez and Akbar 2015; Islam 2015; Alam et al. 2015a, 2015b; Hafez 2016; Akbulut et al. 2017; Moussa and Alhakim 2020) for obtaining the solution of the nonlinear fractional differential equation. In order to understand exp($-\varphi(\xi)$)-expansion method, we assume the following nonlinear fractional partial differential equation in two independent variables x and t of the type

$$Q\left(u, u_x, u_{xx}, u_{xxx}..., D_t^\alpha u, ...\right) = 0, \quad 0 < \alpha \leq 1, \tag{25.1}$$

where $u = u(x, t)$ is an unknown function, and Q is a polynomial of u and its partial fractional derivatives, in which the highest order derivatives and the nonlinear terms are involved.

The following are the main steps in this method.

Step 1. First, the traveling wave variable is considered as (Li and He 2010)

$$u(x, t) = u(\xi), \tag{25.2}$$

$$\xi = kx - \frac{ct^\alpha}{\Gamma(1 + \alpha)}, \tag{25.3}$$

where c and k are nonzero constants that are to be determined later.

Computational Fractional Dynamical Systems: Fractional Differential Equations and Applications, First Edition.
Snehashish Chakraverty, Rajarama Mohan Jena, and Subrat Kumar Jena.
© 2023 John Wiley & Sons, Inc. Published 2023 by John Wiley & Sons, Inc.

Step 2. Using the chain rule concept (Guner et al. 2015) and Eqs. (25.2) and (25.3), we obtain the following results:

$$\frac{\partial^\alpha u}{\partial t^\alpha} = -c\frac{du}{d\xi}, \quad \frac{\partial u}{\partial x} = k\frac{du}{d\xi}, \quad \frac{\partial^2 u}{\partial x^2} = k^2\frac{d^2 u}{d\xi^2}, \quad \frac{\partial^3 u}{\partial x^3} = k^3\frac{d^3 u}{d\xi^3},$$
(25.4)

and so on for other derivatives. Using Eq. (25.4), Eq. (25.1) is reduced to the following nonlinear ordinary differential equation (ODE) form

$$P\left(u, ku', k^2 u'', k^3 u'''..., -cu', ...\right) = 0,$$
(25.5)

where P is a polynomial of $u(\xi)$ and its various derivatives. The prime (′) denotes the derivative with respect to ξ.

Step 3. Equtaion (25.5) is integrated as long as all terms contain derivatives. This procedure is completed when one of the terms has no derivative. The integration constants are assumed to be zero.

Step 4. According to the exp(−φ(ξ))-expansion method, we assume that the solution of Eq. (25.5) can be expressed in the following finite series expansion form (Hafez et al. 2014; Abdelrahman et al. 2015; Hafez and Akbar 2015; Islam 2015; Alam et al. 2015a, 2015b; Hafez 2016; Akbulut et al. 2017; Moussa and Alhakim 2020):

$$u(\xi) = \sum_{i=0}^{M} a_i\left(\exp\left(-\phi(\xi)\right)\right)^i,$$
(25.6)

where $a_i(a_M \neq 0)$ are the constants to be determined later and $\phi(\xi)$ satisfies the following auxiliary ODE (Islam 2015; Hafez 2016; Akbulut et al. 2017)

$$\phi'(\xi) = \exp\left(-\phi(\xi)\right) + \mu \exp\left(\phi(\xi)\right) + \lambda.$$
(25.7)

Step 5. The value of M may be computed by balancing the highest order derivatives with the nonlinear terms arising in Eq. (25.5). More precisely, we define the degree of $u(\xi)$ as $\deg[u(\xi)] = M$, which leads to the degrees of the other expressions as follows (Gepreel and Omran 2012):

$$\deg\left[\frac{d^p u}{d\xi^p}\right] = M + p,$$
$$\deg\left[u^s\left(\frac{d^p u}{d\xi^p}\right)^q\right] = Ms + q(p + M).$$
(25.8)

Accordingly, we can obtain the value of M in Eq. (25.6).

Step 6. One may obtain the solution of auxiliary Eq. (25.7) as follows (Islam 2015; Hafez 2016; Akbulut et al. 2017):
Case 1: (Hyperbolic function solutions)
When $\lambda^2 - 4\mu > 0$ and $\mu \neq 0$,

$$\phi_1(\xi) = \ln\left(-\sqrt{\lambda^2 - 4\mu}\tanh\left(\sqrt{\lambda^2 - 4\mu}/2(\xi + C)\right) - 2\lambda\mu\right).$$
(25.9)

Case 2: (Trigonometric function solutions)
When $\lambda^2 - 4\mu < 0$ and $\mu \neq 0$,

$$\phi_2(\xi) = \ln\left(\frac{\sqrt{4\mu - \lambda^2}\tan\left(\sqrt{4\mu - \lambda^2}/2(\xi + C)\right) - \lambda}{2\mu}\right).$$
(25.10)

Case 3: (Hyperbolic function solutions)
When $\lambda^2 - 4\mu > 0$, $\mu = 0$, and $\lambda \neq 0$,

$$\phi_3(\xi) = -\ln\left(\frac{\lambda}{\cosh(\lambda(\xi + c)) + \sinh(\lambda(\xi + C)) - 1}\right).$$
(25.11)

Case 4: (Rational function solutions)
When $\lambda^2 - 4\mu = 0$, $\mu \neq 0$, and $\lambda \neq 0$,

$$\phi_4(\xi) = \ln\left(-\frac{2(\lambda(\xi + C) + 2)}{\lambda^2(\xi + C)}\right).$$
(25.12)

Case 5

When $\lambda^2 - 4\mu = 0$, $\mu = 0$, and $\lambda = 0$,

$$\phi_5(\xi) = \ln(\xi + C), \tag{25.13}$$

where C is an integration constant.

Step 7. Substituting Eq. (25.6) into Eq. (25.5) using Eq. (25.7) and collecting all terms with the same power of exp $(-\phi(\xi))^i (i = 0, 1, 2, ...,)$ together, we get a polynomial in $\exp(-\phi(\xi))$. Equating each coefficient of this polynomial to zero yields a set of algebraic equations for $a_i (i = 0, 1, 2, ...M)$, k, c, λ, μ. Solving the algebraic equations system, we can construct a variety of exact solutions for Eq. (25.1).

25.3 Numerical Examples

We employ the present method to solve the nonlinear time-fractional KdV equation in Example 25.1 and the mKdV equation in Example 25.2.

Example 25.1 Let us consider the following nonlinear time-fractional KdV equation (Wazwaz 2009)

$$\frac{\partial^\alpha u}{\partial t^\alpha} + 6u\frac{\partial u}{\partial x} + \frac{\partial^3 u}{\partial x^3} = 0, \quad 0 < \alpha \le 1, t > 0. \tag{25.14}$$

Solution

Using the transformation Eqs. (25.2) and (25.3), Eq. (25.14) reduces to

$$-cu' + 6ku\,u' + k^3 u''' = 0, \tag{25.15}$$

where prime denotes the differentiation with respect to ξ. Integrating Eq. (25.15) with respect to ξ and substituting the integration constant to zero, it gives

$$-cu + 3ku^2 + k^3 u'' = 0. \tag{25.16}$$

The highest order linear and nonlinear terms in Eq. (25.16) are u'' and u^2. So balancing the order of u'' and u^2 by using the concept given in Eq. (25.8), we get

$$\begin{aligned} &\deg(u'') = \deg(u^2), \\ &M + 2 = 2M \Rightarrow M = 2. \end{aligned} \tag{25.17}$$

Now, we assume Eq. (25.6) as the solution of Eq. (25.16) at $M = 2$. This may be written as

$$u(\xi) = a_0 + a_1 \exp(-\phi(\xi)) + a_2 \exp(-2\phi(\xi)) \tag{25.18}$$

where a_0, a_1, and a_2 are constants to be calculated. From Eq. (25.18), we have

$$u'(\xi) = -a_1 \exp(-\phi(\xi))\phi'(\xi) - 2a_2 \exp(-2\phi(\xi))\,\phi'(\xi). \tag{25.19}$$

Using Eq. (25.7), Eq. (25.19) can be written as

$$u'(\xi) = -a_1\mu - (2a_2\mu + a_1\lambda)\exp(-\phi(\xi)) - (a_1 + 2a_2\lambda)\exp(-2\phi(\xi)) - 2a_2\exp(-3\phi(\xi)). \tag{25.20}$$

Similarly,

$$\begin{aligned} u''(\xi) = &\,a_1\lambda\mu + 2a_2\mu^2 + (2a_1\mu + a_1\lambda^2 + 6a_2\lambda\mu)\exp(-\phi(\xi)) + (3a_1\lambda + 8a_2\mu + 4a_2\lambda^2) \\ &\exp(-2\phi(\xi)) + (2a_1 + 10a_2\lambda)\exp(-3\phi(\xi)) + 6a_2\exp(-4\phi(\xi)). \end{aligned} \tag{25.21}$$

Substituting Eqs. (25.18) and (25.21) into Eq. (25.16) yields a polynomial in $\exp(-n\phi(\xi))$, $n = 0, 1, 2, ...$ with the parameters $\lambda, \mu, k, c, a_0, a_1$, and a_2 as follows:

$$\begin{aligned} &-c\{a_0 + a_1\exp(-\phi(\xi)) + a_2\exp(-2\phi(\xi))\} + 3k(a_0 + a_1\exp(-\phi(\xi)) + a_2\exp(-2\phi(\xi)))^2 \\ &+ k^3\left\{\begin{aligned} &a_1\lambda\mu + 2a_2\mu^2 + (2a_1\mu + a_1\lambda^2 + 6a_2\lambda\mu)\exp(-\phi(\xi)) + (3a_1\lambda + 8a_2\mu + 4a_2\lambda^2) \\ &\exp(-2\phi(\xi)) + (2a_1 + 10a_2\lambda)\exp(-3\phi(\xi)) + 6a_2\exp(-4\phi(\xi)) \end{aligned}\right\} = 0 \end{aligned} \tag{25.22}$$

Now, collecting the coefficients of $\exp(-n\phi(\xi))$, $n = 0, 1, 2, \dots$ and setting these to zero, we obtain the following system:

$$
\begin{aligned}
&\exp(-4\phi(\xi)): 6k^3 a_2 + 3ka_2^2, \\
&\exp(-3\phi(\xi)): 10k^3 \lambda a_2 + 2k^3 a_1 + 6ka_1 a_2, \\
&\exp(-2\phi(\xi)): 4k^3 \lambda^2 a_2 + 3k^3 \lambda a_1 + 8k^3 \mu\, a_2 + 6ka_0 a_2 + 3ka_1^2 - ca_2, \\
&\exp(-\phi(\xi)): k^3 \lambda^2 a_1 + 6k^3 \lambda\mu\, a_2 + 2k^3 \mu\, a_1 + 6ka_0 a_1 - ca_1, \\
&\text{Constant: } k^3 \lambda\mu a_1 + 2k^3 \mu^2\, a_2 + 3ka_0^2 - ca_0.
\end{aligned}
\tag{25.23}
$$

Solving the above nonlinear system, we obtain different values of c, k, a_0, a_1, a_2 in various cases. Some of them are listed below.

Case 1

$$
a_0 = -\frac{1}{3}k^2(\lambda^2 + 2\mu), \quad a_1 = -2k^2\lambda, \quad a_2 = -2k^2, \quad c = -k^3(\lambda^2 - 4\mu), \quad k = k, \tag{25.24}
$$

where k is a free parameter.

(1.1) (Hyperbolic function solution): Substituting the values of constants given in Eq. (25.24) into Eq. (25.18) and using Eq. (25.9), we obtain

$$
\begin{aligned}
u_{1,1}(\xi) = {}&-\frac{1}{3}k^2(\lambda^2 + 2\mu) - \frac{2k^2\lambda}{\left(-\sqrt{\lambda^2 - 4\mu}\tanh\left(\sqrt{\lambda^2 - 4\mu}/2(\xi + C)\right) - 2\lambda\mu\right)} \\
&- \frac{2k^2}{\left(-\sqrt{\lambda^2 - 4\mu}\tanh\left(\sqrt{\lambda^2 - 4\mu}/2(\xi + C)\right) - 2\lambda\mu\right)^2}.
\end{aligned}
\tag{25.25}
$$

when $\lambda^2 - 4\mu > 0$ and $\mu \neq 0$.

(1.2) (Trigonometric function solution): When $\lambda^2 - 4\mu < 0$ and $\mu \neq 0$, we obtain

$$
\begin{aligned}
u_{1,2}(\xi) = {}&-\frac{1}{3}k^2(\lambda^2 + 2\mu) - \frac{4k^2\mu\lambda}{\left(\sqrt{4\mu - \lambda^2}\tan\left(\sqrt{4\mu - \lambda^2}/2(\xi + C)\right) - \lambda\right)} \\
&- \frac{8\mu^2 k^2}{\left(\sqrt{4\mu - \lambda^2}\tan\left(\sqrt{4\mu - \lambda^2}/2(\xi + C)\right) - \lambda\right)^2}.
\end{aligned}
\tag{25.26}
$$

(1.3) (Hyperbolic function solution): When $\lambda^2 - 4\mu > 0$, $\mu = 0$, and $\lambda \neq 0$, we get

$$
\begin{aligned}
u_{1,3}(\xi) = {}&-\frac{1}{3}k^2(\lambda^2 + 2\mu) - \frac{2k^2\lambda^2}{(\cosh(\lambda(\xi + c)) + \sinh(\lambda(\xi + C)) - 1)} \\
&- \frac{2k^2\lambda^2}{(\cosh(\lambda(\xi + c)) + \sinh(\lambda(\xi + C)) - 1)^2}.
\end{aligned}
\tag{25.27}
$$

(1.4) (Rational function solution): When $\lambda^2 - 4\mu = 0$, $\mu \neq 0$, and $\lambda \neq 0$, we have

$$
u_{1,4}(\xi) = -\frac{1}{3}k^2(\lambda^2 + 2\mu) - \frac{2k^2\lambda^3(\xi + C)}{(-2(\lambda(\xi + C) - 2))} - \frac{2k^2\lambda^4(\xi + C)^2}{(2(\lambda(\xi + C) + 2))^2}.
\tag{25.28}
$$

(1.5) When $\lambda^2 - 4\mu = 0$, $\mu = 0$, and $\lambda = 0$, we obtain

$$
u_{1,5}(\xi) = -\frac{1}{3}k^2(\lambda^2 + 2\mu) - \frac{2k^2\lambda}{(\xi + C)} - \frac{2k^2}{(\xi + C)^2}.
\tag{25.29}
$$

In this case, $\xi = kx + k^3(\lambda^2 - 4\mu)\dfrac{t^\alpha}{\Gamma(1 + \alpha)}$.

Case 2

$$
a_0 = -2k^2\mu, \quad a_1 = -2k^2\lambda, \quad a_2 = -2k^2, \quad c = -k^3(-\lambda^2 + 4\mu), \quad k = k, \tag{25.30}
$$

where k is free parameter.

(2.1) (Hyperbolic function solution): When $\lambda^2 - 4\mu > 0$ and $\mu \neq 0$,

$$u_{2,1}(\xi) = -2k^2\mu - \frac{2k^2\lambda}{\left(-\sqrt{\lambda^2-4\mu}\tanh\left(\sqrt{\lambda^2-4\mu}/2(\xi+C)\right)-2\lambda\mu\right)}$$
$$-\frac{2k^2}{\left(-\sqrt{\lambda^2-4\mu}\tanh\left(\sqrt{\lambda^2-4\mu}/2(\xi+C)\right)-2\lambda\mu\right)^2}.$$

(25.31)

(2.2) (Trigonometric function solution): When $\lambda^2 - 4\mu < 0$ and $\mu \neq 0$,

$$u_{2,2}(\xi) = -2k^2\mu - \frac{4k^2\mu\lambda}{\left(\sqrt{4\mu-\lambda^2}\tan\left(\sqrt{4\mu-\lambda^2}/2(\xi+C)\right)-\lambda\right)}$$
$$-\frac{8\mu^2k^2}{\left(\sqrt{4\mu-\lambda^2}\tan\left(\sqrt{4\mu-\lambda^2}/2(\xi+C)\right)-\lambda\right)^2}.$$

(25.32)

(2.3) (Hyperbolic function solution): When $\lambda^2 - 4\mu > 0$, $\mu = 0$, and $\lambda \neq 0$,

$$u_{2,3}(\xi) = -2k^2\mu - \frac{2k^2\lambda^2}{\left(\cosh\left(\lambda(\xi+c)\right)+\sinh\left(\lambda(\xi+C)\right)-1\right)}$$
$$-\frac{2k^2\lambda^2}{\left(\cosh\left(\lambda(\xi+c)\right)+\sinh\left(\lambda(\xi+C)\right)-1\right)^2}.$$

(25.33)

(2.4) (Rational function solution): When $\lambda^2 - 4\mu = 0$, $\mu \neq 0$, and $\lambda \neq 0$,

$$u_{2,4}(\xi) = -2k^2\mu - \frac{2k^2\lambda^3(\xi+C)}{\left(-2\left(\lambda(\xi+C)-2\right)\right)} - \frac{2k^2\lambda^4(\xi+C)^2}{\left(2\left(\lambda(\xi+C)+2\right)\right)^2}.$$

(25.34)

(2.5) When $\lambda^2 - 4\mu = 0$, $\mu = 0$, and $\lambda = 0$, we obtain

$$u_{2,5}(\xi) = -2k^2\mu - \frac{2k^2\lambda}{(\xi+C)} - \frac{2k^2}{(\xi+C)^2}.$$

(25.35)

In case 2, we have $\xi = kx + k^3\left(-\lambda^2 + 4\mu\right)\dfrac{t^\alpha}{\Gamma(1+\alpha)}$.

All the above Eqs. (25.25)–(25.29) and (25.31)–(25.35) are the exact solutions of the nonlinear time-fractional KdV equation for case 1 and case 2.

Example 25.2 We consider the following nonlinear time-fractional mKdV equation (Wazwaz 2009):

$$\frac{\partial^\alpha u}{\partial t^\alpha} - 6u^2\frac{\partial u}{\partial x} + \frac{\partial^3 u}{\partial x^3} = 0, \qquad 0 < \alpha \leq 1, t > 0.$$

(25.36)

Solution

Using the transformation Eqs. (25.2) and (25.3) in Eq. (25.36), it reduces to

$$-cu' - 6ku^2 u' + k^3 u''' = 0.$$

(25.37)

Integrating Eq. (25.37) with respect to ξ and substituting the integration constant to zero, it gives

$$-cu - 2ku^3 + k^3 u'' = 0.$$

(25.38)

The highest order linear and nonlinear terms in Eq. (25.38) are u'' and u^3. So balancing the order of u'' and u^3 by using the concept in Eq. (25.8), we get

$$\deg(u'') = \deg(u^3),$$
$$M + 2 = 3M \Rightarrow M = 1.$$

(25.39)

Let us assume Eq. (25.6) as the solution of Eq. (25.38) at $M = 1$. This may be written as

$$u(\xi) = a_0 + a_1 \exp\left(-\phi(\xi)\right) \tag{25.40}$$

where a_0 and a_1 are constants. Using Eq. (25.7), from Eq. (25.40), we get

$$u'(\xi) = -a_1\mu - a_1\lambda \exp\left(-\phi(\xi)\right) - a_1 \exp\left(-2\phi(\xi)\right). \tag{25.41}$$

$$u''(\xi) = a_1\lambda\mu + \left(2a_1\mu + a_1\lambda^2\right) \exp\left(-\phi(\xi)\right) + 3a_1\lambda \exp\left(-2\phi(\xi)\right) + 2a_1 \exp\left(-3\phi(\xi)\right). \tag{25.42}$$

Substituting Eqs. (25.40) and (25.42) into Eq. (25.38), we obtain a polynomial in $\exp\left(-n\phi(\xi)\right)$, $n = 0, 1, 2, \dots$ with the parameters λ, μ, k, c, a_0, and a_1 as follows

$$-2ka_0^3 + k^3 a_1\lambda\mu - ca_0 + \left(-6ka_0^2 a_1 + 2k^3 a_1\mu + k^3 a_1\lambda^2 - ca_1\right) \exp\left(-\phi(\xi)\right)$$
$$+ \left(-6ka_0 a_1^2 + 3k^3 a_1\lambda\right) \exp\left(-2\phi(\xi)\right) + \left(-2ka_1^3 + 2k^3 a_1\right) \exp\left(-3\phi(\xi)\right) = 0. \tag{25.43}$$

Collecting the coefficients of $\exp\left(-n\phi(\xi)\right)$, $n = 0, 1, 2, \dots$ and setting these to zero, we obtain the following system:

$$\begin{aligned}
&\exp\left(-3\phi(\xi)\right): 2k^3 a_1 - 2ka_1^3, \\
&\exp\left(-2\phi(\xi)\right): 3k^3\lambda a_1 - 6ka_0 a_1^2, \\
&\exp\left(-\phi(\xi)\right): k^3\lambda^2 a_1 + 2k^3\mu\, a_1 - 6ka_0^2 a_1 - ca_1, \\
&\text{Constant: } k^3\lambda\mu\, a_1 - 2ka_0^3 - ca_0.
\end{aligned} \tag{25.44}$$

Solving the above nonlinear system, we get

Case 1

$$a_0 = -\frac{k\lambda}{2}, \quad a_1 = -k, \quad c = -\frac{k^3}{2}\left(\lambda^2 - 4\mu\right), \quad k = k, \tag{25.45}$$

where k is a free parameter.

(1.1) (Hyperbolic function solution): Substituting Eq. (25.45) into Eq. (25.40) and using Eq. (25.9), we obtain

$$u_{1,1}(\xi) = -\frac{k\lambda}{2} - \frac{k}{\left(-\sqrt{\lambda^2 - 4\mu}\tanh\left(\sqrt{\lambda^2 - 4\mu}/2(\xi + C)\right) - 2\lambda\mu\right)}, \tag{25.46}$$

when $\lambda^2 - 4\mu > 0$ and $\mu \neq 0$.

(1.2) (Trigonometric function solution): When $\lambda^2 - 4\mu < 0$ and $\mu \neq 0$, we obtain

$$u_{1,2}(\xi) = -\frac{k\lambda}{2} - \frac{2\mu k}{\left(\sqrt{4\mu - \lambda^2}\tan\left(\sqrt{4\mu - \lambda^2}/2(\xi + C)\right) - \lambda\right)}. \tag{25.47}$$

(1.3) (Hyperbolic function solution): When $\lambda^2 - 4\mu > 0$, $\mu = 0$, and $\lambda \neq 0$, we get

$$u_{1,3}(\xi) = -\frac{k\lambda}{2} - \frac{k\lambda}{\left(\cosh\left(\lambda(\xi + c)\right) + \sinh\left(\lambda(\xi + C)\right) - 1\right)}. \tag{25.48}$$

(1.4) (Rational function solution): When $\lambda^2 - 4\mu = 0$, $\mu \neq 0$, and $\lambda \neq 0$, we have

$$u_{1,4}(\xi) = -\frac{k\lambda}{2} - \frac{k\lambda^2(\xi + C)}{\left(-2\left(\lambda(\xi + C) - 2\right)\right)}. \tag{25.49}$$

(1.5) When $\lambda^2 - 4\mu = 0$, $\mu = 0$, and $\lambda = 0$, we obtain

$$u_{1,5}(\xi) = -\frac{k\lambda}{2} - \frac{k}{(\xi + C)}. \tag{25.50}$$

In this case, $\xi = kx + \dfrac{k^3}{2}\left(\lambda^2 - 4\mu\right)\dfrac{t^\alpha}{\Gamma(1 + \alpha)}$.

Case 2

$$a_0 = \frac{k\lambda}{2}, \, a_1 = k, \, c = -\frac{k^3}{2}\left(\lambda^2 - 4\mu\right), \, k = k,$$ (25.51)

where k is a free parameter.

(2.1) (Hyperbolic function solution): When $\lambda^2 - 4\mu > 0$ and $\mu \neq 0$,

$$u_{2,1}(\xi) = \frac{k\lambda}{2} + \frac{k}{\left(-\sqrt{\lambda^2 - 4\mu}\tanh\left(\sqrt{\lambda^2 - 4\mu}/2(\xi + C)\right) - 2\lambda\mu\right)}.$$ (25.52)

(2.2) (Trigonometric function solution): When $\lambda^2 - 4\mu < 0$ and $\mu \neq 0$,

$$u_{2,2}(\xi) = \frac{k\lambda}{2} + \frac{2\mu k}{\left(\sqrt{4\mu - \lambda^2}\tan\left(\sqrt{4\mu - \lambda^2}/2(\xi + C)\right) - \lambda\right)}.$$ (25.53)

(2.3) (Hyperbolic function solution): When $\lambda^2 - 4\mu > 0$, $\mu = 0$, and $\lambda \neq 0$,

$$u_{2,3}(\xi) = \frac{k\lambda}{2} + \frac{k\lambda}{\left(\cosh\left(\lambda(\xi + c)\right) + \sinh\left(\lambda(\xi + C)\right) - 1\right)}.$$ (25.54)

(2.4) (Rational function solution): When $\lambda^2 - 4\mu = 0$, $\mu \neq 0$, and $\lambda \neq 0$,

$$u_{2,4}(\xi) = \frac{k\lambda}{2} + \frac{k\lambda^2(\xi + C)}{\left(-2\left(\lambda(\xi + C) - 2\right)\right)}.$$ (25.55)

(2.5) When $\lambda^2 - 4\mu = 0$, $\mu = 0$, and $\lambda = 0$, we obtain

$$u_{2,5}(\xi) = \frac{k\lambda}{2} + \frac{k}{(\xi + C)}.$$ (25.56)

Here, $\xi = kx + \frac{k^3}{2}\left(\lambda^2 - 4\mu\right)\frac{t^\alpha}{\Gamma(1 + \alpha)}$.

All the above Eqs. (25.46)–(25.50) and (25.52)–(25.56) are the exact solutions of the nonlinear time-fractional mKdV equation for different cases.

References

Abdelrahman, M.A.E., Zahran, E.H.M., and Khater, M.M.A. (2015). The exp(−ϕ (ξ))-expansion method and its application for solving nonlinear evolution equations. *International Journal of Modern Nonlinear Theory and Application* 4: 37–47.

Akbulut, A., Kaplan, M., and Tascan, F. (2017). The investigation of exact solutions of nonlinear partial differential equations by using exp(-Φ(ξ)) method. *Optik* 132: 382–387.

Alam, M.N., Hafez, M.G., Akbar, M.A., and Roshid, H.O. (2015a). Exact traveling wave solutions to the (3 + 1)-dimensional mKdV–ZK and the (2 + 1)-dimensional Burgers equations via exp(-Φ(ξ))-expansion method. *Alexandria Engineering Journal* 54: 635–644.

Alam, M.N., Hafez, M.G., Akbar, M.A., and Roshid, H.O. (2015b). Exact solutions to the (2 + 1)-dimensional boussinesq equation via exp(Φ(η))-expansion method. *Journal of Scientific Research* 7 (3): 1–10.

Gepreel, K.A. and Omran, S. (2012). Exact solutions for nonlinear partial fractional differential equations. *Chinese Physics B* 21 (11): 110204.

Guner, O., Bekir, A., and Cevikel, A.C. (2015). A variety of exact solutions for the time fractional Cahn–Allen equation. *Eur. Phys. J. Plus* 130: 146 (1–13).

Hafez, M.G. (2016). Exact solutions to the (3 + 1)-dimensional coupled Klein–Gordon–Zakharov equation using exp (-Φ(ξ))-expansion method. *Alexandria Engineering Journal* 55: 1635–1645.

Hafez, M.G. and Akbar, M.A. (2015). New exact traveling wave solutions to the (1 + 1)-dimensional Klein–Gordon–Zakharov equation for wave propagation in plasma using the exp(-Φ(ξ))-expansion method. *Propulsion and Power Research* 4 (1): 31–39.

Hafez, M.G., Kauser, M.A., and Akter, M.T. (2014). Some new exact traveling wave solutions for the Zhiber–Shabat equation. *British Journal of Mathematics & Computer Science* 4 (18): 2582–2593.

Islam, S.M.R. (2015). Applications of the exp(-Φ(ξ))-expansion method to find exact traveling wave solutions of the Benney–Luke equation in mathematical physics. *American Journal of Applied Mathematics* 3 (3): 100–105.

Li, Z.B. and He, J.H. (2010). Fractional complex transformation for fractional differential equations. *Computer and Mathematics with Applications* 15: 970–973.

Moussa, A.A. and Alhakim, L.A. (2020). Fractional exp(-Φ(ξ))-expansion method and its application to space-time nonlinear fractional equations. *Australian Journal of Mathematical Analysis and Applications* 17 (2): 1–15.

Wazwaz, A.M. (2009). *Partial Differential Equations and Solitary Waves Theory*, Nonlinear Physical Science. Berlin, Heidelberg: Springer.

Index